Creating the American West

CREATING THE AMERICAN WEST
Boundaries and Borderlands

Derek R. Everett

University of Oklahoma : Norman

Also by Derek R. Everett
The Colorado State Capitol: History, Politics, Preservation (Boulder, Colo., 2005)

Publication of this book is made possible through the generosity of Edith Kinney Gaylord.

Chapter 3 is based on Derek R. Everett's article "On the Extreme Frontier: Crafting the Western Arkansas Boundary," which was published in *Arkansas Historical Quarterly* 67:1 (Spring 2008), pp. 1–26. Material in chapter 4 is adapted from Everett's article "To Shed Our Blood for Our Beloved Territory: The Iowa-Missouri Borderland," which was published in *The Annals of Iowa* 67 (Fall 2008), pp. 269–97.

Library of Congress Cataloging-in-Publication Data
Everett, Derek R.
  Creating the American West : boundaries and borderlands / Derek R. Everett.
    pages cm
  Includes bibliographical references and index.
  ISBN 978-0-8061-4446-7 (hardcover : alk. paper) 1. United States—Boundaries—History. 2. West (U.S.)—Boundaries—History. 3. United States—Territorial expansion. I. Title.
  E179.5.E95 2014
  978—dc23
                    2013039702

The paper in this book meets the guidelines for permanence and durability of the Committee on Production Guidelines for Book Longevity of the Council on Library Resources, Inc. ∞

Copyright © 2014 by the University of Oklahoma Press, Norman, Publishing Division of the University. Manufactured in the U.S.A.

All rights reserved. No part of this publication may be reproduced, stored in a retrieval system, or transmitted, in any form or by any means, electronic, mechanical, photocopying, recording, or otherwise—except as permitted under Section 107 or 108 of the United States Copyright Act—without the prior written permission of the University of Oklahoma Press. To request permission to reproduce selections from this book, write to Permissions, University of Oklahoma Press, 2800 Venture Drive, Norman OK 73069, or email rights.oupress@ou.edu.

1 2 3 4 5 6 7 8 9 10

For Heather, with all my heart

> We go to gain a little patch of ground
> That hath in it no profit but the name.

—Norwegian captain, in Shakespeare's *Hamlet*,
Act IV, Scene 4

Contents

List of Illustrations	xi
Acknowledgments	xiii
Introduction: The Significance of State Boundaries	3
1. A Little Patch of Ground: The Precedents for Western State Boundaries	19
2. The Fairest Portion of Our Union: Early Boundaries in the Trans-Mississippi West	45
3. On the Extreme Frontier: The Western Arkansas Boundary	71
4. Blood Will Be Shed: The Missouri–Iowa Boundary	95
5. Nature Has Marked out the Boundaries: The Oregon Country Boundaries	121
6. A State Bordering upon Anarchy: The California–Nevada Boundary	143
7. Two Distinct Civilizations: The New Mexico–Colorado Boundary	167
8. Let Us Divide: The Dakota Boundaries	191
Conclusion: A Broader View of Borderlands	215
Appendix: "The Honey War"	229
Notes	231
Bibliography	277
Index	293

Illustrations

FIGURES

The author at the Colorado-Nebraska-Wyoming junction	xiv
Four Corners	2
The junction of Connecticut, Massachusetts, and Rhode Island	18
State highway signs near Maysville, Arkansas	46
The original "hinge" in the western Arkansas line	70
Missouri's original northwestern corner	94
The Oregon-Washington boundary at the Columbia River	120
Bisected swimming pool at the Cal-Neva Casino Resort	142
Fence following the New Mexico–Colorado boundary	166
Marker and fence along North Dakota–South Dakota boundary	190
A man and his ass on New Mexico–Arizona boundary	214
A man and his ass in the "border town" of Texarkana	225

MAPS

John Wesley Powell's 1890 proposal for states corresponding to drainage basins	13
English "Sea to Sea" charter grants	22
Land claims in the trans-Appalachian West	27
The Ordinance of 1784	32
The first trans-Mississippi boundaries, 1803–1804	48
Proposal for western boundaries extrapolated from the *St. Louis Enquirer*, 1819	53
National Intelligencer proposal for western boundaries, 1819	55
Early proposals to modify state boundaries, 1831–57	59
Western Arkansas, 1808–23	73
Western Arkansas, 1824–28	81
Northern Missouri lines, 1808–24	97

Competing boundaries of Albert Miller Lea's 1839 report	104
The Honey War, 1839	109
The Oregon Country	123
California-Nevada boundary region	145
New Mexico–Colorado boundary region	169
The Northern Great Plains	193

Acknowledgments

In a sense, this project began with my research into the history of Colorado's state capitol, as I pondered how the Centennial State's arbitrary boundaries lump together diverse populations and environments. Similar lines have created regional identities out of otherwise anonymous land across the country. I wanted to know how state boundaries came to be, who participated in that process, and what motivated them to leave us with the familiar shapes that make up our country. Guided by the kind and supportive wisdom of my adviser, Elliott West, I pondered these questions while pursuing my doctorate at the University of Arkansas. Throughout the process I learned a great deal about the evolution of the United States and especially the trans-Mississippi West in the nineteenth century. The experience increased my admiration for this patchwork country on a more intimate level. Above all, I had fun. Granted, there were moments when I wondered who would save the petty, fanatic, conniving, silly people of the past from themselves. Virginians staking a claim to the Louisiana Purchase, the comic warriors of Missouri and Iowa, drunken correspondents in the Sierra Nevada, and dispute over Dakota Territory's prairie chicken law all come to mind.

Many friends and colleagues helped me shape this work. At the University of Arkansas, I received assistance from the Department of History, the Fulbright College of Arts and Sciences, the Graduate School, and the staff of the Mullins and Young Law Libraries. I thank in particular the unparalleled (and unmeridianed) Elliott West, as well as Lynda Coon, Beth Juhl, Dan Sutherland, Jeannie Whayne, and Patrick Williams. I am indebted to my fellow disciples Julie Courtwright and Jason Pierce, and everyone else who slogged through the trenches together, especially Aneilya Barnes, Matt and Tammy Byron, and Gene Vinzant. At Metropolitan State University of Denver, I appreciate the input of Tom Altherr, Vince de Baca, Jim Drake, Kim Klimek, Todd Laugen, Steve Leonard, Andrea Maestrejuan, and John Monnett, as well as the staff at the Auraria Library. I extend similar thanks to Mark Fiege, Robert Gudmestad, Adrian Howkins, Diane

The author at the monument marking the junction of Colorado, Nebraska, and Wyoming

Margolf, Jared Orsi, and the Morgan Library staff at Colorado State University. The folks at the University of Oklahoma Press (in particular Chuck Rankin, Jay Dew, Emily Jerman, and Amy Hernandez), freelance copy editor Chris Dodge, cartographer Tom Jonas, and my anonymous reviewers made the book in your hands possible. All photographs and postcard reproductions in this book come from my personal collection. For their early encouragement and insightful critiques, I express special gratitude to Stephen Aron and the late David Weber.

For inexhaustible confidence and support over the years, I thank my parents, Dave and Sandy Everett; my grandparents, Claire Everett and Don and Glenita Emarine; and my parents-in-law, Brad and Marva Craig. I dedicate this book to my wife, Heather, whose encouragement ensured its completion. She understood my long absences from her side as I toiled in my basement office and even sent me there when my enthusiasm flagged. I have no idea what I did to deserve her love, but I seek only to earn it and return it every day. In the midst of this project Louisa Claire arrived and brought her parents more joy and unconditional affection than they ever thought possible. I look forward to exploring life with my beautiful girls. In the spirit of this book, I'll evoke Johnny Cash: because they're mine, I walk the line.

Countless times over the course of this project, friends and colleagues delighted in referring to me as a "borderline historian." I am proud to say that they were right.

Creating the American West

Postcard of the Four Corners monument, mid-twentieth century

Introduction
The Significance of State Boundaries

Boundaries pervade American life. Lines imposed on the landscape dictate which candidates we can vote for, what schools our children attend, how much tax we must pay, whether we can buy fireworks or alcohol, where our responsibility for mowing grass or shoveling snow ceases and our neighbors' begins, and much more. Even time itself is harnessed by unseen barriers into Eastern, Central, Mountain, and Pacific zones. People in the United States interact with countless boundaries every day, as they have for a long time. As a Missouri newspaper declared in 1836, "anything relative to the boundaries of a country, or even a plantation are always so considered by the discerning of every age and nation."[1] The creation of state boundaries in particular represents an essential aspect of the country's history. By drawing lines in the trans-Mississippi West throughout the nineteenth century, Americans imposed a specific form of political organization inherited from the older states of the Union. State boundaries both divide and unite, providing a structure to help manage the disparate populations and environments that constitute the United States. For countless people then and now, these lines matter.

Yet these ubiquitous, invisible lines have long been overlooked by historians and the public at large. To many modern Americans, western state boundaries in particular represent little more than goals on cross-country road trips, the completion of a journey across one giant rectangle and the start of another. Such attitudes echo the limited scholarly interest in states and their boundaries since the late nineteenth century. The major events that established the external limits of the United States—the American Revolution, the Louisiana Purchase, Texas's annexation, the acquisition of the Oregon Country, the war with Mexico, and so on—appear in most narrative histories of the country. By contrast, historians have accepted the existence of states with the same sort of manifest destiny that fueled territorial annexation. It is as if national borders evolved but state boundaries had always existed, awaiting their discovery by intrepid pioneers. "Of course that's where North Dakota should be," one almost hears, or "Where else would we put Nevada?" or perhaps even "Because God intended *that* to be Wyoming and *that* to be Colorado."

This passive mind-set obscures a dynamic story. Transcontinental expansion in the nineteenth century did not create the American West. Any country can alter its borders to claim more land, more resources, more people. It was the act of drawing lines to shape polities modeled upon the eastern states that made the West truly American. Political limits restructured vast tracts acquired from international competitors and populated by diverse cultures. Throughout the nineteenth century the United States supplanted American Indian societies and Euro-American empires by restructuring an extensive region in its own image. The chapters to come will assert the essential function of internal lines in American history generally and in the trans-Mississippi West in particular. The creation and evolution of western state boundaries represented the single most important method for transforming a naturally anonymous landscape into a functional part of the United States.

During the mid-nineteenth century, scholars emphasized this state structure as they focused upon the political development of the country, appropriate considering the state-federal tension of the Civil War era.[2] But a new generation of historians who grew up during the Gilded Age moved away from state-centered narratives. Historian Frederick Jackson Turner credited maps that emphasized natural features and omitted "any names or political boundaries" with inspiring his frontier thesis of 1893.[3] Four years later, Woodrow Wilson accepted Turner's approach as the best one for understanding American history. Wilson noted with despair that throughout the nineteenth century "new regions have been suffered to become new States," which he thought "whimsically enough formed." He went on: "We have joined mining communities with agricultural, the mountain with the plain, the ranch with the farm, and have left the making of uniform rules to the sagacity and practical habit of neighbors ill at ease with one another."[4] Turner insisted upon this notion again in 1901, when he bemoaned the academic practice of "fixing our attention too exclusively upon the artificial boundary lines of the States."[5] The study of these lines has endured in state historical journals, a handful of boundary-specific monographs, and encyclopedic works of varying quality.[6] Yet general American historiography since the Progressive Era has shrugged off state-centered research as provincial and passé.

More than a century of scholarly neglect does an injustice to an essential aspect of the story of the United States. At the end of the eighteenth century, Americans possessed a shaky independence and a new, relatively untried constitution. In the first half of the nineteenth century they incorporated land beyond the country's original western border through purchase, negotiation, and conquest. The federal government needed to control these regions effectively to demonstrate to itself and other countries its viability. Among the methods used to effect this transformation, perhaps the most dramatic and long-lasting was extending lines beyond the Mississippi River to create polities on the model of the original states. Imposing political organization on these tracts constituted a reworking of the landscape into something distinctly American. In the pages that follow, "American" refers to all things European American, including political institutions modeled on the first states, those along the Atlantic coast. "Boundary" and "border" describe the invisible lines between states and countries, respectively. In the trans-Mississippi West, state boundaries were imposed in sync with the country's chosen system of republican democracy. Inserting political divides into the West thus reshaped it as a truly American region. Although the process aided in the destruction of American Indian independence, it reinforced the survival and potential for growth of the federal experiment. As Representative James E. Belser of Alabama argued in 1845, "the Union was strengthened and preserved by expanding its blessings."[7] In addition, the West offered Washington, D.C., an area it could control—in theory at least—in contrast to the challenges posed by other sections of the country. State-making in the West turned miscellaneous land claims into effective polities and affixed the label "American" onto formerly foreign soil.

Drawing state lines instigated widespread passionate debate—in Washington, D.C., among settlers who moved into the lands the boundaries divided, and across the country generally over the nineteenth century. In the chapters that follow, stories about state boundaries drawn from accounts in newspapers hundreds or even thousands of miles away from the new states bear this out repeatedly. One might expect coverage of the congressional debates that helped shape these lines to appear in newspapers of the nation's capital. But what interest might the residents of Pennsylvania or South Carolina have had in a boundary dispute between Missouri and Iowa? Who in Kansas City or Milwaukee would have bothered to read about the line separating New Mexico and Colorado? Why would Illinoisans or Oregonians have cared about the North Dakota–South Dakota divide? And what accounts for the myriad western boundary stories that appeared in New York City newspapers? Nineteenth-century press attention to events and people half a continent away suggests that Americans of the day followed the stories of state lines with interest. In the heyday of boundary-making, as the nation as a whole grappled with the debate between state and federal authority, the origins of fellow members of the Union proved compelling. Boundary-making shaped not only new polities in the West but the future of the country as a whole.

Western state boundaries were being drawn at the same time that cartography was blossoming as a profession. In *Mapping the Nation*, historian Susan Schulten outlines this new field and its political and social consequences. She demonstrates that after the late-eighteenth-century revolutions in North America and Europe, Western cultures grew eager to chart their borders and make clear to the world their sphere of influence. Geographic comprehension and representation went hand-in-hand with this expression of authority, hence the proliferation of maps and atlases—both historical and modern—in the United States in particular. Emma Willard's foundational 1828 public school textbook included maps that depicted state lines overlaid upon lands of pre-contact native cultures, suggesting states' foreordained nature. Similarly, Rufus Blanchard's 1876 "Historical Map of the United States" included faint state boundaries but downplayed national borders, again placing the origin of states ahead of the country. The intent of incorporating state lines might have been inspired by a simple need to provide familiar references for readers, but their presence asserts their primacy to the audience. Through maps like these, Americans perceived stability in their institutions. In Schulten's words: "The quest for government control transformed national space by placing a premium on the articulation of territory."[8] This expression of power depended upon state boundaries and national borders alike, at times the former more than the latter. After all, for much of the nineteenth century the United States operated more as a collection of parts than a cohesive unit. The proliferation of mapmaking coincided with Americans' nineteenth-century obsession with creating new states, especially in their vast western claims.

Americans who came to occupy the West shared a desire to transform this immense region by re-creating the political institutions they knew east of the Mississippi River. On a western tour in the 1860s Albert D. Richardson described this phenomenon: "Making governments and building towns are the natural employments of the migratory Yankee. He takes to them as instinctively as a young duck to water. Congregate a hundred Americans any where beyond the settlements, and they immediately lay out a city, frame a State Constitution, and apply for admission into the Union, while twenty-five of them become candidates for the United States Senate."[9]

Such endeavors demanded the surveying and enforcement of state boundaries. U.S. Representative Thomas Patterson of Colorado gave voice to this desire when he suggested in 1878 that only the "official ascertainment and proper marking of the lines separating these great political divisions of our country" could protect personal rights and local sovereignty.[10] Across the West, transplanted Americans demanded the same rights and responsibilities they had known in their former homes. U.S. Representative James B. Belford of Colorado stated in 1883 that "while the East is manufacturing boots and shoes and baskets the people of the West have widened the pathway of the pioneer into the highway of empire and are manufacturing States and commonwealths."[11] The *New York World* observed similarly in 1877 that "western states and territories occupy vast tracts, empires in themselves, and their rapid settlement will yet see divisions and subdivisions, changes that . . . may blot out all the territories and cut them up into well-peopled states."[12] This divide-and-conquer system supplanted preexisting native societies and imperial claims of other countries and marked an essential aspect of the growth of the United States.

Even as lines intended to define authority, state boundaries inspired occasional challenges to that authority. After the annexation of Texas in 1845, for example, the Sabine River—part of the former border between the Spanish empire and the United States—evolved into a boundary between Texas and Louisiana. Determining the exact division—where each state's authority began and ended—proved contentious. In 1848, both Louisiana and Texas demanded that the federal government specify which state controlled the Sabine River "in order that crimes and offences committed thereupon should be redressed in a speedy and convenient manner."[13] Historian Matthew A. Byron has explored this problem with reference to the nationwide phenomenon of dueling. He suggests that nearly 20 percent of all nineteenth-century duelists crossed state or territorial boundaries to satisfy their honor. By doing so they evaded the laws of their own states and committed crimes in other ones, polities unlikely to seek extradition once they returned home. Land parcels in one state that a changed river's course fused to another state also provided havens for duelists and other criminals.[14] Often in the nineteenth century, state boundaries reflected both the strength and weakness of relatively new legal and political institutions.

As with many issues explored throughout this book, the question of the authority represented by these invisible lines endures as more than a quaint relic of days gone by. Indeed, state boundaries have both facilitated and complicated the enforcement of law and order ever since their creation. One western state in particular inspires concern in the early twenty-first century. When Montanans revised their constitution in 1972, inexplicably they left out their state's boundaries. Almost forty years later, Cathy Hackett, a state Republican Party activist, attempted to rectify what she considered a potential legal loophole. Hackett feared that the lack of physical definition placed "state sovereignty in jeopardy" and declared, "You need the boundaries for Montana law to have jurisdiction." Without the lines identified in Montana's charter, Hackett feared that criminals could evade prosecution on any charge. Unable to muster enough support for an amendment reinstating the lines, however, Hackett withdrew her proposal shortly before the 2010 election.[15] Montana thus remains a limitless state, perhaps an inspiration for its temporary abolition of daytime speed limits in the late 1990s.

Boundaries and borders alike also play important roles in creating identities for the people living between them. They help define a group and how it differs from others, inspiring both a sense of belonging and uniqueness. Some invisible lines on the North American continent make one group Americans, another Canadians, and yet another Mexicans. Historians Jeremy Adelman and Stephen Aron view this consequence skeptically: "If borders appeared juridically to divide North American people, they also inscribed in notions of citizenship new and *exclusivist* meanings. They defined not only external sovereignty but also internal membership in the political communities of North America."[16] But the process of creating identity does not rest solely with international borders. Other lines make some people Arizonans, others Coloradans, and still others Missourians. Identity provided by state boundaries is an odd twist on the notion of state sovereignty cherished in the early nineteenth century and still demanded today, considering that the boundaries exist due to state-making actions of the federal government.

The lines that embrace states and engender identities across the country also define valued local institutions and symbols and give rise to exclusivism. As historian Carl Abbott observes, even modern transnational forces "have not eliminated the desire to define ourselves in terms of smaller communities and distinct places."[17] This process of identity creation takes place in many ways. The shape of one's state provides an immediately recognizable representation of each political community. Perhaps the most familiar western outline belongs to Texas, an admired icon that appears everywhere from belt buckles to ashtrays, from earrings to novelty pens. At the dedication of a new Colorado history center in 2011, officials stated that the building's design drew inspiration from the state's shape, joking that its rectangularity made things "much easier than if Colorado were the shape of, say, Florida."[18] Boundaries divide the member states of the Union and create fifty distinct identities in addition to the national one.

Of all the questions raised during boundary-making debates in the nineteenth, few appeared more often or with more vigor than the question of whether to use geometric or geographic lines, ones corresponding to parallels and meridians or ones fashioned by natural features such as rivers and mountains. The system inherited from British colonial days involved a mix of the two, often anchoring such barriers on geographic features with geometric divides extending from them into the uncharted hinterland. The confederation government established the nation's preference for geometry in the 1780s through three ordinances—explained more fully in chapter 1—that required straight lines to reshape land beyond the Appalachian Mountains. The Land Ordinance of 1785, for example, depending on geometry over geography in defining private property (not political authority), helped to justify straight-line boundaries. Scholars have different views on the Land Ordinance's effect on this debate. One thorough treatise on the subject comes from William D. Pattison, who calls the rectilinear system "a striking example of geometry triumphant over physical geography."[19] Andro Linklater suggests that the ordinance "was to leave its mark on almost every acre of real estate, every farm and every city block west of the Appalachians."[20] Less enthusiastically, the authors of the *Atlas of the New West* lament: "Most attempts to make sensible political boundaries in the West have failed in the face of an implacable grid laid out by the first land surveys."[21] Yet only one state boundary—that dividing the Dakotas, detailed in chapter 8—follows a land survey line. Although vital to the story of American expansionism, private property lines and political boundaries have distinct stories.

The geometry-versus-geography debate percolated during almost every proposal for a new line to be drawn in the western United States. These debates eventually created a political web of forty-eight boundaries common to neighboring states in the trans-Mississippi West. (The number climbs to fifty if one counts the points that touch at Four Corners.) Yet only part of one line corresponds to a chain of mountains—the Continental Divide and Bitterroot Range between Idaho and Montana. Just over a third incorporate rivers in part or in whole. The rest consist entirely of geometric lines. In some cases, inaccurate surveys created deviations from the lines as described by law. Regardless, to many Americans then and now these boundaries seem straight, well-defined, and tidy.

The triumph of geometry over geography, plowing political divides through otherwise cohesive landscapes and populations, was met with scorn numerous times in the nineteenth century. The territorial governor of Nevada appealed unsuccessfully to the California legislature in 1862 to move their shared line to the Sierra Nevada crest. "Nations have ever sought to discard mere straight lines," he declared, "striving for more prominent objects, such as a sea, river, or chain of mountains—great landmarks planted by the hand of Nature to designate the boundaries between States and Empires."[22] The Silver State's legislature pleaded again nine years later that "naturally and geographically defined lines are generally preferable, and should be adopted rather than artificial ones."[23] Both arguments fell on deaf ears in Sacramento. Once established, even a problematic geometric division proved difficult to alter.

The geometric order widely imposed within the country encounters disapproval from most scholars who pay even limited attention to state boundary-making. Writing in the 1880s, British historian James Bryce declared that state lines "are for the most part not natural boundaries fixed by mountain ranges, nor even historical boundaries due to a series of events, but purely artificial boundaries determined by an authority which carved the national territory into strips of convenient size, as a building company lays out its suburban lots."[24] This attitude prevails among recent scholars of the plethora of political rectangles in the West. As geographer Malcolm G. Comeaux observes, "The United States Congress had a predilection for symmetry when drawing Western boundaries."[25] Straight lines appeared much cleaner on the nation's maps than meandering geographic divides. Historical geographer D. W. Meinig argues further that "the long-standing preference for straight-line geometric boundaries ordered this wonderfully variegated land into a simple set of political boxes. And for all the efforts of the cartographers and geographers of the time, it was the names affixed to these huge, more or less rectangular units that would provide the most influential guide for ready assimilation of this Far West into American minds."[26]

I admit to surrendering to the geographic lobby myself. My book on the Colorado State Capitol notes the consequences of two lines each of latitude and longitude: "Geographically, there is no sensible reason for the state of Colorado to exist. . . . [T]he simple rectangle that demarcates Colorado's boundaries affords practically nothing . . . capable of bringing this disparate region into a single political entity."[27]

Scholars should not embrace geography over geometry so readily, however. The natural features promoted by many sources as proper political divisions often prove just as problematic, if not more so. Rivers, for example, make terrible boundaries. A glimpse at a map of the Mississippi River below Cairo, Illinois, with dozens of small parcels sliced away from one state and attached physically if not legally to another, bears this out. By the late nineteenth century the Missouri River's peregrinations led several states to ask Congress and the courts to adjust their divisions dependent upon it accordingly. A portion of Nemaha County, Nebraska, found itself on the Missouri side of the river in the mid-1870s, and Missouri sought to adopt the wayward Cornhuskers into their state.[28] Several years later the governments of Nebraska and Iowa requested authority to trade parcels of land along the river that the meandering Missouri cut off over the course of several decades. A Nebraska senator submitted a resolution to that effect noting "serious disputes concerning the ownership of such lands and . . . the defeat of justice in civil and criminal cases on account of questions of jurisdiction."[29] Only the most immovable channels make sensible boundaries—Hell's Canyon on the Snake River, the Columbia River gorge, and Boulder Canyon on the Colorado River, for instance. The constantly changing nature of rivers often makes their use as political lines a headache.

The logic that rejects rivers as useful boundaries might encourage one to adopt mountain ranges and watersheds instead. Westerners with their perennial concerns about water rights could see the appeal of a political community organized around a drainage area. This attitude found its greatest support in the spring of 1890, during hearings of the House Select Committee on Irrigation. John Wesley Powell, the famed scientist and explorer of the late nineteenth century, presented information to Congress on the challenges facing arid lands in the West. In his testimony, Powell identified geometric boundaries as yet another complication to managing water resources in the region. "The present State lines and present county lines," he observed, "were not laid out with the end in view of securing a homogenous body of people, a people having one common interest in one county or one State government. If this country had been divided into counties and States by river basins, that difficulty would have been avoided." To rectify this unnatural reality, Powell hinted at remaking the West by revising state boundaries to the edges of drainage basins. To that end, he mapped a proposition to transform the region then dominated by geometry into nearly two dozen geographic polities.[30] Powell's map represented one of several emphasizing nature and deemphasizing politics that helped inspire Turner's frontier thesis three years later.

Powell's suggestion, as sensible geographically is it may have been, never caught on in the West. For one thing, altering boundaries requires the consent of state and federal governments alike. By the summer of 1890, eighteen of the present twenty-two political communities west of the Mississippi River had achieved statehood, and one struggles to imagine all of them consenting to his plan by dismantling themselves. Powell's proposal thus remained an academic one, fuel for those who criticized western state boundaries but of little practical value. In particular, the difficulty of marking geographic limits—so important in shaping property rights and legal jurisdiction—undermined his concept. The features that split basins from one another present serious problems as boundaries. As a California state senator grumbled in 1863, in response to Nevada Territory's insistence on the Sierra Nevada crest as their common limit: "Adopt the summit of the Sierras as the dividing line and there would be no time within the next 6,000 years when we could quite agree where our jurisdiction terminated."[31] Imagine a survey team marking a boundary from peak to peak along the crest of a mountain range, battered by extreme weather while straining to determine the exact course of the divide. Even with topographical maps and satellite images, modern surveyors would struggle to identify the exact cleft between river systems in the mountains. Extending this problem onto the broad plains, where vagueness of

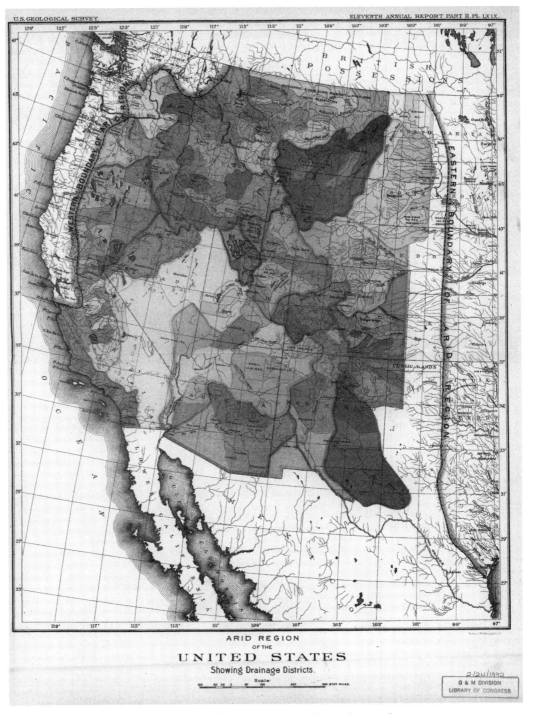

John Wesley Powell's 1890 proposal for states corresponding to drainage basins

such divisions would render definitions of land ownership and law and order nearly impossible, further illustrates the limits of Powell's scheme. Small wonder, then, that surveyors marked the only western boundary that follows either a mountain range or watershed divide—most of the line between Idaho and Montana—only after completing their work along every other western state line.[32]

Some westerners viewed natural features as unifying rather than dividing factors when considering their future states. The utility of rivers as either political boundaries or centers of states, an essential question in the wake of the Louisiana Purchase, appears in chapter 2. As American settlement infiltrated the Rockies, Sierra Nevada, Cascades, and other ranges, the question of their place in western polities emerged. William Gilpin, a soldier and scout who served as Colorado's first territorial governor, advocated mountains at the center of western states rather than at the states' edges. Such an approach would prevent disputes over profitable mining claims. In his address to the first territorial legislature in 1861, Gilpin described Colorado as a "commonwealth of the primeval mountains."[33] A British visitor to the territory later that decade encountered Gilpin on a stagecoach and asked a traveling companion about their fellow passenger. The friend illuminated Gilpin's logic for placing mountains at the center of western states: "He's eloquent about making the Great Sierra unite the two halves of the continent instead of dividing them."[34] New England newspaperman Samuel Bowles echoed this sentiment in 1869, comparing the Keystone State of Pennsylvania with Colorado, "the key-stone in the grand continental formation." Bowles touted Colorado's location, perched astride "the back-bone, the stiffening of the Republic."[35] Embracing mountain ranges and rivers within political communities, and insisting upon geometry over geography overall, reflected a stereotypically American attitude—that humans could master the landscape.

Geometric boundaries emerged most often due to incomplete knowledge of western geography, a reality for which they receive a fair amount of criticism. Yet we should not deride nineteenth-century federal officials or American settlers and politicians in western communities for acting through ignorance. They faced what they considered an immediate need to organize expanding populations in the trans-Mississippi West. This growth demanded the stability of legal and political systems known in older states. In 1884, one U.S. senator criticized the proposal to divide Dakota Territory geometrically rather than following a geographic feature. His colleague Benjamin Harrison of Indiana refuted this: "The Senator is too familiar with the map not to know that artificial lines are the frequent, and even the ordinary, boundaries of States in this country. His reason will show him that it is the best boundary, because it is certain and unfluctuating."[36] Geometric boundaries proved both expedient and effective for carving up the West into nascent polities capable of operating on an equal basis with the older members of the Union.

In 1850 the federal government provided perhaps the most dramatic example of faith in geometry to divide its territory, when it established the American prime meridian. Measured from the U.S. Naval Observatory in Washington, D.C., this patriotic method for calculating longitude reflected the nation's self-confidence inspired by its recent continental expansion. Within the United States this new meridian—located just a few miles to the west of a Greenwich line—served as the focal point for astronomical measurements, especially those made by surveyors and boundary-makers. As a result, more than a dozen trans-Mississippi state lines declared within the next few decades consisted in whole or part of meridians "west of Washington." The longest of these include the boundaries splitting Colorado and New Mexico from Utah and Arizona—thus contributing half of the lines responsible for Four Corners—and the Dakotas and Nebraska from Montana and Wyoming. This competing set of longitudinal lines led to occasional conflicts between states, such as an overlap created by California's state constitution that declared a Greenwich line in contrast to Nevada's organic act passed by Congress with a Washington meridian, a topic discussed in chapter 6. Even though the federal government resumed using Greenwich in 1912, its flirtation with an American longitudinal system reflected the country's unwavering faith in making its own straight lines.[37]

Nineteenth-century westerners themselves preferred horizontal rectangles when shaping their states. Extensive debates over dividing Dakota Territory offered several good examples of this attitude. As Dakota's surveyor general argued in 1878: "To have homogenous peoples and States, climate, topography, and manifest tendencies should be considered; there should be about double the longitude than there is latitude. The division by an east and west line on or near the forty-sixth parallel is in happy harmony with every one of the laws natural, commercial, and social."[38] A local newspaper seconded his opinion, noting the horizontality of most overland travel routes. The press declared that "if there is to be any affirmation of sentiment or affinity of interest among the people of any state or territory, it should embrace as few degrees of latitude and as many of longitude as circumstances of the case will admit."[39] Some eastern politicians accepted arguments like these. Referring to Oregon, an Illinois representative argued in 1876 that "while natural boundaries are desirable, they are not of paramount consideration. Compactness and regularity of form, extent of territorial area, and some regard for latitude and longitude, are the controlling considerations in fixing the boundary of a county or State."[40] Such clearly defined limits seemed logical for a country dedicated to property rights and land ownership, which coincided with and encouraged the contemporary flourishing of cartography. Many Americans involved in this process advocated linear boundaries more often than vague lines following geographic features. After all, political communities are artificial, not natural. It seems appropriate that the lines that shape them should be artificial as well.

Produced half a century ago, the chauvinistic epic *How the West Was Won* commenced: "This land has a name today, and is marked on maps. But the names and the marks and the land all had to be won, won from nature and from primitive man."[41] While this book eschews such arrogant ethnocentrism, it acknowledges the basic argument. *Creating the American West* demonstrates the significance of imposing state boundaries west of the Mississippi River and the consequences for the region and the nation as a whole. The first two chapters focus on the origins of boundary-making in the United States from the colonial and confederation eras through the Louisiana Purchase, the first inspiration for transforming land beyond the country's original borders into functional units of the federal republic. From those foundations, the narrative shifts to a series of case studies, selected from a plethora of possibilities. These appear in roughly chronological order based upon the most contentious era for each one. Early divisions in the Louisiana Purchase provide structure for chapters 3 and 4, including the boundaries between Arkansas and its neighboring native cultures and Missouri and Iowa.[42] Chapters 5 and 6 move toward the Pacific coast during the mid-nineteenth century, with emphasis on the lines splitting the Oregon Country and the states of California and Nevada. The ethnic and political consequences of the boundary between New Mexico and Colorado appear in chapter 7. Chapter 8 narrates the evolution of the line splitting the Dakotas, the last great internal American boundary battle. The conclusion elaborates upon ways in which state boundaries can contribute toward new interpretations of borderlands history.

The conclusion, which views state lines through the lens of borderlands, a traditionally international story, might surprise some scholars as an effort to link supposedly disparate subjects. Yet as the pages that follow bear out repeatedly, internal American boundaries mirror the characteristics of contested international zones across the continent and across the centuries. The lines have affected in countless ways how local and national officials exercised control where jurisdictions met. Residents of places bisected by state boundaries have experienced many of the same opportunities and obstacles known to those who have lived in other borderlands, both as the lines evolved in the nineteenth century and into the present. Their marginal situation allowed them to play territories and states off one another. Perched on the edges of authority, at times these borderline denizens skirted the law, and at other times, when they sought its protection, they railed against its weak enforcement. The process of defining state boundaries complicated property rights across the continent, as polities and residents clashed over their position. Some Americans who felt ignored by their territory or state even considered secession to recast their divided region as its own state. The lives of those whose homes stood at the limits of new western polities are threads in complicated tapestry, and the role state boundary-making could play in redefining borderlands scholarship completes the picture in the book's conclusion.

There would be no United States of America without the individual members of the Union. The lines that shape its parts represent a political framework upon which the structure and function of the country depends. They create regional identities, embrace unique institutions, customs, and histories, and both divide and unite a vast nation. By studying these lines and their myriad broader consequences, historians can overcome an unwarranted neglect dating back more than a century. Of all the divisions within the United States, none stand more important than state boundaries. In the trans-Mississippi region, these imposed lines carved diverse landscapes and populations into polities on the model of older states. They re-created political institutions on formerly foreign soil, and in doing so they supplanted American Indian and European imperial authorities. Most importantly, state boundaries constructed a distinctly American West.

Granite post marking the junction of Massachusetts, Rhode Island, and Connecticut

CHAPTER 1

A Little Patch of Ground
The Precedents for Western State Boundaries

Once upon a time, while on a diplomatic errand for his usurping uncle, the prince of Denmark happened upon the army of Norway. This angst-riddled young man, Hamlet, asked an officer of the infantry their destination. The Norwegian captain responded that they were bound for Poland "to gain a little patch of ground that hath in it no profit but the name." The officer considered the land unsuitable for farming and doubted that it possessed any valuable natural resources, yet the desire to expand Norway's influence trumped the tract's limitations.[1] This interaction from one of the world's most famous works of literature hints at a common theme in history—the desire of cultures and countries to take territory from others and make it their own. Two centuries after William Shakespeare penned the tale of the melancholy Dane, the United States emulated his Norwegians by expanding their land claims through purchase, negotiation, and conquest. Throughout western regions Americans imposed boundaries to create new political communities, restructuring territory acquired from foreign powers in their own way. This process built upon British imperial organization and was codified through some of the most influential political acts in early U.S. history. Each new "little patch of ground" annexed and assimilated by Americans reflected their expansive national vision as well as their peculiar methods of territorial organization.

From Americans looking beyond the Appalachian Mountains to Norwegians marching through northern Europe and beyond, the process of taking and controlling land echoes throughout the pages of world history. The Assyrians, for example, introduced a system of provincial authority to facilitate governing an empire that stretched from Egypt to Persia in the eighth century BC.[2] Countless examples of internal organization followed the Assyrians' model over the centuries. From organizers of Roman provinces to Islamic emirates to Anglo-Saxon shires, diverse groups recognized the value of controlling large tracts by dividing them into smaller, more efficient units. This phenomenon flourished during the early modern era with the rise of the nation-state. Europeans carried the notion with them through exploration and conquest, particularly during their several centuries of colonizing efforts in North America. Policy makers conceived of the continent as a blank slate upon which to impose communities like those in the homeland, hence "New England," "New France," "New Spain," "New Netherlands," "New Sweden," and so on. As historian Alan Taylor argues, "the remaking of the Americas was a team effort . . . led and partially managed (but never fully controlled) by European people."[3] Indeed, native cultures influenced in myriad ways the political destiny of North America, even if European notions of political authority dismissed the preexisting yet effective systems used by many American Indian cultures, as would the United States in years to come.[4]

By the early seventeenth century three major imperial players orchestrated Europe's efforts to dominate North America, and each approached the land that constitutes present-day United States differently. The Spanish, for example, viewed their northern frontier at the time as little more than a buffer for their pillaged holdings in Mexico. Meanwhile, the French had established an economic presence in the St. Lawrence River valley, but few from their homeland wanted to settle in it. In both the Spanish and French claims, American Indians remained the dominant players for many years. Wedged between the vast Spanish and French empires in North America, the English proved more interested in what is now the United States than the Spanish and more effective at peopling it than the French, establishing colonies along the Atlantic coast in the seventeenth and eighteenth centuries, often with little detailed knowledge of local geography. Thus their boundaries depended less on the landscape than on a general desire to define far-flung communities of English colonists. After the American Revolution, as the colonies became a union of states, Anglo Americans sought to re-create their forms of government beyond the Appalachians. As a result, English precedents have resonated across the United States for well over two hundred years.

The first English attempt to set up a colony in North America reflected well the desire to establish a presence without much understanding of geographic reality. It also demonstrated England's desire to amass as much land as possible to compete with its imperial challengers, especially the powerful Spanish who dominated the western hemisphere throughout the sixteenth century. In 1584, as tensions with Spain escalated rapidly, Queen Elizabeth I made England's first public declaration for a North American colony. Her instructions to Walter Raleigh specified no bounds for it, however. Instead she charged him to "discouer, search, finde out, and view such remote, heathen and barbarous lands, countries, and territories, not actually possessed of any Christian Prince, nor inhabited by Christian people" on the other side of the Atlantic Ocean. Once he accomplished that sweeping task, Raleigh should set about laying the foundations for an English colony.[5] Following these vague guidelines, Raleigh helped orchestrate the expedition that located a settlement on Roanoke Island in present-day North Carolina in 1585. Yet the Roanoke colony failed in less than five years, the victim of competition with Spain and unfamiliarity with the demands of colonization.[6] After the mysterious disappearance of Roanoke's colonists, England ceased its North American efforts for almost twenty years.

At the dawn of the seventeenth century, the English government reassessed its approach to colonization. By 1606 it decided to try again, with slightly more specific guidelines, including physical limits for the two colonies the English hoped to establish. The 1606 charter approved by the new king, James I, called for one settlement to be located between the thirty-forth and forty-first parallels and another between the thirty-eighth and forty-fifth parallels. It identified the overlap between the two colony zones as a neutral ground to limit competition between the transplanted English.[7] The document paved the way for Jamestown and Plymouth, England's first successful forays at a permanent American presence. Yet both actually commenced under revised charters issued in 1609, when both companies organizing colonization efforts received "Sea to Sea" charters, reflecting England's grandiose vision as well as ignorance about the amount of land it purported to bestow. While the Pilgrims who came to Plymouth made little effort to enforce the expansive claim, their southern neighbors proved a different story. James bestowed upon the managers of the Virginia Company—the London-based firm organizing the Jamestown effort—an enormous tract of land over which he had no real control. The king's 1609 clarification included the present-day United States north of a line extending from Wilmington, North Carolina, to Santa Barbara, California, and below a northwesterly running line that included almost

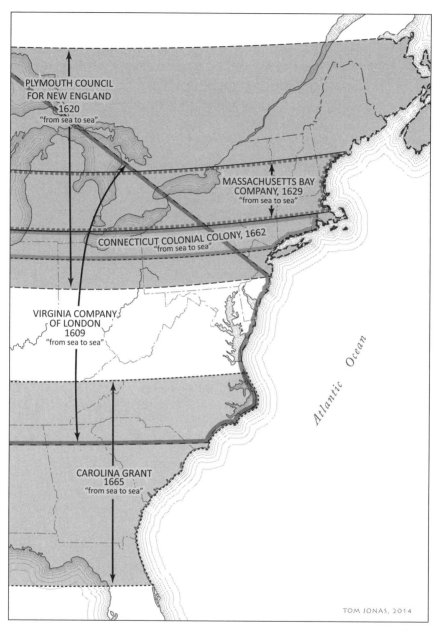

English "Sea to Sea" colonial grants in the seventeenth and eighteenth centuries

all land west of the Great Lakes, most of western Canada, and all of Alaska, a far greater largesse than he decreed in 1606.[8] Although no one expected Virginia to dominate all that land immediately, nor even understood just how much territory the 1609 charter covered, the document illustrated England's designs on North America.

For centuries afterward, these "Sea to Sea" grants led to bizarre disputes as subsequent states tried to enforce their extraordinary claims, even long after their governments repudiated those parts of the charters. Of particular concern during and after the American Revolution was the fact that these documents produced overlapping jurisdictions west of the Appalachians, over which several states at once tried to enforce their authority. The first U.S. government settled most of these contested claims in the 1780s, a subject discussed later in this chapter. But for some states their expansive origins refused to fade into obscurity. Virginia authorities, for example, used their old charter to declare their ownership of the northern Louisiana Purchase in 1819. Although the state had ceded its trans-Appalachian lands to the national government decades earlier, Virginia denied that it had ever abandoned its claims beyond the Mississippi River. The Louisiana Purchase thus resurrected Virginia's 1609 charter, as if the federal government had rescued a portion of the state from a French exile. This argument found little support, however, and Virginia abandoned its efforts to enforce the claim.[9] In addition, in 1936 a former Virginia governor hearkened back to the 1606 charter by noting that the Pilgrims—whose American origins also dated back to that document—had landed in 1620 in what they considered "northern Virginia." He proclaimed with an odd interpretation of geography: "This means that not only was Plymouth Rock and Boston in Virginia, but every state in the present union except North Dakota and Washington on the north and Louisiana on the south!"[10] Texans may have a modern reputation for thinking big, but Virginians have been doing so for more than two centuries longer.

As the English government organized new colonies in North America throughout the seventeenth century, it introduced a slew of boundaries. Most blended geographic and geometric elements starting at well-known features, including capes and rivers, and extended into the backcountry via straight lines. The 1609 bounds for Virginia included points of latitude and longitude as well as geographic features on the Atlantic coast. The boundaries between later colonies developed similarly. The 1629 charter of Massachusetts Bay defined a bewildering set of "Lymitts" stretching several

"English Myles" from a series of rivers near the coast.[11] Maryland, established in 1632, boasted a southern boundary that followed the Potomac River to its source, and from there turned due north to the fortieth parallel, then returned back toward the coast, carving its panhandle in the Appalachian Mountains.[12] This aberration offers an example of the unexpected ways in which boundary-making with little knowledge of geography could shape the lives of generations to come. Yet England's system offered a practical solution to the challenge of political organization, a seemingly immediate need as its North American presence grew ever larger.

Each new colony created by the English government affected the shape of the older ones by carving them into smaller and more manageable units. In 1664, the Duke of York allowed two noblemen to break off part of the land claimed by New York to found New Jersey. They bounded the new colony between the Delaware and Hudson rivers, contained to the north by a geometric division between two points on those flows.[13] A year later, the revised charter of Carolina extended from sea to sea between the latitudes of twenty-nine degrees and thirty-six degrees thirty minutes.[14] The latter line replaced the southern boundary of Virginia as defined by the 1609 charter. In the wake of the American Revolution, the process of modifying limits between former colonies devolved to the states themselves. Thomas Jefferson observed in his *Notes on the State of Virginia*, written in the 1780s, that well-informed conventions between Virginia and its neighbors honed what he considered vague and unsatisfactory lines imposed upon them by their colonial masters.[15] The obsession with defining and clarifying boundaries, whether for English colonies or American states, reflected the vital role land ownership played under both authorities. From farms to states, citizens sought to know just where they stood.

The creation of one late-seventeenth-century English colony led to one of the most famous lines ever imposed on North America. Pennsylvania's 1681 charter overlapped the claims of both New York and Maryland, resulting in decades of territorial disputes between the three polities. The tension with Maryland in particular lasted for eighty years and complicated property rights in the contested zone. During the 1760s the colonies settled the matter with an impartial survey of their divide conducted by two respected British scholars, Charles Mason and Jeremiah Dixon. Their expedition set out in 1763, as the Great War for Empire ended and Pontiac's Rebellion began. Five years of intermittent mapping followed, with the effort slowed by the threat of American Indian resistance. By 1768

Mason and Dixon completed the first scientific survey of a major North American boundary, the line separating Maryland from Pennsylvania and Delaware (then considered part of Pennsylvania). The surveyors marked the boundary with impressive crown stones carved with the family crests of Maryland's Calverts and Pennsylvania's Penns. Boasting a professional study beyond any other such boundary line in North America, this line took on a life of its own. Over the years it morphed into a symbolic divide between the northern and southern colonies and eventually states. Few internal boundaries in American history can tout as resonant a legacy as the Mason-Dixon line.[16]

As Mason and Dixon started their work in 1763, the British government introduced a new kind of boundary to North America.[17] After the Great War for Empire—also known as the French and Indian War or Seven Years' War—the French surrendered their imperial tracts in Canada and the Mississippi River valley and the Spanish lost Florida. The British now claimed land stretching from the Atlantic Ocean west to the Mississippi and from the Gulf of Mexico north to the frozen tundra. To maintain the delicate peace between backcountry settlements and native cultures living beyond the Appalachians, the government of King George III created the Royal Proclamation Line. It decreed an invisible barrier along the Appalachian crest reinforced by military patrols to keep colonists and native societies separate and at peace. A few clandestine settlements in present-day western Pennsylvania, Kentucky, and Tennessee challenged this royal decree, and British officers and native leaders fumed at the presence of some fifty thousand trespassers by 1774. By illegalizing land speculation and preventing these squatters from holding legal title to their land, however, it retarded even greater numbers from crossing the mountains until the outbreak of the American Revolution shifted British priorities.[18] The 1763 divide added another element to the North American notions of boundaries, helping keep the peace between troublesome groups at the edge of established authority. Later generations of Americans adopted this principle when organizing their own distant West.

An important shift in self-conception took place in British North America during and after the Great War for Empire. As historian James D. Drake describes, colonists along the Atlantic coast viewed themselves increasingly as residents of a distinct, self-sufficient continent ("continent" was then a relatively new term in geography), an attitude that conflicted with their second-class status below those living in Great Britain. Colonists' attachment to their continent and their colony was at loggerheads with various postwar proposals by British intellectuals to merge the patchwork of polities into a more logical structure. The increasing use of the term "continental" in the mid-eighteenth century reflected this American sensibility, as in the names of the Continental Congress, which organized in 1774 to give aggrieved colonists a voice, and the Continental Army it established to protect their interests. As they pondered a break with the British, some Americans looked forward to a time when a confederation of like-minded polities including their coastal colonies might cover North America. The landmass on which they lived provided an argument for independence and was essential in Americans' visions of their future. In Drake's words, "the notion of the continent as an ally made a difference, it helped give rise to the United States."[19] Represented by a plural noun, "United States," the new country would depend both on its unity and its component parts.

By the revolutionary decade of the 1770s a diverse system based on geography and geometry separated the Atlantic colonies of British North America from each other and from the trans-Appalachian West. Attempting to harness the energy of colonial frustration, in the summer of 1776 the Second Continental Congress authorized a committee to propose some sort of provisional government while another crafted the Declaration of Independence. In August the first committee presented the fruits of their labor, the Articles of Confederation. A little over a year later, in November 1777, Congress accepted the articles, which focused upon wartime matters and foreign policy while leaving most authority to the states. Recognizing the potential for competition between the member states of the new country, the articles also tackled thorny problems left over from the days of British rule. In Article IX the document provided for settling the overlapping land claims of various colonial charters over the years. Congress declared itself "the last resort on appeal in all disputes and differences now subsisting or [that] may hereafter arise between two or more States concerning boundary, jurisdiction or any other causes whatever."[20] This clause proved one of the most challenging for the confederation government to enforce. With so much territory, influence, and prestige at stake, many states with claims beyond the Appalachians resisted surrendering anything to their neighbors or the new national administration.

The land question proved so contentious that it delayed the confederation government from taking full effect for four years. Although the

Overlapping claims to the trans-Appalachian West in the early 1780s

Continental Congress accepted the plan in 1777, three holdout states—Maryland, Delaware, and New Jersey—refused to ratify the articles until states with large trans-Appalachian claims handed over those tracts to Congress. These three polities lacked similar charters and feared perpetual domination by the larger states. Two in particular earned the most enmity—Virginia, with its expansive "Sea to Sea" grant, and New York, which claimed everything between the Tennessee River and the Great Lakes. Leaders of both states considered such lands as potential rewards for soldiers in the revolutionary armies, thinking they "ought to be the common interest of the United States, they being defended at the common expense."[21] Regardless, Virginia, New York, and other states proved resistant to surrendering their far-flung tracts.

Wedged between the powerhouses of New York and Pennsylvania, New Jersey leaders spoke loudly in favor of sharing the trans-Appalachian region among all the states. In the summer of 1778 New Jersey's legislature memorialized Congress to hold the area "in trust for the use and benefit of the United States" without regard to colonial grants. The New Jersey assembly also feared that states without western claims might find themselves overburdened by debt and their economies in shambles without land to sell after the war. New Jersey's government pointed out another challenge facing the confederation's members—the many poorly marked state boundaries close to the Atlantic coast. It noted the "salutary effects" of better-defined political lines, including "preventing jealousies as well as controversies, and promoting harmony and confidence among the states."[22] The professionally surveyed Mason-Dixon line represented the exception rather than the rule for such boundaries. New Jersey argued that state governments and landowners far and wide would likely view more formal studies of their limits favorably. Yet Congress dismissed New Jersey's pleas, finding little interest in the proposals of a state that refused to join the confederation.

Consequences of vague state divisions dating from this revolutionary era resonate into the twenty-first century. The line between the Carolinas, for example, fuels at least one perennial debate. South Carolina claims Andrew Jackson as its native son, and a monument quoting his will, affirming that belief, stands in a state park at his birthplace. At the time of Jackson's birth in 1767, however, incomplete and inaccurate surveys left the site's location in dispute. Thus North Carolina considers Jackson its own as well and includes him in a monument to the state's native-born presidents on the capitol grounds in Raleigh.[23] High school football teams in neighboring Union County, North Carolina, and Lancaster County, South Carolina, turned the competing Jackson claims into a gridiron contest in 1979. As a North Carolina state archivist quipped: "I'm not aware of any other time in American history when the outcome of a football game has affected the accuracy of a historical event."[24] The northern team won the "Old Hickory Football Classic" by a score of 36-6 in front of four thousand spectators, earning Union County the right to display a small bust of

Jackson in its courthouse. Although the crowd focused upon the fun of the game more than the birthplace dispute, officials on both sides appreciated the levity in what has often been a cantankerous argument. A South Carolina state historian considered the matter "too good a controversy to bury. I can't imagine either state conceding to the other."[25] Settling anything relating to political boundaries remains a challenge, as many states have demonstrated over the decades. From the trivia of a presidential birthplace to questions of law enforcement and property rights, the significance of state boundaries echoes.

Disputes that seem quaint or amusing to modern audiences proved bitterly contentious in the late eighteenth century. To the chagrin of the Continental Congress, the three recalcitrant states prevented the confederation from exercising full national authority as the war with Great Britain raged on. Such division undermined the image of American unity the new government sought to foster. After a little more than a year, however, two of the intransigent states joined the confederation: in the winter of 1778–79, New Jersey and Delaware judged weak national voices better than none at all and accepted the articles. Maryland, without any claim beyond its awkward panhandle in the Appalachians, remained the sole holdout. It refused to ratify the articles until resolution of the territorial issue, with particular resentment aimed at the two largest states. The states with the largest claims then made their own compromises. Bowing to popular sentiment New York ceded its western tracts to the national authority in 1780. Support built in Virginia for a similar gesture, and a year later it declared itself ready to surrender its claim to territory north of the Ohio River. Satisfied at long last that the most valuable western land belonged to the country as a whole, Maryland's legislature acceded to the compact. On March 1, 1781, Congress declared the formal establishment of the confederation government.[26] Within eight months a combined French and American force defeated the British at Yorktown, not far from the ruins of Jamestown, and fighting in the American Revolution came to an end as well.

With independence secured on the battlefield, diplomats argued over the logistics necessary to complete the political divorce. The Continental Congress had produced its own vision of American borders in 1779: the Mississippi River to the west and Florida to the south, and extending far north into present-day Ontario. These sweeping demands met with surprise among the British, French, and Spanish alike. For a time the British seemed amenable to the American vision, but victories against the Spanish and French in the early 1780s stiffened their backbone against their rebellious former colonies.²⁷ British and American representatives from both sides worked out a provisional treaty in Paris in 1782, and they signed a formal peace on September 3, 1783. According to the document a series of geographic and geometric borders contained the confederation generally between the Atlantic Ocean and the Mississippi River to the east and west and British Canada and Spanish Florida to the north and south.²⁸ While some Americans rejoiced at their grandiose new size, others recognized myriad problems inherent in controlling that much territory. The British rewarded their independent kin with a Pandora's box sure to keep them busy for decades to come.

After to the Treaty of Paris, American land claims stretched much farther than their actual influence. Powerful native cultures, suspicions of U.S. intentions in their homelands, represented one major challenge. Another came from the states themselves, as they struggled to organize the trans-Appalachian West in a system harmonious with the states along the Atlantic coast. Ten days after the delegates in Paris accepted the treaty and long before it arrived in Philadelphia, the western land issue percolated in the Continental Congress yet again. Maryland delegate James McHenry—the namesake of Baltimore's famous fort—asked his colleagues for a committee to study transforming the region into "one or more convenient and independent states." Such new polities "will tend to increase the general happiness of mankind, by rendering the purchase of land easy, and the possession of liberty permanent," McHenry suggested. The New Hampshire and New Jersey delegations supported the Marylander's proposal but all the other representations rejected it. Then the largest state offered a proposal of its own. Virginia delegates encouraged Congress to organize the West into states approximately ten thousand to twenty-three thousand miles square, corresponding roughly to present-day Vermont and West Virginia, respectively.²⁹ This idea found little favor as well, although within a few months Congress gave a similar suggestion a more positive hearing. For the time being, though, the West remained unorganized.

Details regarding the largest western land cessions took several years to settle. In the early months of 1784, Congress worked out the transfers from New York and Virginia of "waste and unappropriated Lands in the Western Country . . . for the common Benefit of the Union," with Virginia issuing the final deed for land north of the Ohio River on March 1.³⁰ Yet significant questions remained, including how to organize land held in trust to the entire country, how to disperse it to settlers, and how to establish law and order in an area difficult to reach from the Atlantic coast. On the same day that Virginia made formal its land cession, a congressional committee of three delegates—David Howell of Rhode Island, Thomas Jefferson of Virginia, and Jeremiah Townley of Maryland—released its suggestions for transforming the trans-Appalachian West into states.[31] This committee produced the first expression of a distinctly American way of organizing vast tracts into future states of the Union.

Of the three delegates, Jefferson's name resonates loudest through the centuries. He played a major role in shaping the polities west of the Appalachians for most of his career. Enough of a pragmatist to realize that Richmond, Philadelphia, and other capitals could not exercise effective control beyond the mountains, he hinted at a series of new western states in a proposed constitution for Virginia in 1776. Jefferson's draft insisted that any new polities beyond the Appalachians, organized in land his state still claimed at that time, "shall be free and independant [sic] of this colony and of all the world."[32] During the Revolutionary War he concocted schemes for creating new states in the Ohio River valley, including one that called for six polities shaped by convoluted geometric boundaries linking rivers with other natural features.[33] As Virginia's governor in 1781 he approved the land cession, which he helped formalize three years later when he sat in the Continental Congress once again. There Jefferson replaced his wartime notions of a patchwork of potential states with a more orderly proposal that historian Julian P. Boyd calls "the foundation stone of American territorial policy."[34] The committee called for a geometric division of the territory between the Appalachians and the Mississippi River. A meridian line through the falls of the Ohio River at present-day Louisville, Kentucky, would split the region in two. Boundaries spaced every two degrees of latitude carved it into more than a dozen new political communities between the Great Lakes and Spanish Florida.[35] The geometry of the committee's plan reflected a desire to organize the land as soon as possible. It also indicated the limited American knowledge of the region's geography, a trait shared by those who created the British colonies and one that would often emerge as the United States shaped the trans-Mississippi West in the next century.

The tidy, straight boundaries the committee proposed mirrored the practicality and orderliness so praised by Enlightenment philosophers of the eighteenth century, of whom Jefferson ranked as one of the most

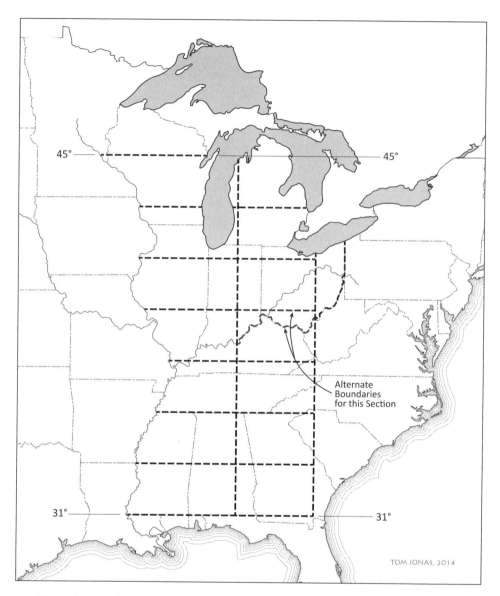

The Ordinance of 1784

prominent in the United States. In another nod to intellectualism, the new polities shaped by these efficient lines received classical-sounding names ranging from "Cherronesus" and "Assenisipia" to "Metropotamia" and "Pelisipia." Other names proved prophetic—"Michigania" and "Illinoia," for example—while two evoked the recent history of the new country:

"Saratoga" and "Washington."[36] Such plans assumed that Connecticut and Massachusetts would eventually surrender their claims north of the Ohio River, a process underway at the time. With Virginia, North Carolina, and Georgia retaining claims over the lands to their west for the moment, the committee did not invent names for the southern tiers of states. According to the 1784 plan the national government would manage each parcel as a territory until it contained twenty thousand free American settlers, after which time it would adopt a constitution and join the confederation on an equal footing with the thirteen original members. The committee's plan banned slavery from the entire trans-Appalachian West and guaranteed the permanency of each new state's boundaries unless altered by both Congress and the state affected.[37] The opinions of American Indian societies who actually controlled this region at the time received no consideration.

Congress held its final debate on the committee's proposals on April 23, 1784. During the discussion the body rejected two amendments giving the confederation government nearly dictatorial powers over each territory before its statehood.[38] It supported the practical and logical structure of parallels and meridians to create manageable states west of the Appalachian Mountains. Congress also approved of the provision granting it control over distribution of western land without interference from the territories or even later sovereign states. The body made a few alterations to the committee's proposal, such as making the Ohio River above its fall line at present-day Louisville a future state boundary in place of the nearby thirty-ninth parallel. It demanded that the new states establish republican forms of government and remain forever loyal to the confederation, but it rescinded the anti-slavery provision. Every state delegation except South Carolina's voted in favor of its adoption. This Ordinance of 1784 made Congress the ultimate authority over matters outside the boundaries of the original states. Excepting its power to distribute land, the document provided for a rather liberal system of local control and equality after statehood. This reflected the overriding faith in state authority that served as a hallmark of the confederation period. Congressional insistence on retaining control of the land that made up these states, however, bespoke the later rise of federal dominance.[39]

With a method in hand to create future states, Congress pondered how to encourage settlement west of the Appalachians. Even more than in state boundary-making, the distribution of western land depended upon the triumph of geometry over geography. A reflection of the practicality praised by Enlightenment philosophy, the former drew upon the utility of straight political lines, a practice that Americans inherited from the English. The latter, however, transformed that basic solution into a requirement. In an effort to stop squatters and speculators from overrunning the region, Congress approved a policy for distributing land legally to settlers in 1785. Jefferson's committee had the previous year suggested a rectilinear system for surveying and selling western land using "hundreds" of ten miles square. Congress amended the notion into the Land Ordinance of 1785, transforming the "hundreds" into units six miles square, called townships. The ordinance designated the thirty-six square miles therein as sections, which could be further divided into half-sections, quarter-sections, and so on. Congress instructed surveyors to plat the landscape thoroughly before its sale.[40] The knowledge surveyors thus accrued of water and salt springs, mineral resources, and agricultural potential provided invaluable assistance to Americans crossing the Appalachians in search of a new home.

The tracts carved up by the Land Ordinance commenced where the western boundary of Pennsylvania crossed the Ohio River. From there, over the course of the next century, surveys extended these squares across the continent, excepting only Texas, Kentucky, Tennessee, and parts of Ohio.[41] In contrast with privately owned parcels of every conceivable shape on the Atlantic coast, another English legacy, the land system that transformed much of North America in the years to come demonstrated an abiding faith in geometry. The Land Ordinance staked a distinctly American claim to the landscape, one far more regimented than any other colonial system on the continent or perhaps in the world. Only the occasional surveying error or accommodation for the curvature of the earth deviates the grid system from its intended orderliness. Whether or not they realize it, most Americans living west of the Appalachian Mountains today interact with the effects of the Land Ordinance on perhaps a daily basis. Land ownership, roads and highways, city and county governments—all these and more from the Ohio River to the Pacific Ocean bear the trademark squares often recognizable from an airplane high above the country. In the words of western suburbanite D. J. Waldie: "Before they put a grid over it, and restrained the ground from indifference, any place was as good as any other."[42] The Land Ordinance enforced geometry's primacy over geography in Americans' conceptions of controlling the land they claimed. Similarly, most boundaries drawn to encompass states over the next century depended upon straight lines far more than natural features. Geometry thus started as an expedient for the English but morphed into a necessity for the Americans.

By the mid-1780s, Congress had thus established a system of land ownership and state organization to encourage growth and stability in the country's western two-thirds. The body still struggled to convince Massachusetts, Connecticut, North Carolina, and Georgia to repudiate their "Sea to Sea" charters, awarded between 1629 and 1732 with a somewhat more limited scope than those granted to Virginia and Plymouth in 1609. These negotiations continued into the early nineteenth century.[43] With the larger and more controversial cessions of Virginia and New York out of the way, however, the door seemed open to the heart of the trans-Appalachian region, specifically the alluring Ohio River valley. Yet not everyone in Congress approved of the two ordinances intended to distribute land and organize states in the West. Dissatisfied members of Congress lobbied to amend both ordinances almost immediately. Virginia delegate James Monroe sent reports of these critiques to Jefferson, by then serving the confederation as minister to France. Writing in early 1786, Monroe expressed doubts about the viability of Jefferson's 1784 ordinance. He feared that much of the land north of the Ohio River would not suit American farmers, especially near the Great Lakes. In addition, Monroe declared that much of what is today the state of Illinois "consists of extensive plains [which] have not had from appearance and will not have a single bush on them, for ages." Such land could never support adequate numbers of American settlers and this might prevent many of the proposed states from joining the Confederation. Monroe hinted that Congress might of necessity abandon his state-making ideas in favor of larger polities more likely to embrace a population sufficient to warrant their membership in the national government.[44]

Discussions over revising the proposed trans-Appalachian state boundaries percolated in Congress through the spring and summer of 1786. By mid-May Monroe related to Jefferson a suggested new approach to political organization north of the Ohio. A congressional report argued in favor of somewhere between two and five states carved from the region, and it hinted at a more forceful pre-statehood authority in the nascent polities. Each would receive officials including a governor, executive council, and judges to guide it toward joining the confederation. Monroe compared this less democratic approach to organizing new states with British colonial policies that incited the revolution. Regardless, this suggestion heralded the political adolescence known as the "territory" through which most future states would pass. The proposed new ordinance also raised the population requirement for statehood from twenty-five thousand to a level equal with the smallest state at the time, Delaware with thirty-seven thousand inhabitants.[45] Within two months, Congress shifted away from amending the 1784 ordinance to replacing it entirely. Monroe noted a particular interest in raising the population threshold for states even higher, up to the equivalent of one-thirteenth of all free Americans. Such a level would surpass 180,000 people easily, at least nine times greater than the 1784 requirement, and a threshold held at the time by less than half of the thirteen states. It would likely put statehood anywhere west of the Appalachians out of reach for decades to come. Such a policy reflected the jealousy with which the original states clung to their influence. Yet Monroe feared that it might force western Americans to rebel against the confederation and perhaps join forces with the British, who retained military outposts and alliances with native cultures in the region even after recognizing independence.[46]

The criticism of Jefferson's 1784 ordinance bore fruit in 1787, when the confederation government revamped the way it viewed the political future of the trans-Appalachian region. The Northwest Ordinance of 1787 succeeded the Ordinance of 1784 as the formula for state-making in the West. More than twice as long as its predecessor, it emerged out of the less democratic vision of admitting new states that Monroe chronicled throughout 1786. The Northwest Ordinance called for provisional governments to shepherd the region north of the Ohio River through the transition to statehood. It provided for a local legislature, a governor and judges appointed from the national capital, legal and property rights, and religious freedom. The ordinance also included territorial boundaries of geography (rivers) and geometry (lines drawn from river confluences and lakes), starting with three polities that correspond to present-day Illinois, Indiana, and Ohio. Congress divided up these territories and envisioned perhaps two more carved from the region's northernmost reaches in years to come. In this sense, Congress emphasized the adaptability of boundaries to changing circumstances in 1787 in contrast to its 1784 structure of prefabricated states to admit as their population allowed. In addition the Northwest Ordinance recognized the need to treat with American Indians to achieve full legal control over the region. The law pegged the number of citizens required for a territory to apply for statehood at sixty thousand, greater than the free population of Delaware, Rhode Island, or Georgia at the time. It also made Congress the arbiter in permitting new states to join rather than accepting any polity that met the requirements, as in the 1784 regulations. The final article of the 1787 document prohibited slavery in this new Northwest Territory, resurrecting an abandoned concept from 1784. Supporters of the 1787 ordinance thus sought to ensure the perceived status of the Northwest as a haven for free, industrious Americans, while constraining western populations in an effort to ensure the prolonged dominance of the Atlantic coast.[47]

The Northwest Ordinance affected American history across time and place. As the most comprehensive law ever approved regarding the admission of new states it served as a starting point for later generations of politicians to debate the fate of other western lands. Following the Louisiana Purchase of 1803 some members of Congress referenced it as a useful model for organizing the Territory of Louisiana north of thirty-three degrees, the present Louisiana-Arkansas boundary.[48] The debates over the 1820 Missouri Compromise also cited the Northwest Ordinance with surprising consistency. Most references to it pointed not to the legal details of state-making but instead to its prohibition of slavery in the region. Politicians from free states used the 1787 law as an argument for banning slavery north of some invisible divide within the Louisiana Purchase. Senator James Burrill, Jr., of Rhode Island, for example, suggested that since Indiana Territory had governed the land north of the thirty-third parallel for about a year before the creation of Missouri Territory in 1805, and the Northwest Ordinance prohibited slavery in Indiana, the same rule applied to all land north of the state of Louisiana and west of the Mississippi River.[49] Such interpretations stretched the imagination and led to intense arguments in Congress.

In contrast to their "free soil" counterparts, slave state representatives in the early nineteenth century argued that the "celebrated ordinance"—in the phraseology of Senator Freeman Walker of Georgia—did not apply to the Louisiana Purchase.[50] Senator Nicholas van Dyke of Delaware echoed this sentiment. Van Dyke dismissed the idea of forcing Northwest Ordinance principles, especially the slavery barrier, on the land west of the Mississippi: "Different, in all respects, is the case of Missouri: a part of a territory acquired by treaty from a foreign Power—never subject to the ordinance of 1787—involuntary servitude existed there at the time of cession, and still exists—the people object to this restriction—insist upon their rights under the treaty, and deny your power to impose such a condition. Under circumstances so entirely dissimilar, the Northwestern States furnish not even the frail authority of precedent to bind Missouri."[51] The right of Congress to impose limitations on slavery served as a central point of contention during the entire Missouri debate. But as one northern politician noted: "The very assignment of boundaries is in the nature of a condition" placed on American territory.[52] Just because a political division might be invisible to the naked eye did not make it inconsequential. Quite the reverse: boundary-making represents one of the federal government's most lasting and resonant legacies.

By the 1850s the importance of the Northwest Ordinance waned in favor of more recent territorial precedents, especially the Missouri Compromise of 1820. The U.S. acquisition of northern Mexico in 1848 through the Treaty of Guadalupe Hidalgo raised the question of extending that compromise line, a geometric division across the entire trans-Mississippi West between slave and free regions. For many Americans the Missouri bargain supplanted the 1787 law as the preeminent precedent for boundary-making and political organization, especially in the trans-Mississippi West. Nonetheless, the ordinance did not fade away completely. Its precedents for territorial tutelage and a certain population required in order to qualify for statehood remained in effect for many years to come. An 1877 memorial to Congress from the legislature of Dakota Territory, for example, referred to "the benign provisions of that historic enactment" that continued to influence territorial policy ninety years after the Northwest Ordinance was enacted.[53] Its insurance of equal rights and responsibilities for states regardless of their age, incorporated from the Ordinance of 1784, earned the Northwest Ordinance a place among the most important and effective laws ever approved for the United States. In the words of historian Catherine Drinker Bowen, it ranks as "the third great document of American history, after the Declaration of Independence and the Constitution."[54] The Northwest Ordinance also illustrated a shift in American sovereignty away from the people, whose voice resonated the loudest under Jefferson's 1784 plan, and toward the national government, which ever after would dominate territories and choose subjectively which ones it thought worthy to join the Union.

Contemporary events would reinforce the country's gradual transition from state to federal control. In the summer of 1787, while the confederation's legislature met in New York City and adopted measures like the

Northwest Ordinance, another body assembled at the Pennsylvania statehouse in Philadelphia. Meeting in the same room where eleven years earlier the Second Continental Congress had declared independence from Great Britain, the group convened with the blessings of the confederation to propose amendments to its articles. Sweeping legislation like the three western ordinances aside, the confederation struggled to coordinate thirteen states with their own jealous interests. In the end this secretive meeting in Philadelphia proposed an entirely new national authority, a more powerful federal government outlined in the U.S. Constitution. Along with many other concerns, the delegates to the Constitutional Convention pondered how to reinforce the confederation measures intended to strengthen control over the trans-Appalachian West. For this clandestine assemblage, promoting prosperity nationwide involved not only a new political system but encouraging its spread via future states in the western domain as well.

Throughout the Constitutional Convention, delegates spoke of creating and admitting new states. On May 29, 1787, four days after the body commenced its work, Edmund J. Randolph of Virginia outlined various issues facing the fledgling country and pointed to the method of welcoming new states as one of many questions before the convention.[55] Over the following weeks this issue cropped up several times, with some delegates proposing not only a formula for admitting new polities but also a complete redrawing of preexisting state boundaries inherited from the British. Such proposals drew upon the attitudes of influential figures like Thomas Paine, whose writings during the revolution tried to downplay colonial or state loyalty and foster a more united affection in American hearts.[56] In the 1787 convention, George Read of Delaware suggested abolishing the lines altogether to make one cohesive country. By contrast William Paterson of New Jersey proposed a redrawing of the boundaries to make thirteen political communities of roughly equal population in order to maintain state equality.[57] Delegate and convention scribe James Madison considered "a general amalgamation and repartition of the States to be practicable" but preferred a union of preexisting members to the complexity of creating a national government and thirteen state governments at the same time.[58] States and the boundaries that encompassed them remained at the forefront of delegates' minds, notwithstanding Paine's efforts to downplay such local sentiments.

With the existing state lines safe for the moment, the convention turned to the matter of the trans-Appalachian West in July 1787, just as the confederation government put the finishing touches on the Northwest Ordinance. The question of new states' equal status with the original thirteen provided much fodder for debate. Gouverneur Morris of Pennsylvania wanted the coastal polities to maintain forever their domination of the country: "The new States will know less of the public interest than [the Atlantic States]; will have an interest in many respects different; in particular will be little scrupulous of involving the community in wars the burdens and operations of which would fall chiefly on the maritime States. Provision ought, therefore, to be made to prevent the maritime States from being hereafter outvoted by them."

Morris doubted that his proposal would excite much anger if settlers bound for those future states west of the Appalachians were to know their subservient status from the beginning. "The busy haunts of men," he argued, "not the remote wilderness, was the proper school of political talents."[59] Morris went so far as to belittle state organization as a whole, suggesting that a new national government should draw its support from the general population rather than depend upon petty, squabbling former colonies.[60] Elbridge Gerry of Massachusetts suggested a limit to the number of new states west of the mountains and expressed concern about supposedly unpredictable immigrant groups tending toward that region.[61] The larger, established states of 1787 feared competition from the future generations that might come to occupy the western half of the country.

Over the objections of Morris and Gerry, the U.S. Constitution as adopted treated states equally in principle regardless of age.[62] Article IV, section 3 created a federal foundation upon which to build future states across North America and beyond. It provided for new states to join the Union with the permission of Congress, as long as a territory did not come from preexisting states that divided or united themselves without federal approval. The article also ensured the federal government's supremacy over all land in the United States not yet a sovereign state itself as well, thus ensuring its power to both shape and admit new members of the Union.[63] Nowhere else in the Constitution did boundary-making appear, either for the country as a whole or for its member states. It lacked even a general description of the borders that shaped the country and gave jurisdiction to the government. Apparently the charter's authors took for granted that none of the Union's members, or indeed the Union itself, existed without lines to contain them. Mentioning their existence thus must have seemed redundant, even though almost all state constitutions written since then include their individual boundaries. For more than a century, Congress and the states used Article IV, section 3 and the confederation's ordinances to reshape vast stretches of North America. The Constitution's provisions in this matter have worked so efficiently that they have never required amending.[64]

The federal government as organized under the new constitution continued the western work instigated by the confederation. As the first secretary of state, Jefferson claimed the management and organization of the trans-Appalachian region for his department in 1789.[65] The first session of the new Congress affirmed the Northwest Ordinance as the method by which it would admit new states north of the Ohio River. As a top-down approach compared to the more democratic 1784 measure, the Northwest Ordinance meshed well with the efforts to promote a more powerful federal authority through the Constitution.[66] The federal legislature also approved two acts that defined further the organization of land beyond the mountains. In mid-April 1790, members of Congress approved a cession of land from western North Carolina, essentially present-day Tennessee, to add to the tracts held in trust for the country. Among its various provisions, this act specified that "no regulations made or to be made by Congress shall tend to emancipate slaves."[67] Several weeks later President George Washington signed another bill relating to the region. It replicated the territorial system of 1787 for the land south of the Ohio River. Although a federal law rather than a confederation one, it received a moniker appropriate to the latter—the Southwest Ordinance. Far shorter than its counterpart for the Northwest Territory, this ordinance differed in its permission of slavery. It did not permit slavery expressly but instead required the law to match "the conditions expressed" in the North Carolina cession statute.[68] Amid the myriad issues facing the nascent federal government, it recognized the necessity of planning for its future by planning for more members of the Union to emerge in the trans-Appalachian West.

For decades following the adoption of the Constitution, state and federal governments struggled to determine which would exercise the most power, one of several points of contention that would inspire the Civil War. The Constitution gave national officials more authority than had the Articles of Confederation, and adherents of the new system sought from the beginning to assert federal authority. Perhaps inevitably, states and their boundaries offered a tempting target. During the first Congress, meeting in New York City, Senator William Maclay of Pennsylvania recorded a proposal that would reinvent the component states of the Union. His colleague Rufus King of New York, a Federalist senator who believed in national power over the states, offered a dramatic suggestion on February 11, 1791. King called upon his fellow senators to enable President Washington to divide the nation into six regional government districts. These units, supposedly a more efficient method of local control, would bear no similarity whatsoever to state lines as they then stood. Maclay reported that King's idea consisted of two units northeast of the Hudson River, two between that flow and the Potomac River, and two south of the Potomac. In true Federalist style, King declared that the new national authority "had no right to pay any more attention to the State boundaries than to the boundaries of the Cham of Tartary." Maclay, whose sympathies tended toward limits on federal power, expressed fear about King's notion: "Annihilation of State government is undoubtedly the object of these people." After a brief debate the Senate rejected King's proposal.[69] The lines of existing states survived, as the Constitution guaranteed. The federal government's power to create boundaries for future members of the Union, however, gave a century's worth of elected officials a political playground in which to exercise their imaginations.

Less than a month after the Senate debated King's proposal to wipe states from the U.S. map, the fourteenth joined the Union, Vermont. The next year witnessed the first trans-Appalachian applicant to seek admission as an equal polity. Virginia's western reaches had lobbied for recognition as their own state as early as 1780, proposing to encompass land on both sides of the Ohio River.[70] After Virginia ceded its claims on the north bank in 1784, rumblings of statehood continued on the south side. The effort succeeded when Kentucky received its promotion in the summer of 1792. Four years later the western cession of North Carolina joined the Union as the state of Tennessee. The Washington administration thus demonstrated that new polities could join the old on a basis that was at least theoretically equal.

Revolutionary war veterans and their families made up a large portion of the settlers in the first two trans-Appalachian states, encouraged in large part by land bounties awarded to those who fought against the British.[71] The settlers pushing inland from the Atlantic coast possessed "weak loyalties to provincial or state governments, and they had greater reason to think of themselves as Americans," in the words of historian James D. Drake.[72] Yet these migrants carried with them the desire for government to protect their property and rights. Inevitably, they understood established state (and formerly colonial) authority far better than the new federal one, so many sought to create states through the ordinances set up in the 1780s. In so doing, Kentuckians and Tennesseans established a century-long precedent. Well into the late nineteenth century, state-making was habitual for Americans moving into land outside the original thirteen states of the Union. They wanted that sense of security offered by boundaries that placed them in a defined, effective, regional political community. For the first century of U.S. history after independence, that meant *states*. As historian Forrest McDonald points out, the eventual triumph of the federal government after the Civil War did not whitewash the decades of history that preceded it. Even the British recognized individually the thirteen polities that it surrendered in the Treaty of Paris, not the United States as a whole.[73] States represented the building blocks of the country and fostered the notion that the more states admitted to the Union, the stronger the whole and its component parts.

At the dawn of the nineteenth century a fourth post-revolution state earned a star on the national banner. The first polity carved from the Northwest Territory, Ohio joined the Union early in 1803. Like its neighbors to the south of its namesake river, many of Ohio's early American settlers had served in the military during the revolution. They came from older states along the Atlantic coast, especially nearby regions of Pennsylvania and Virginia. Following the model of Kentucky and Tennessee, these settlers sought statehood to secure their livelihoods legally. As would later advocates of statehood, nascent Ohioans found their fortunes buffeted by political shenanigans at the highest levels, with Federalists hesitant to allow another backcountry state into the club and their opponents eager to embrace an agrarian commonwealth in the West.[74] With Thomas Jefferson's election as president in 1800 and the beginning of the end for the Federalist Party, the door opened for Ohio and other trans-Appalachian states to join their peers in the Union in the early nineteenth century.

Few sources outside the Buckeye State rank Ohio's statehood as the most important event for the United States in 1803. Instead, the country's first acquisition of foreign territory stands squarely in the limelight. That year the country began a half-century of expansion through the processes of purchase, negotiation, and conquest. Whatever excitement it may have engendered, the Louisiana Purchase of 1803 also inspired numerous questions, its political future one of the most pressing. Yet both the confederation and early federal state-making policies applied only to the land contained within the borders of the United States at the end of the Revolutionary War. Whether these precedents would suffice to incorporate formerly foreign land into the federal republic remained uncertain. In the years to come the extension of state boundaries across the Mississippi River helped determine the fate of the country as a whole and reconstructed vast tracts along the model of eastern states. The system of organization inherited from the British, codified by the confederation, and modified to fit the needs of the new republic now faced its most important challenge, one that expanded in complexity as the country itself grew larger throughout the nineteenth century.

CHAPTER 2

The Fairest Portion of Our Union
Early Boundaries in the Trans-Mississippi West

The Louisiana Purchase of 1803 doubled the size of the United States by extending its dominion westward across the Mississippi River. The young republic now claimed a diverse, complicated region populated by scores of American Indian societies as well as French and Spanish colonists. But to many, the vast territory now labeled on maps as part of the United States did not yet seem *American*. To that end, the federal government stamped its distinct form of authority on the purchased lands. Building upon the precedents of British colonial and American policies, and in conjunction with settlers crossing the Mississippi, officials imposed boundaries on the region in the early nineteenth century. Their actions provided the framework for incorporating almost all additions of land into the federal system. As an Arkansan claimed in 1822, the Louisiana Purchase represented "the fairest portion of our Union—a tract of country, out of which infant States and Territories would soon rear their heads, and with emulative pride, claim protection at the hands of the general government."[1] State boundaries introduced U.S. institutions into formerly foreign land for the first time and proved essential in its transformation into distinctly American territory. This first official extension of American institutions beyond the country's original boundaries shaped the future of the Louisiana Purchase and established precedents for a continental nation.

The Louisiana Purchase offered new challenges for American policy makers. Ordinances in 1787 and 1790 laid the foundations for territorial and state governments between the Appalachian Mountains and the Mississippi River. Yet extending those provisions onto what had been foreign soil and over a foreign population proved problematic. For decades after the Louisiana Purchase, officials in Washington, D.C., and American

Nonsensical set of state highway signs near Maysville, Arkansas

settlers living in the purchased land engaged in an extensive debate over how to blend it into the nation. These boundary debates took on national significance as the lines they led to demarcated not only the limits of territories and states but also the extent of slavery, best viewed through the state boundaries upon which the Missouri Compromise depended. Historian Peter Kastor argues that doubling the country's land claims in 1803 forced Americans to consider both how to manage the immense tract and how it affected their national identity. He argues that "the Louisiana Purchase eventually reinforced union. In the process, it was Louisiana that helped Americanize the United States."[2] But Americanization took place both in the region and the country as a whole. The imposition of territorial and state boundaries to organize effective political communities west of the Mississippi River offers the clearest example of this transformation in the Louisiana Purchase and beyond. Limits of territories and states represented more than administrative necessity in the West.

The early-nineteenth-century process of imposing boundaries in the Louisiana Purchase inspired wide-ranging debate about the future of both the region and the nation. In 1803, American diplomats in France anticipated the question of how the United States would manage the Louisiana Purchase by placing a provision ensuring its inhabitants a political future amid the protections of their religious freedom, property rights, and colonial records. Article III of the cession treaty stated in part: "The inhabitants of the ceded territory shall be incorporated in the Union of the United State[s], and admitted as soon as possible, according to the principles of the Federal Constitution, to the enjoyment of all the rights, advantages, and immunities, of citizens of the United States."[3] President Thomas Jefferson echoed this call in his annual message to Congress that fall. He praised the American economic and political control over the Mississippi River valley and saw in the region "a wide spread for the blessings of freedom and equal laws."[4] Jefferson encouraged Congress to act speedily to fuse the purchased land into the Union. Some members, however, doubted the likelihood of a smooth incorporation due to Louisiana's ethnic and religious diversity. As Representative William Eustis of Massachusetts suggested, "the people of this country [Louisiana] differ materially from the citizens of the United States. I speak of the character of the people at the present time."[5] The perceived need to Americanize the Louisiana Purchase—to transform it into something similar in structure and function to the eastern states—dominated discussions about the region. Over the next few decades, two connected forces aided this change: increasing migration into the Purchase from eastern states and the establishment of political communities by drawing boundaries in the immense territory.

Within months of the treaty's signing Congress answered Jefferson's call for political organization by inserting the first boundary into the Purchase, a line dividing the new Territory of Orleans to the south from the enormous District of Louisiana to the north. Politicians long deliberated where to draw the line, from as far south as the thirty-first parallel (the present-day latitudinal Louisiana-Mississippi line) to as far north as the thirty-fourth parallel (just north of the mouth of the Arkansas River). By considering only geometric boundaries, Congress reinforced the well-established American notion of imposing new authority on an area via artificial lines rather than natural divisions. By the spring of 1804, congressional members settled on the thirty-third parallel as the dividing line. Most congressmen expected the District of Louisiana to the north to remain an American Indian reserve. As such, they initially did not appoint a territorial government to prepare it for statehood as they had done in the trans-Appalachian West and present-day Louisiana. Instead, it came

Extending state boundaries into the Louisiana Purchase, 1803–1804

temporarily under the jurisdiction of Indiana Territory until federal officials decided what to do with it.[6] Regardless, by selecting the thirty-third parallel west of the Mississippi River as a political boundary, Congress extended American institutions onto formerly foreign soil for the first time. Through this mundane action, the federal government proved the American experiment transplantable.

Few Euro-Americans living in the Louisiana Purchase shared the satisfaction of Congress and the president at the first line drawn beyond the country's original borders. By the end of 1804, Congress had received several petitions from influential citizens throughout the Purchase's sparse settlements objecting to its division. Some southern Louisianans, preferring immediate statehood, chafed at their territorial status and considered it a violation of treaty stipulations. One petition doubted the sincerity of Congress to treat the newly acquired lands with respect: "Do political axioms on the Atlantic become problems when transferred to the shores of the Mississippi? [O]r are the unfortunate inhabitants of these regions the only people who are excluded from those equal rights acknowledged in your declaration of independence, repeated in the different State constitutions, and ratified by that of which we claim to be a member?"[7] To protect their unique interests within the federal system, settlers insisted upon their equal status to those living east of the Mississippi River.

Petitioners throughout the Purchase focused much of their anger on the use of the thirty-third parallel to divide Orleans and Louisiana. One rejected the use of American precedents for boundary-making, dismissing the Northwest Ordinance and other such measures as unsuitable for lands acquired from other countries. Considering its distinct population and geographical needs, Louisiana deserved a brand-new plan.[8] Another petition received by Congress included concerns from settlers on both sides of the line. Some in the Territory of Orleans lamented the loss of their "Louisiana" moniker. Others north of the boundary wondered if the federal government had authorized the division "to prolong our state of political tutelage" by isolating Louisianans from the region's more established population near New Orleans.[9] Some of the Purchase's residents looked upon Congress establishing any kind of Orleans-Louisiana boundary with suspicion, fearing their region would become little more than a plaything for eastern politicians: "[I]f Congress had a right to divide Louisiana into two Territories last year, they may claim next year the right to divide it into four, into eight Territories.... Your petitioners ... see no end to the oppression likely to result from such a precedent; and ill-fated Louisiana is condemned to drag along for ages the fetters of an endless territorial infancy."[10]

Regardless of the protests, the thirty-third parallel boundary struck eastern politicians as practical and useful, and the line stayed put. Possessing more familiarity with the American boundary-making process than Purchase residents, the federal government believed it knew better how to shape the trans-Mississippi West.

Those living south of thirty-three degrees did not resent the federal government's actions as much as their unorganized compatriots upriver did, however. Fearing a future for their region as a vast native reserve rather than farms and plantations, some disaffected residents north of the parallel took action. In 1804 a group of firebrands in present-day Missouri and Arkansas declared themselves the legislature of the District of Louisiana. In a petition to Congress they contrasted the experience of other states-in-waiting—"the open, disinterested countenance of a fond adoptive mother exhibited to our sister territories"—with their own status under "the stern, distrustful look of a severe, imperious master."[11] In response Congress reorganized the District of Louisiana into the Territory of Louisiana in March 1805 and appointed officials to oversee its political affairs from St. Louis. Yet Congress did not permit residents of the new territory to elect their own local representative body, a right granted to most other such polities throughout the nineteenth century.[12] Their suspicion of the new extralegal legislative body echoed earlier dismissals of self-organized territories and states like Transylvania (in present-day eastern Kentucky) and Franklin (today's eastern Tennessee) in the trans-Appalachian West.[13] If the federal officials realized how many times this pattern would repeat itself in western history they might have reconsidered their attitude.

By the beginning of the War of 1812 the Territory of Orleans achieved statehood as Louisiana, and Euro-Americans throughout the rest of the Purchase interacted with the territorial government centered in St. Louis. But the steady increase in settlement from east of the Mississippi River demanded plans for the political future of the rest of the Purchase. By the war's end, with the trans-Appalachian region cleared of British interference and powerful American Indian groups defeated in battle, even more Americans crossed the river to find agricultural and economic opportunities. To maintain order, protect property rights, and exercise their franchise, these settlers wanted territories and states of their own, and that meant drawing new lines in the trans-Mississippi West. Explorers, farmers, merchants, politicians, and countless others of varied ancestry contributed to this process from the War of 1812 through the rest of the nineteenth century. They built on the tentative precedents established by the early denizens of the Louisiana Purchase. Yet, regardless of their background or diversity, those involved in the process participated in a common task—the extension of established American institutions into land acquired from other countries and cultures. Boundary-making represented an essential early step toward that goal, providing a framework upon which to build new political communities.

Among the various challenges to drawing political lines within the Louisiana Purchase, a lack of good information about the landscape represented perhaps the most daunting. Jefferson expressed concern about this dearth of knowledge shortly after receiving the treaty from France: "It is, indeed, probable that surveys have never been made upon so extensive a scale as to afford the means of laying down the various regions of a country, which, in some of its parts, appears to have been but imperfectly explored."[14] Long worried about this ignorance, Jefferson had lobbied for scientific surveys of the trans-Mississippi region since the end of the American Revolution, two decades before the United States acquired Louisiana.[15] Many members of Congress fretted similarly over how to control Louisiana when so few Americans knew much about it. As one stated: "It is highly desirable that this extensive region should be visited, in some parts at least, by intelligent men."[16] With legislative support, Jefferson authorized a series of expeditions to begin the process of comprehending the territory. Over the next decade and a half, teams led by Dunbar and Hunter, Freeman and Custis, Lewis and Clark, Pike, and Long provided some details about the landscape, particularly along its major rivers. Even after these efforts, however, much of the trans-Mississippi map remained blank. As the *St. Louis Enquirer*, edited by avid expansionist and future senator Thomas Hart Benton, declared in 1819: "The formation of states west of the Mississippi will require an intimate acquaintance with the geography of the country from the right bank of the river to the foot of the Rocky mountains."[17] Most early boundaries beyond the river, however, developed through a combination of geographic guesswork and geometric compromise.

Boundary-making precedents established in the Louisiana Purchase resonated across the continent throughout the nineteenth century. As the region's best-understood natural features, rivers played an essential role in debates over how to divide it. Jefferson recognized the vitality of rivers in determining the future of the Purchase: "Many of the present establishments are separated from each other by immense and trackless deserts, having no communication with each other by land.... This is particularly the case on the west of the Mississippi, where the communication is kept up only by water between the capital and the distant settlements; three months being required to convey intelligence from the one to the other by the Mississippi."[18]

In the days before road networks stretched across the landscape, rivers served as highways of exploration, settlement, and commerce. Using them as a starting point for boundary discussions thus struck many as sensible. Early boundary proposals tended to focus on one of three riverine notions: states centered upon major tributaries of the Mississippi, states separated by those tributaries, or states that possessed a Mississippi frontage without consideration of those tributaries. Judging by newspaper accounts, most residents of the Louisiana Purchase favored the first of the three. They encouraged the creation of polities with rivers at their heart. Benton's *St. Louis Enquirer* suggested that "the centre of each state should be on a river" like the Missouri, Arkansas, and St. Peter (later renamed the Minnesota) "and the frontiers in the naked plains" that lay between.[19] These states would stretch west from the Mississippi two to three hundred miles. Beyond them would follow a tier of other states centered on the Kansas, Platte, and Yellowstone and the upper reaches of the Missouri and Arkansas Rivers. The *Enquirer* expected its river-centered plan to create out of the Purchase no less than twenty-six states roughly the same size of the older members of the Union.[20]

River-centered states offered a sensible political organization for the time. With most American settlers establishing homes and trade networks along rivers, they represented the arteries through which the lifeblood of new communities flowed. In the vast tracts between these rivers, straight lines would cut through land then unknown and undesired, compartmentalizing the more valued stretches along the river banks. As the *Enquirer* was one of the most prominent newspapers of the region, its support for such a scheme proved invaluable. In 1819 it suggested how such a method of state construction would create cohesive political communities: "By this means the centres of the states would always be in the part of the country which admitted of population, agriculture and commerce, that is to say on the borders of the rivers, where wood is found; and the extremes, or ends of the states would be in the vast plains found between the rivers, and

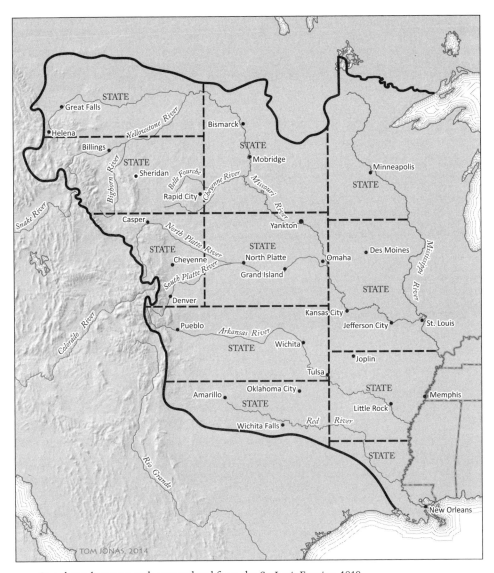

State boundary proposals extrapolated from the *St. Louis Enquirer*, 1819

uninhabitable for want of wood and water. Such a system would line the banks of the Missouri and other great rivers with a succession of great cities, capitals of states, rising above one another from the centre of the valley of the Mississippi to the foot of the Shining [Rocky] Mountains."[21]

A statehood petition sent to Congress by settlers along the Missouri River that same year followed the *Enquirer*'s model. Attempting to subvert arguments that their proposed lines would create an excessively large state, the petitioners claimed: "The districts of country that are fertile and susceptible of settlement are small, and are detached and separated from each other at great distances by immense plains and barren tracts, which must for ages remain waste and uninhabited."[22] These nascent Missourians expected their state's frontiers to remain forever a no-man's-land between agrarian regions along major rivers and drew their lines accordingly.

Not everyone living in the upper Louisiana Purchase favored states divided by geometry with rivers at the center, however. Some suggested using rivers as natural boundaries between future states rather than their centers. They noted the precedent of an early division of Missouri Territory—organized in 1812 out of the old District of Louisiana—into five counties separated principally by rivers.[23] Some inhabitants of Cape Girardeau, more than a hundred miles downriver from St. Louis, called for a state wedged between the thirty-sixth parallel and the Missouri River, with Cape Girardeau as the capital. These boundaries blended geometry and geography to shape what residents of that town considered a governable polity. The *Enquirer* lampooned the proposal, however, arguing against creating two states out of one unified region with two governments charging twice the taxes.[24] It also noted haughtily: "In laying off a state or a county there is a natural feeling which displays itself in the human heart. Every one wishes to be in the centre; no one is willing to be on the outline. . . . Of course there are but few who can be pleased, and many to be dissatisfied."[25]

Regardless of the *Enquirer*'s derision, the notion of using western rivers as boundaries cropped up occasionally in the mid-nineteenth century. Several polities sought to shift their federally bestowed geometric lines to nearby rivers. Kansas Territory asked Congress in 1859 to move its northern boundary from the fortieth parallel to the Platte River, which would have gobbled a large slice of southern Nebraska Territory. The Kansas legislature offered few arguments for their request but suggested that the population south of the Platte preferred their proposal.[26] Six years later Utah Territory made a similar request to take the land north of the Colorado River in northwestern Arizona Territory. Such a shift would have opened up commercial access to the southern farming and mining districts of Utah "and more ready means of import to encourage the reclamation of the desert."[27] The federal government acceded to neither of these requests, likely practical decisions. The mercurial character of the Platte would have caused Kansas and Nebraska countless jurisdictional problems, while dividing the Grand Canyon between two territories would have

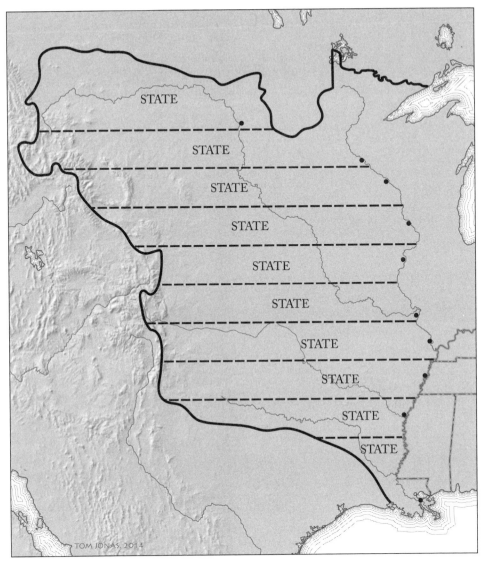

State boundary proposal made by Washington, D.C.'s *National Intelligencer*, 1819

posed logistical challenges of its own. The skepticism about using rivers as boundaries expressed by the *Enquirer* in 1819 resonated across the West.

The farther from the Louisiana Purchase one traveled, the less important decision makers considered the Mississippi's tributaries when pondering potential boundaries. The Washington *National Intelligencer*, the capital's predominant nineteenth-century newspaper, rejected the use of tributaries to unite or divide. Instead, the *Intelligencer* proposed a series of elongated states shaped like Tennessee on the western side of the Mississippi River, stretching west to the Rockies. This, it was said, would prevent the domination of that vital flow by three or four immense states. The *Intelligencer* contrasted its boundary proposal with that promoted by the *Enquirer*: "In the one case, there would be a single line of states west of the Mississippi, all having the benefit of a front on that great channel, possessing equal geographical advantages, stretching their jurisdiction to the western limits of the Union; all of them populous, wealthy, and powerful. In the other case . . . there will be two lines of states west of the Mississippi; one of them lying immediately upon the river, vigorous and opulent; the other line, behind them (should the remaining territory ever become states) far in the west, cut off from the river, remote and feeble with a foreign nation on their confines."[28]

The *Intelligencer* recognized the essential nature of the Mississippi in its proposal, but proved ignorant of the importance of its large western tributaries. It also failed to consider the possibility of the U.S. borders expanding beyond their then-terminus at the Continental Divide. In a sense the *Intelligencer* echoed the Ordinance of 1784 by arguing for prefabricated states to admit as their populations increased rather than the Northwest Ordinance model that modified state lines to accommodate the evolution of local communities and national desires.

The influence of the *Intelligencer* in this and many other national political matters emerged in congressional debates. Some members of Congress expressed similar doubts about the need to consider tributaries when carving up the remaining Purchase lands. During the Missouri statehood debate in 1820, for example, Senator Samuel Dana of Connecticut worried that the large size proposed by Missourians would give them "a position commanding all the passes into the interior."[29] Keeping in mind the interest of fairness to Missouri's future neighbors, some congressmen feared making that state the valve that would regulate the flow of settlement and commerce up its namesake river to the land beyond. For politicians from free states, undoubtedly Missouri's status as a potential slave state factored into this attitude. What the residents of the proposed state saw as their blessed position as the gateway to the West, some in Congress viewed as potential dangerously excessive authority over the country's farthest reaches. Regardless, both groups recognized the long-term effects of boundary decisions.

Many residents of the Louisiana Purchase resented the notion that those living far from their homes might decide the region's political fate.

Benton's *Enquirer*, unsurprisingly, reacted strongly against the *Intelligencer*'s skinny-state proposal. The newspaper presented its readers with a potential image of their state as "900 miles long, 120 miles wide, a superficial content of . . . 100,000 square miles, only a few habitable spots in it where creeks and rivers were crossed, and the people mocked with the illusion of a front upon the Mississippi, to arrive at which they must travel over land several hundred miles between two narrow and parallel lines." The St. Louis newspaper moaned with a sense of forlorn defeat: "The wishes of the Missouri people are therefore to have no weight in the division of their own country. Atlantic gentlemen will cut it up for them. Fresh proof of the systematic contempt with which the people of the west are treated."[30] The *Enquirer*'s attitude echoed the sentiments of many westerners over the course of the nineteenth century, as countless other communities chafed against a perceived lack of self-determination.

In the end, federal officials chose to listen to the locals and organized the first tier of trans-Mississippi states north of Louisiana with rivers and riverine communities at their center, with geometric divisions slicing through less accessible land to the north and south. The voices of those living in the regions to be divided found willing ears in Washington. Even so, the lines reflected limited geographic understanding, even among those living in the upper Louisiana Purchase, particularly in the case of Missouri. After Congress carved Arkansas Territory from the southern reaches of Missouri Territory in March 1819, some citizens of Missouri sent a petition to Washington praying for statehood. Their boundary proposals blended geography and geometry to separate the new state from the rest of the Purchase. The petition called for boundaries that incorporated many rivers and their confluences linked by latitudinal lines.[31] This convoluted system would have given Missouri considerably more land to the west and north, and slightly to the south, than it possesses at present. Most importantly the proposal retained the Missouri River at the center of the state and sliced geometric divisions through places little understood or desired at the time by American settlers. Neither the federal government nor local residents possessed a monopoly on geographic ignorance.

In at least one apocryphal instance, a popular story if difficult to prove, a private citizen took advantage of the amorphous nature of boundary-making for his own benefit. John Hardeman Walker, a landowner along the west bank of the Mississippi south of its confluence with the Ohio, refused to flee the area after the 1811 New Madrid earthquakes. In fact he acquired many of his neighbors' abandoned farms and plantations and established a sizeable fiefdom. Walker believed that a sovereign state government offered more protection for his property than a federally managed territory. To that end in the late 1810s he lobbied politicians to include his extensive spread in the soon-to-be state of Missouri rather than Arkansas Territory. When the federal government established the latter in 1819, its northeastern boundary corresponded to his property between the Mississippi and St. Francis Rivers and the thirty-sixth and thirty-six thirty parallels. Thus, according to legend, Walker's political and economic influence secured the unusual jog in the line and created the "boot heel" of southeastern Missouri, although little evidence survives to back up the story.[32]

As time passed and settlement increased in the trans-Mississippi West, pushing inland from the largest tributaries toward the invisible lines that purported to divide territories and states, new boundary demands arose. Newcomers sought to alter lines to correspond with the greater knowledge of the region's geography that came with the passage of time. Through such revisions they hoped to create what they considered more efficient and appropriate boundaries. Vague lines drawn across heretofore unknown stretches of land far from major rivers no longer met the needs of those living in the Far West. Once again rivers fueled these discussions, this time the tributaries of the Missouri River in particular. American settlers demanded control over these vital local waterways, much smaller than those they fed into yet no less important for the people living along them.

Time and again, Missouri represented a boundary-making proving ground. Several times after achieving statehood in 1821, Missouri politicians requested that Congress extend their boundaries westward for economic and ethnic reasons. These proposals met with mixed results. A decade after statehood, for example, Missouri noted that "many inconveniences have already arisen, and others are expected to arise, from the improvident manner in which certain parts of the boundaries of this State have been designated." Beginning in 1831, Missouri politicians requested to move the state's northwestern corner. Originally it consisted of a line extending a hundred miles due north from the confluence of the Kansas and Missouri Rivers (at present-day Kansas City) and then turned ninety degrees to the east. In the 1830s, Missourians sought to revise this by shifting their geometric northwestern limit westward to the Missouri River channel.[33] This region lacked significant American settlement in the early 1820s but had flourished as an agricultural destination in the decade since. Press and government sources referred to it as the Platte Purchase in reference to a

Proposed alterations to early trans-Mississippi state boundaries, 1831–57

river flowing through the triangular territory (not to be confused with the larger Platte River in present-day Nebraska).

The federal government provided a strong motivation for Missouri's efforts to expand its northwestern boundary to its namesake river. The Indian Removal Act in 1830 threatened the state's aims regarding the Platte Purchase. Congress and President Andrew Jackson threatened to cut off access to the larger river for northwestern Missouri by designating federal land west of the state line as an Indian reserve. Missourians needed quick action to stave off defeat. Petitions sent to Congress to bolster the state's claim to the Platte Purchase portrayed Native groups living in the area already as "a poor, drunken, miserable set of beings, dwindling away to nothing, quarrelling among themselves, killing each other, and in constant broils with their white neighbors."[34] Increasing numbers of American squatters in the region made a tense situation worse, especially as the trespassers railed against supposed depredations they endured at the hands of American Indians. Missourians argued that transferring the region to their state was the only way to keep the peace and provide for American settlement at the same time.[35] Petitioners clamored for Congress to shift the line and recognize "the true boundaries of the State."[36] By the middle of the 1830s, therefore, Missourians abandoned their faith in rivers as the center of geometric states and demanded instead that Missouri's domain extend to the upper Missouri River. Political theory crumbled when contrasted with an opportunity for more land.

To prevent trans-Appalachian Native groups from constraining Missouri, its politicians appealed passionately in favor of annexing the Platte Purchase. They warned of economic and ethnic catastrophe without the transfer. State leaders described their western boundary drawn years earlier as having been made in "ignorance of the geography of the country."[37] Such an inconvenience should not constrain Missouri forever, they insisted. As geographical information evolved, so too should state lines. Congress bowed to the vocal demands of Missourians in the summer of 1836 by approving the extension of their northwest line to the river.[38] Shortly thereafter, agent William Clark negotiated with Native groups living in the region to surrender their claims for land and assistance to move west of the Missouri River. The acquisition of the Platte Purchase led to a rush of settlement into the region as settlers claimed valuable land near the Missouri River and organized county governments.[39] The land transfer reflected federal faith in rivers as barriers between Native cultures and American settlers, hydrological hedges separating intractable neighbors.[40] But the new geographic line offered problems of its own. In the late 1850s, for example, Missouri and Kansas bickered over control of an island in the river until federal officials decided the case in Missouri's favor.[41] Both geographic and geometric divisions posed challenges.

Eventually Missourians pushed their boundary demands too far. In 1850 the state legislature requested an extension of the southwestern boundary to the Neosho and Spring Rivers. Using tactics similar to those that worked for the Platte Purchase, the petition called those flows "the only natural outlet to market" for farmers in the area. Missouri politicians argued paradoxically that the region offered little to the Senecas and Quapaws living there at the time, yet touted the area's agricultural potential and its sources of coal and marble. If given to enterprising Americans rather than the commercially enfeebled Natives, "a city would soon be built up where a wilderness now stands" and the heretofore geometric corner of southwestern Missouri could flourish. The legislature suggested buying off the Senecas and Quapaws and asked congressional approval to begin negotiations with the tribes for the land. Congress, however, believed Missouri—then the second-largest state in area after Texas—claimed enough land already and refused this shift of its lines.[42]

Later trans-Mississippi polities followed Missouri's lead in insisting upon the flexibility of their boundaries. Iowans made similar demands that Congress redraw Iowa's western limit. Originally an enormous territory between the Mississippi and Missouri Rivers up to the British Canadian border, Iowa was pared by federal officials into a more manageable size in the 1840s to prepare it for statehood. Blending the ideas of river-centered and river-bounded states, in 1845 Congress centered Iowa on the west bank of the Mississippi River north of Missouri with several tributaries cutting diagonally through it. The confluence of the St. Peter (now Minnesota) and Blue Earth Rivers—near present-day Mankato, Minnesota—represented Iowa's new northwestern corner. From there geometric boundaries extended south to the Missouri line and east to the Mississippi River. Congress considered these limits sensible for a state delineated by "the boundaries of nature" slicing through inconsequential lands far from the riverine population.[43] But the nascent Hawkeyes disagreed. In a rare example of western settlers postponing their desire for statehood, Iowans rejected the boundaries bestowed upon them by Congress. Instead they demanded both eastern and western access to rivers, as had their neighbors in Missouri. Such limits offered easier access for agricultural commerce and facilitated governing the western reaches of the territory.[44]

When Congress returned to the matter of Iowa in 1846 some members sympathized with the riverine boundary argument. Representative Stephen Douglas of Illinois, whose interest in western territories would increase steadily, called lines proposed by Congress "most unnatural" ones that left awkward tracts to the west and north for future states.[45] Indeed, many members of Congress looked to the boundaries of Iowa and other western lands with an eye toward the future shape of the entire region.

Iowa's delegate to Congress, Augustus C. Dodge, pleaded with his colleagues for a state bounded to the east and west by major rivers.[46] The territory's leading newspaper, the *Burlington Hawk-Eye*, also desired access to the Missouri River rather than a long Mississippi River frontage: "What is the consequence of this greedy grasping? We retain a portion of this wilderness of pestilent lagoons and friable sand stone and have a Chinese Wall of boundary between us and the Missouri River."[47] Once again the protests of local residents helped sway federal opinion. In late 1846 Congress altered Iowa's lines to those desired by its citizens, and statehood followed by the year's end.[48]

Like Missourians, Iowans also tried to shift their lines after achieving statehood. Throughout the 1850s Iowa's legislature sent petitions to Congress requesting an extension of the northwestern corner of Iowa beyond the Big Sioux River to the Missouri River, "the most natural and appropriate western boundary" for the state.[49] Iowa politicians pointed to the impossibility of navigation on the silt-laden Big Sioux and argued the logic of the navigable Missouri River for their state's entire western boundary. The process of building bridges across murky flows like the Big Sioux, "as dangerous in crossing to the traveller as to the buffalo," might proceed easier if in the jurisdiction of one state instead of two, it was asserted.[50] The proposal would have given Iowa a northwestern panhandle out of present-day South Dakota but Congress ignored the notion and left Iowa's boundary where it was. Win or lose in these battles with Congress, westerners often looked at their boundaries as malleable things, evolving as their geographic knowledge improved and population shifted.

The creation and evolution of territorial and state boundaries demanded a great deal of attention from diverse groups of people, and the importance of those lines extended far beyond local ramifications. In the early nineteenth century, local and national leaders viewed the fate of the Louisiana Purchase and the fate of the country as linked inextricably. Thus at times the lines separating these new political communities acquired unexpected significance, taking on new meanings for the nation at large. One trans-Mississippi boundary in particular accrued an unparalleled national importance, the Missouri-Arkansas line.

Many early-nineteenth-century settlers along the lower Missouri River brought slaves with them, establishing plantation agriculture there. As national politicians pondered boundaries for a state of Missouri in the late 1810s, they debated the future of slavery in that territory and the rest of the Purchase lands. As one proslavery contributor to the *St. Louis Enquirer* cautioned, politicians opposed to slavery might "[d]ivide and govern" Missouri. He went on: "If the slaves of Missouri cannot be expelled to make room for a New England population, then let Missouri itself be destroyed. Let her territory be cut down and split up into ribbons and garters; let her soil be severed into thin slices, so that neither weight nor strength shall remain in the disjointed parts."[51] Both sides of this incendiary debate realized that the precedents involved in state-making in the Louisiana Purchase extended beyond the principles behind drawing boundaries. Indeed, when Congress took up the issue of Missouri statehood in early 1820, the fate of its boundaries and its slaves—not to mention slavery throughout the remaining Purchase lands—became one and the same issue.

The battle over slavery's future in the Louisiana Purchase involved a part of the country located farther from the Purchase than almost any other. In years shortly after the War of 1812, new states entered the Union on a yearly basis—Indiana (1816), Mississippi (1817), Illinois (1818), and Alabama (1819)—maintaining a balance between slave and free states in the Senate. By the end of the decade, southern politicians, outnumbered in the House of Representatives since 1789, came to view that balance as vital to their interests. The latest threat to this political equilibrium came from York County, Massachusetts, a remnant of the immense colonial land grant of Massachusetts Bay. It applied for statehood under the moniker "Maine" in the late 1810s. To balance the admission of Maine, southern politicians demanded a slave state, and increasingly populous Missouri seemed a likely candidate. Sectional attitudes flared as a result, with free-state politicians bemoaning the perceived hijacking of Maine's statehood to serve the interests of the South.[52] In this atmosphere of crackling tension nationwide, boundaries in the Louisiana Purchase stood ready to play a major role.

As the debate dragged on, some in Congress despaired of ever finding a solution, while others asked their colleagues whether the trans-Mississippi region merited all the fuss and bother. Senator Harrison G. Otis of Massachusetts wondered "if it would have been happier for us if the Mississippi had been a torrent of burning lava, impassable as the lake which separates the evil from the good, and the regions beyond it destined to be covered forever with breaks and jungles, and the impenetrable haunts of the wolf and the panther."[53] Senator James Barbour of Virginia cautioned his colleagues not to let the issue fester until western settlers took matters into their own hands. Americans living beyond the Mississippi, he claimed, would not linger patiently in colonialism forever. Barbour declared that "those hardy sons of the West" drew inspiration from the spirit of 1776, and he warned of a revolutionary upheaval in the Louisiana Purchase if the federal government dallied in creating political communities.[54] The pressure to draw boundaries to shape new states beyond the Mississippi would grow stronger, not weaker, as the years passed, and Congress needed to prepare itself for the inevitable.

While politicians harangued in the legislative chambers over the fate of Missouri, the inequity of tying Maine to a national problem it did not deserve, and the extension of a controversial institution, slavery, into the West, a solution emerged. On January 18, 1820, relatively early in the debate, Senator Jesse B. Thomas of Illinois proposed to contain slavery by using a series of state boundaries. He proposed a bill that hearkened back to the Northwest Ordinance by suggesting that while slavery would be allowed in the proposed state of Missouri, the older territorial policy should, "to all intents and purposes, be deemed and held applicable to, and shall have full force and effect in and over all the territory belonging to the United States" north of thirty-six degrees thirty minutes, the northern Arkansas Territory boundary, and outside Missouri's state lines.[55] Thomas's bill was trundled off to a committee while congressmen considered the broader implications of western state-making.

Thomas's proposal followed others that suggested limiting slavery in the Louisiana Purchase by geometric lines. In 1811, for example, Representative Jonathan Roberts of Pennsylvania suggested a barrier to the institution north of a parallel drawn through the confluence of the Mississippi and Ohio Rivers, but the idea fizzled with the country preoccupied about the impending war with the British.[56] By the decade's end, however, the introduction of territorial boundaries within the Purchase gave Congress a framework upon which to shape both future states and the future of western slavery. Three days after Thomas introduced his bill, the Senate appointed a select committee of five members—Thomas, James Burrill, Jr., of Rhode Island, Richard M. Johnson of Kentucky, William A. Palmer of Vermont, and James Pleasants of Virginia—to consider state boundaries as barriers for slavery in the West.[57] Over the course of two weeks the committee altered Thomas's proposal from dependence upon state lines specifically, to a revision banning slavery in "that tract of country ceded by France to the United States, under the name of Louisiana, which lies north of thirty-six degrees and thirty minutes north latitude" but permitting it in the new state of Missouri.[58] Thus the line that sliced through the Ozarks to divide Missouri and Arkansas came to represent even more to Americans at large.

This proposal for dividing slavery's future via the Missouri-Arkansas line (excepting Missouri) provided Congress a springboard from which to launch a discussion about slavery in general. Some southern politicians, for instance, approved of the committee's notion as tacit support of slavery south of thirty-six thirty degrees.[59] One even suggested pressing the divider north to the fortieth parallel—the present Kansas-Nebraska boundary—to provide a little more territory for the institution. Although this measure fell to defeat, so too did several attempts to prohibit slavery everywhere west of the Mississippi River except Louisiana, the proposed state of Missouri, and Arkansas Territory east of the ninety-fourth meridian (the present geometric division separating Louisiana and Arkansas from Texas north of the Sabine River).[60]

Missourians did not intend their desire for state organization to create such a national ruckus, nor did they expect their southern boundary to attract such widespread attention. The arguments over a line that indicated initially only where their authority stopped and their neighbor's started overwhelmed the line's more humble origins. Some congressmen worried about the national repercussions of creating significant fractures along the political divides between members of the Union. Looking forward to a time when American claims might span the continent, Representative James Stevens of Connecticut remarked: "The south line of Pennsylvania State and the Ohio waters now form the boundary line between the two parties [slave and free states]. If you continue that line by the 36° 30' of north latitude, to the Pacific ocean, I fear it will not prove a pacific measure." Another member of the House stated: "Conscientious scruples are not bounded by the Mississippi, or limited to the latitude of 36 degrees 30 minutes."[61] The repercussions of adapting western state boundaries to national policy proposals resonated loudly and widely.

Throughout the compromise debates numerous politicians made reference to the long-term consequences of their deliberations. For many, the proposal affected not only the future of Missouri but of the nation too. Northern congressmen considered the stakes of Missouri's statehood debate to be astronomical. Senator David Morril of New Hampshire warned that the country's and the world's future generations "will have occasion to look back upon the measures of this Congress with joy or sorrow, delight or regret, perhaps, to the last period of time."⁶² Senator Prentiss Mellin of Massachusetts observed similarly that "we are not legislating for a year, nor for the period of our own lives, but for centuries to come."⁶³ As the debate drew to a close, Representative Charles Kinsey of New Jersey remarked prophetically: "We have arrived at an awful period in the history of our empire, when it behooves every member of this House now to pause and consider that on the next step we take depends the fate of unborn millions. I firmly believe that on the question now before us rests the highest interests of the whole human family. Now, sir, is to be tested, whether this grand and hitherto successful experiment of free government is to continue, or, after more than forty years enjoyment of the choicest blessings of Heaven under its administration, we are to break asunder on a dispute concerning the division of territory."⁶⁴ Kinsey reminded his colleagues that when they stripped away all the emotion surrounding the question of slavery's future in the West, the heart of the matter remained—a boundary helping to shape two future members of the Union like all the others across the country, yet one upon which they sought to place a burden heavier than any other such line carried.

By early March 1820 both houses of Congress agreed to the Thomas proposal of using state boundaries as slavery boundaries. The Senate approved by a vote of 34–10, and the House of Representatives consented with a vote of 134–42, thus paving the way for Maine's and Missouri's statehood and a geometric limit on the expansion of slavery in the West—the famed Missouri Compromise.⁶⁵ It demonstrated the wide-ranging utility Americans ascribed to their state lines. For expediency's sake geometric state boundaries rather than natural barriers provided the primary structure upon which politicians grafted questions of national importance. Whereas the famous Mason-Dixon line separating Virginia and Maryland from Pennsylvania provided only a symbolic divide for slavery, in 1820 the Missouri-Arkansas line emerged as the official barrier between free soil and land open to slavery west of the Mississippi. The line that meant one thing to those living near it meant something else entirely to the nation at large. Both interpretations proved just as valid and important.

In the years immediately following the Missouri Compromise many Americans looked upon the 1820 agreement as a stabilizing force. For people across the United States the line represented a pledge for the country's future, although certain groups and regions interpreted it differently. Its significance only increased with the rapid territorial growth of the 1840s.

Five years after Texas's annexation and statehood in 1845, for example, the state's wild claim to land stretching into present-day Wyoming fell victim to the Compromise of 1850. A provision to amend Texas's northern boundary to fit the Compromise line lay within the various pieces of legislation passed to calm sectional frustrations. Even after thirty years the Missouri-Arkansas line retained its importance. President James Polk reflected this attitude in late 1848 when he informed Congress of his decision to approve a territorial government in Oregon while leaving the Mexican Cession lands yet in doubt. Polk noted that he felt comfortable approving Oregon's organization considering that territory's position entirely north of thirty-six thirty. Unlike the lands taken from Mexico that overlapped that invisible line, the president knew that he need not worry about passionate debates over the future of slavery within the Oregon Country.[66]

By the mid-1850s, however, a new generation of politicians supplanted the Missouri Compromise with political machinations to manage slavery's spread into the West, making that region the opening battleground for the sectional crisis of the 1860s. The Compromise of 1850 and the Kansas-Nebraska Act, both of which allowed in theory the expansion of slavery into western territories regardless of their relation to the line, weakened the 1820 deal. The *Dred Scott v. Sanford* decision in 1857 eliminated it completely. Yet some states raised its specter with either reverence or revulsion in each subsequent debate. In the 1850s the legislatures of Connecticut, Massachusetts, Michigan, New Hampshire, New York, Ohio, and Rhode Island submitted numerous resolutions to Washington, D.C., opposing any alteration to the Missouri Compromise.[67] Maine, which entered the Union as part of the 1820 agreement, flooded Congress with resolutions from 1854 through 1858, demanding the geometric containment of slavery.[68] Many northerners believed that Missouri's southern boundary served an almost sacred role in defining the region to its north as forever free soil.

Southern states likewise used the Missouri-Arkansas boundary for their own cause, viewing it initially as a sacrosanct pledge to preserve and extend slavery anywhere south of thirty-six degrees thirty minutes. Within months of the capture of Mexico City in 1847, for example, the Missouri legislature informed its representatives in Washington to vote "in accordance with the provisions and the spirit" of the compromise that relied upon their southern boundary. Missouri leaders encouraged the continued use of that line "in relation to the organization of new Territories or States, out of the territory now belonging to the United States, or which may hereafter be acquired either by purchase, by treaty, or conquest."[69] When the Wilmot Proviso threatened to bar slavery from the land seized from Mexico, the Virginia legislature responded that the Missouri Compromise guaranteed the institution south of thirty-six degrees thirty minutes.[70] After the war with Mexico, however, this attitude changed. Southern states now decried the 1820 deal as federal interference in the expansion of American institutions. In Arkansas, legislators endorsed the repeal of the limit on slavery represented by their northern boundary, a notion echoed by their colleagues in Tennessee and Texas.[71] Missourians dismissed similarly the federal constraints on slavery that had figured so prominently in their own application for statehood, "even if such act ever did impose any obligation upon the slave-holding States."[72] For southern politicians the line that started as an insurance policy for slavery evolved into an intolerable constraint.

Even as the nation stumbled toward civil war, the power of the Missouri Compromise line proved overwhelming. Eight days after South Carolina declared itself out of the Union, politicians in Washington turned once more to the Compromise in an effort to save the crumbling nation. Senator Henry M. Rice of Minnesota submitted a resolution that observed that "the Territories of the United States, and the question of the admission of new States into the Union have caused most, if not all, the agitation of the question of slavery." To halt the nation's disintegration, Rice proposed immediately making two massive states out of all remaining territories, with the land north of thirty-six thirty becoming the state of Washington, and that to the south the state of Jefferson. The senator proposed dividing these enormous states into smaller units once their populations increased and spread.[73] While the proposal did not succeed, it demonstrated that in a desperate hour American leaders looked once more to the Missouri-Arkansas boundary and the unparalleled national significance placed upon it in an effort to save the Union. The perceived importance of the line separating these two states pervaded national decision-making in the first half of the nineteenth century.[74]

The significance of the Louisiana Purchase cannot be understood without considering the importance of extending state boundaries into it. As Senator John Breckinridge of Kentucky observed in 1803, the territory "will, beyond all question, be settled fully at a period not very remote. It is equally certain, that it will be settled by [A]mericans."[75] American settlers moving into the Purchase brought with them not only dreams of prosperity but also notions of politics, economics, and society that they used to transform the area. By demanding lines that, in their opinion, harnessed the vast landscape effectively, these newcomers transformed a region claimed variously by the French and Spanish and occupied by Native cultures, making it effectively a part of the United States. In this process the federal government played an essential role, aided and sometimes prodded by American settlers. The imposition of boundaries into the Louisiana Purchase extended a young but strengthening political system into previously foreign land for the first time. It also helped transform the area into something that seemed to many people more American in terms of its settlers and the institutions they demanded.

Forming territories and states in the region that some came to view as "the fairest portion of our Union" extended the American system of local government beyond the original limits of the United States. This process added new stars to the banner and created a sense of national preservation. Boundaries came to reflect the future of the Louisiana Purchase and, in the case of the Missouri Compromise, the fate of the country itself. Historian Peter Kastor suggests that "[t]he incorporation of Louisiana had defined an American nation," but this process of transformation worked both ways.[76] As Woodrow Wilson hypothesized in 1897, "the Louisiana purchase opened the continent to the planting of States, and took the process of nationalization out of the hands of the original 'partners.'"[77] Just as Louisiana altered the national consciousness and shifted American attention westward, the Purchase itself became American through the process of drawing lines beyond the Mississippi River, establishing political communities on the model of the original members of the Union. In the words of a mid-nineteenth-century Nebraska booster, by acquiring Louisiana "our young territory became American domain—open to American enterprise, American freedom, and American settlers."[78] By imposing boundaries in the Louisiana Purchase, the United States instigated the century-long process of drawing lines that shaped newly acquired regions, creating states in the image of older ones, and claiming authority over diverse cultures and environments. The next six chapters investigate this phenomenon through a series of examples that span the trans-Mississippi West and the course of the nineteenth century.

Stone marker at the original "hinge" of the western Arkansas line, on the grounds of the Fort Smith National Historic Site

CHAPTER 3

On the Extreme Frontier
The Western Arkansas Boundary

Perched along many state lines are communities and features with names inspired by the boundary. Close to the Mason-Dixon line separating Maryland and Delaware, for example, stand the towns of Marydel, Delmar, and Mardela Springs. West of the Mississippi River this phenomenon creates Kanorado, Monida Pass, Ucolo, the MonDak Heritage Center, and the Cal-Nev-Ari Casino. Boundary Peak and the municipalities of Stateline and Bordertown perch along the California-Nevada divide, while Texas shares Texico, Texhoma, Texline, Lake Texoma, and Texarkana with its neighbors. Two portmanteaus along the Arkansas-Oklahoma boundary remind modern audiences of the vibrant history of that particular line. Most residents of the town of Arkoma live within a few blocks of State Line Road, while the crossroads of Arkinda hearkens back to the nineteenth century when American settlers and American Indians shaped their regional claims into governable territories. The Arkansas River, flanked to the north by the Ozarks and in the south by the Ouachitas, served as the focal point for this struggle between diverse groups with disparate visions of their future "on the extreme frontier," as one faction described the contested region.[1]

Americans expanded their presence in the Louisiana Purchase by moving up the Arkansas River in the early nineteenth century. As they did, the settlers imposed their political institutions and therefore their perception of effective control on the land. But they encountered unexpected competition in their boundary-making from American Indians. By that time some factions of Native people in the Deep South had adapted to American concepts of land ownership. They recognized a thoughtful manipulation of such lines as their only hope of retaining some of their former independence. Having lived near Euro-Americans for generations the Choctaws and Cherokees—the native groups with the most expansive claims in 1820s Arkansas Territory—knew the importance their competitors placed upon invisible divisions of land.[2] Both cultures proved just as adept at shaping boundaries as American settlers. The Choctaws and Cherokees secured a line considerably more favorable to their own interests than early American residents of Arkansas imagined possible, much to their frustration.

Shortly after completion of the Louisiana Purchase in 1803, President Thomas Jefferson proposed using some of the acquired land as a refuge for American Indians then occupying valuable land in the southern states and territories, thus presaging mandatory removal by almost three decades.[3] To that end, the federal government negotiated with Native cultures already living in the region to accommodate all concerned. One of the earliest agreements involved a powerful trans-Mississippi society, the Osages. In 1808 they surrendered their vast claim to land between the Missouri and Arkansas Rivers east of a north–south line running through Fort Clark on the Missouri (see chapter 4).[4] Treaties in 1818 and 1825 finalized their pullout from Arkansas. Meanwhile the Cherokees, who had hunted in present-day Arkansas since the end of the Revolutionary War, received permission from Jefferson to relocate some of their population there permanently in 1809, from their homelands in the Southeast.[5] After several clashes with American settlers in the area, the Cherokees negotiated for a diamond-shaped tract between the Arkansas and White Rivers in 1817. The following year the Quapaws, a smaller group long dwelling in the region, secured a reserve on the south side of the Arkansas near its mouth.[6] American settlers from subsistence farmers to town builders and plantation owners filled in the gaps between these defined tracts. By the end of the 1810s, therefore, a myriad of lines defined the territory of diverse neighbors living along the Arkansas River.

Political organization for American settlers moving into the region followed almost immediately. In 1819 Congress drew a line from the Mississippi River at thirty-six degrees westward to the St. Francis River, up that flow to thirty-six degrees thirty minutes, and thence west to the nation's western boundary on the high plains to carve off the Arkansas Territory.[7] This new territory encompassed all of the modern state of Arkansas and most of present-day Oklahoma minus the panhandle and a strip along the northern boundary. After the Missouri Compromise of 1820 many

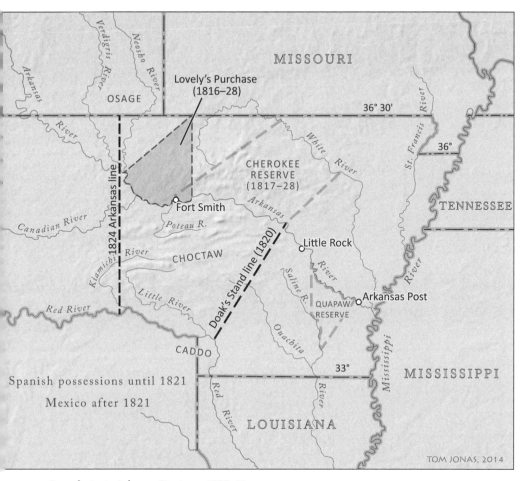

Boundaries in Arkansas Territory, 1808–23

Americans looked to the elongated territory's future as a province of slavery. Those moving into the region, however, focused more attention on its fate as a refuge for eastern Native cultures. Shortly after the organization of Arkansas as a territory, its American residents petitioned President James Monroe about their unmarked northern and southern boundaries. Arkansans needed to know "the extent of country over which our laws are to govern" to ensure the success of their infant political community.[8]

The question of Arkansas Territory's authority involved not only its neighboring states but also the Native cultures living within it. At times these groups clashed, often the result of pressure from the encroachment of American settlement. In late 1820, for example, a fight between Osages and Cherokees took place on the Poteau River near Fort Smith, an isolated military outpost established in 1817. This tension threatened to escalate into a war involving at least a half-dozen Native societies in the area. The *Arkansas Gazette*, the most influential newspaper in the territory, worried about the fate of American settlers near the contested region. The *Gazette* expressed particular concern for Americans living in far-flung Lovely's Purchase, a tract of land deep in the contested zone.[9] It consisted of a triangle extending northeastward from the Arkansas and Neosho Rivers in present-day northeastern Oklahoma and northwestern Arkansas. Agent William Lovely had declared the area a buffer zone between the Osages and Cherokees in 1816. According to the agreement, both groups could hunt in the Purchase but neither had claim to the land itself. Without permanent Native competition an illegal influx of American settlers followed in the late 1810s, thus troubling Lovely's Purchase in a manner not befitting its name.[10] After the 1820 battle, Arkansas territorial governor James Miller parleyed between the Osages and Cherokees to contain the conflict for the time being. The arrival of ever more trespassers, however, kept tensions high.[11]

American settlers, some with slaves, pushed steadily up the Arkansas River throughout the 1820s. Farms, plantations, and towns filled the river valley and spread toward the Ozark and Ouachita Mountains. This trend inspired the territorial legislature to move the capital from Arkansas Post to Little Rock, a hundred miles up the river, in October 1820.[12] At the same time more American Indians emigrated from the Deep South to escape the pressures of American encroachment there, leading to concern among the other newcomers for the future of their vast new territory. The *Gazette* envisioned an onslaught of "those poor deluded wretches" transforming Arkansas into a national dumping ground for displaced cultures.[13] Concocted by people striving to fend off such a fate, in late 1820 several proposals emerged in the territorial press to create a new western boundary for the territory. Up the Arkansas River and its tributaries beyond this new line could dwell all of the local and relocated Native groups. Such an invisible boundary defended by military posts would provide protection from "the tomahawk or scalping knife" as the *Gazette* stated indelicately.[14] Within a year and a half of attaining territorial status, therefore, Arkansans promoted boundaries as political barriers to separate their promising new homes from what they viewed as indigent nuisances and interlopers.

Coincidentally, a contemporary development east of the Mississippi River provided the first step toward just such a boundary. To accommodate American expansion in Mississippi, Choctaw bands living there signed the Treaty of Doak's Stand on October 18, 1820. Negotiated by Andrew Jackson and Thomas Hinds, this agreement granted the Choctaws an elongated land claim between the Red, Canadian, and Arkansas Rivers in then-southwestern Arkansas Territory. A line running northeastward from the Red to the Arkansas upstream from Little Rock separated Choctaw and American claims. Some—but not all—Choctaws in Mississippi would move beyond that line to the new reserve and surrender a significant portion—but not all—of their ancestral lands in Mississippi. The federal government viewed Doak's Stand as a time-buying measure until American and Native societies could live near one another peaceably. As Article 4 of the treaty stated, the new Choctaw cession "shall remain without alteration, until the period at which said nation shall become so civilized and enlightened, as to be made citizens of the United States."[15]

From the point of view of the federal government, Doak's Stand provided the best of both worlds—valuable land in Mississippi and time for the Choctaws to "improve" themselves. American settlers and politicians in Arkansas Territory, however, viewed the matter quite differently. The first whisper of news about the treaty arrived there in early December 1820. It included an unsubstantiated rumor that displeased Choctaws had responded to the agreement by beheading one of their leaders who agreed to it.[16] After the treaty's details appeared in print later that month, the *Gazette* proclaimed the territory's vitriolic opposition. According to the newspaper, the Choctaws surrendered six million acres in Mississippi in exchange for approximately fifteen million acres in Arkansas. The lack of clear knowledge about the rivers involved, however, might mean the loss of even more land. To make matters worse, the paper said, some three thousand Americans—and an unknown number of slaves—currently lived in the parcel granted to the Choctaws. The *Gazette* demanded federal assistance for those settlers to move out of their "Botany Bay" of a territory before a wave of Native migrants arrived.[17]

Over the next few weeks the *Gazette* continued its tirade against "this ridiculous treaty." From the newspaper's perspective the treaty traded only a little land east of the Mississippi River for a lot to the west of it and benefited but a few plantation owners in Mississippi. Doak's Stand thus represented "a death blow to the territory." The newspaper suggested that the treaty aroused intense reactions among the American population of Arkansas because "it not only affects their rights, their liberty, and their property; but it endangers their peace, their safety, their happiness, and their lives. All is at stake."[18] According to the *Gazette*, indignant Arkansans felt as though the federal government had stolen their future and handed it over to the Choctaws now in possession of valuable land on both sides of the Mississippi.

Other newspapers around the United States also noted the treaty but with less passion than in Arkansas. The *National Intelligencer* pointed out that the treaty required Senate confirmation before taking effect. Until that time it represented "but the project of a Treaty," to which the *Arkansas Gazette* retorted: "Yes, it is but a project of a Treaty, and a *wild* one indeed."[19] After the *St. Louis Enquirer* paid the treaty scant attention, the *Gazette* complained that no one outside Arkansas Territory understood their concerns.[20] The *Gazette* found cause to cheer a lengthy letter published in the *Louisiana Advertiser*, however. Its author explored the possible consequences of the treaty for the entire trans-Mississippi region. From the correspondent's point of view, the Choctaws, Cherokees, and Quapaws possessed the best land in Arkansas already. The "swampy and barren" tracts remaining for American settlers meanwhile offered poor prospects for a strong future state. The *Advertiser*'s correspondent also described Arkansas as "a flank guard" to help defend Louisiana and the country as a whole from invasion by the Spanish or Native cultures to the west. Therefore, it was asserted, the federal government must not parcel off the territory for land that might shelter future rebels against American authority.[21]

Arkansas politicians did not stand by idly while the westernmost inhabitants of their territory lost their property. Governor Miller heard of the deal while at Fort Smith negotiating peace after the recent clash between the Cherokees and Osages. Immediately he composed a letter to President Monroe and the Senate pleading against the treaty.[22] The *Gazette* identified Louisiana's state legislature as a potential ally as well. According to the *Gazette* a member of the body had composed a memorial against Doak's Stand, arguing that "it would check the progress of population and improvement in the west. The memorial also suggested that the Choctaw agreement contradicted the Louisiana Purchase treaty, "which held out a promise that the whole extent of the province should in time be admitted to all the advantages of self government, and to a full participation in the blessings of the Federal System."[23] Apparently at least one member of the Louisiana legislature also disputed the notion that the Choctaws might civilize enough to constitute a population worthy of statehood some day.

As the treaty came up before the U.S. Senate in late 1820, Arkansans waited anxiously for the news of its fate. In the meantime the *Gazette* prognosticated about the future for those settlers most affected. It suggested that those displaced would "remove to the Spanish province of Texas, and seek that protection under a foreign monarch, which is denied to them in their native country."[24] Unsurprisingly, the news of the ratification of Doak's Stand meant despair for many Arkansans.[25] Territorial delegate James W. Bates protested the decision in Congress and in letters to executive officials but to no avail.[26] Back home no consolation seemed adequate. When Mississippi's *Port Gibson Correspondent* informed the *Gazette* that the Choctaws "are a civil race of Indians, who have always been friendly to the U. States," the Arkansas paper retorted: "The Editor of the Correspondent is as barren of wit as Spitsbergen is destitute of vegetation."[27] The *Gazette* believed that if Mississippians found such nobility in the Choctaws they should keep them on their side of the river.[28]

After Senate ratification of Doak's Stand, Arkansas politicians and settlers started looking for ways to amend the situation in their favor. Clinging to faith in political boundaries as peacemaking barriers, some interested parties set about trying to push the Choctaw reserve farther west. In the meantime the *Gazette* encouraged displaced settlers not to flee to Texas. Instead it suggested they wait patiently for federal officials to come to their senses and revise the treaty.[29] In March 1821 the Monroe administration, apparently heeding these voices, instructed officials in Arkansas Territory to encourage the Choctaws to settle as far west as possible and to prevent any more American settlers from entering the contested area. Secretary of War John C. Calhoun expected an updated treaty with a new dividing line to follow shortly thereafter.[30] By the end of the year the *Gazette* took solace in reports that relatively few Choctaws had made the trip up the Arkansas River to their new reserve.[31] Their small numbers reflected disputes among the Choctaws themselves. One tribal leader proposed accepting a boundary farther from American settlers while another considered it "improper to divide a family or a nation" with separate reserves in Mississippi and Arkansas.[32] Although intended to resolve tension in the Deep South, Doak's Stand inspired further arguments among all involved.

By the summer of 1822 the proposals for revising the Choctaw line morphed into crafting a new western boundary for Arkansas Territory itself. At that time it stretched across the southern plains to the U.S. border with Mexico hundreds of miles to the west, at the present-day Texas panhandle. Major William Bradford, the commander of Fort Smith, took time from the ongoing Osage-Cherokee negotiations to propose a new territorial boundary. He suggested a straight line running from the southwestern corner of Missouri to the confluence of the Arkansas and Canadian Rivers, about forty miles upstream from the fort. From there Bradford's line extended due south to the international border on the Red River. He considered his proposal a sensible way to provide "sufficient territory for a respectable state."[33] Shortly thereafter some members of Congress considered crafting a new territorial boundary stretching from the southwestern corner of Missouri due south to the Red River. This proposal would require the Choctaws to move their reserve's eastern line to the new boundary. With little incentive to surrender thousands of square miles of their reserve, however, the Choctaws refused the proposal.[34] For the time being, Arkansas's western line stayed put.

The Choctaw reserve filled its share of column inches in the early 1820s but other Native cultures and American settlers created controversies in the region. For many Arkansans the Cherokees represented a paired threat with the Choctaws, the former with a diamond-shaped reserve covering the central Ozark Mountains north of the Arkansas River since 1817. The Cherokee tract operated much the same as the Choctaw reserve, as an outlet for American Indians escaping the competition for their lands east of the Mississippi. After the Treaty of Doak's Stand, local politician Matthew Lyon compared the two native groups relocating in Arkansas Territory. Both possessed land claims east and west of the Mississippi, and Lyon expected that American settlers in Arkansas would come to view their new Native neighbors as troublesome pests.[35] Lyon and others observed that the Cherokee reserve's creation in 1817 had displaced many American settlers north of the Arkansas River. Many had since moved from there south and west onto land now belonging to the Choctaws, from which they would have to leave yet again.[36] From the American perspective, the presence of the Choctaws and Cherokees alike retarded the flourishing of Arkansas Territory.

To the west of the Cherokee reserve sat Lovely's Purchase, the extralegal neutral zone created between Cherokee and Osage claims in 1816. Americans viewed this land as empty, and by the early 1820s some three thousand interlopers had built farms and small towns in it. Articles touting the Purchase appeared in the *Arkansas Gazette*, always eager to promote the Americanization of western Arkansas.[37] Acting territorial governor Robert Crittenden begged federal authorities not to infringe on the development of the region's "garden spot" by evicting its industrious if illegal occupiers.[38] A letter to the *Gazette* in late 1822 argued that the Cherokees lacked any real claim to the land west of their 1817 reserve. It suggested that the Cherokees "wish now to procure a legal right to the land. They are sensible of its great value and importance to this Territory, and know that ere long the whites must have it, and are grasping at it with an avidity which is only equalled by the enormity of their former demands and their known insolence. They are a restless, dissatisfied, avaricious people."[39] Such a claim could have been made just as effectively against the American population that the *Gazette* correspondent represented, a coincidence apparently lost. In any case, the federal government supported the Cherokee claim to hunt in Lovely's Purchase, but did little to evict the Americans trespassing there. Arkansans felt justified in claiming the land for themselves, believing that they deserved to control *something* along their western edge before Native groups took it all.[40]

With so many groups competing for control along the Arkansas River, the challenge of enforcing the laws of conflicting authorities complicated matters even further. The case of Tom Graves, a wealthy and influential Cherokee leader in Arkansas, offers an excellent example of this convoluted situation. In early 1823 the territorial court charged Graves with the murder of an Osage woman in Crawford County, established north of Fort Smith where the lines between Choctaw, Cherokee, Osage, and American control blurred. Although Graves did not deny the act, his lawyer argued that the deed took place on land ceded to the Choctaws by Doak's Stand. According to an 1817 federal law, U.S. courts did not have the jurisdiction to judge offenses committed by Natives on their lands. The prosecution, meanwhile, argued that since so few Choctaws lived in the reserve they did not control it and it should not be considered Native land. More American settlers than Choctaws lived there still, which to the prosecutor meant that the laws of the federal government remained in force regardless of the 1820 treaty. In the end the court acknowledged the reality of a continued non-Native presence in the reserve but it declared the 1817 law in force since the land remained Choctaw property technically. The court therefore freed the Cherokee Graves, a man who benefited from the disputed nature of western Arkansas.[41]

Shortly after the Graves incident an election for territorial delegate focused upon the problem of defining the boundary between American and Native settlement. In a circular printed in the *Arkansas Gazette*, candidate Henry W. Conway listed as his primary goals defining the territorial boundaries, extinguishing Native land claims in Arkansas, securing Lovely's Purchase, and defending the borders militarily. His competitor, Major Bradford of Fort Smith, made almost identical promises. Bradford also proposed a geometric western line from the southwestern corner of Missouri roughly southwest to the Red River, which echoed much of his boundary proposal of a year earlier.[42] During the campaign the *Gazette* printed rumors of proposed revisions to the Choctaw and Cherokee holdings that would benefit the American presence. It published a letter by Missouri's Senator Thomas Hart Benton pledging congressional action on those points as well.[43] In 1823 white Arkansans felt more confident of their impending triumph over their Native neighbors than they had in years and elected Conway to Congress to lobby on their behalf.

Federal officials turned unprecedented attention to western Arkansas in 1824. Reports of the confusing, overlapping claims of American settlers and Native cultures demanded investigation. Secretary Calhoun called upon Congress to survey the contested eastern boundary of the Choctaw reserve to clarify the matter for the region's inhabitants.[44] In the meantime Benton and others in Congress sought to redraw the territorial line regardless of the Choctaw boundary. They proposed a geometric western limit following the thirty-six thirty line due west for forty miles from the southwestern corner of Missouri and running from there due south to the international border on the Red River. The line included most of Lovely's Purchase and a large portion of the eastern Choctaw reserve. Benton argued that such a boundary "will give to the future state of Arkansas that power and magnitude, to which, as a frontier state, in relation both to a foreign nation and numerous Indian tribes, it will be justly entitled."[45] His fellow congressmen agreed and President Monroe signed the bill on May 26, 1824. When the news arrived in Arkansas a few weeks later, it brought jubilation to the Americans. Choctaws and Cherokees, however, viewed the new boundary with disappointment. They expected to lose part or all of their reserves to correspond to the new line.[46] The U.S. Army made preparations for the changes to come that spring. With the territorial boundary now forty miles to its west and reserve lines expected to follow

THE WESTERN ARKANSAS BOUNDARY 81

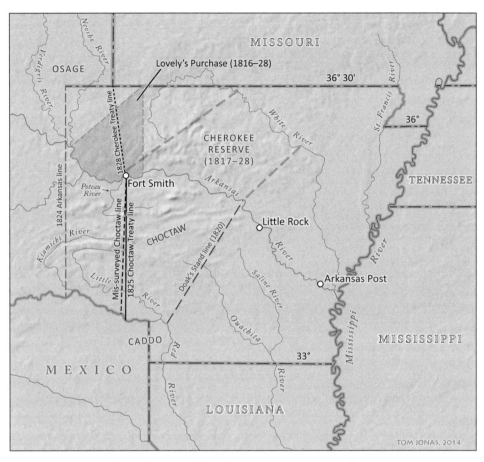

Boundaries in Arkansas Territory, 1824–28

suit, Fort Smith closed. The troops there relocated to a post on the Neosho River, Fort Gibson, just west of the new line.[47]

The year 1824 witnessed other changes in the imposed divisions of Arkansas Territory as well. American settlers looked forward to the official survey and marking of their new western boundary. Additionally, the federal government complied with a memorial of the territorial legislature made the previous year. It negotiated with the Quapaws to cede their reserve south of the Arkansas River below Little Rock, surrounded on all sides by American farms and plantations. Land agents divided the parcel for quick sale to land-hungry settlers eager to fill the area.[48] Yet the fate of Lovely's Purchase remained in question as surveyors marked the western limit of the 1817 Cherokee reserve. Americans resented the expense of the boundary survey for a culture they considered destined to leave that land in short order. Many Cherokees opposed the work as well, doubting the impartiality of the surveyors. They petitioned President Monroe, proclaiming their "universal dissatisfaction and disgust" with the property division.[49]

While the Quapaws headed southwest into Texas exile, a delegation of Choctaw leaders traveled east to Washington, D.C., in an effort to settle their reserve's boundary once and for all. In the negotiations, Calhoun hoped they would accept the new western limit of Arkansas as the eastern limit of their territory. He suggested: "There is no probability that any State or Territory will be erected to the west of Arkansas Territory, which would leave the Choctaws, who might choose to emigrate, a quiet and undisturbed possession of the immense country."[50] But the Choctaws proved far shrewder than Calhoun expected. Their representatives objected to losing so much fertile ground, especially since Doak's Stand emphasized the intent to make them "civilized" American farmers. A compromise proposal came from the Choctaw delegation calling for a line run due south from the southwestern corner of Missouri to the Red River. This line would push the new territorial limit east forty miles and preserve that land for native groups on both sides of the Arkansas River. Calhoun and other federal officials spent the last two months of 1824 pleading with the delegation to change their minds.[51] Yet the Choctaws understood the importance Americans ascribed to these invisible lines and proved just as determined to shape them to their own advantage.

When they realized that the Choctaws would not acquiesce, Arkansas politicians pleaded their case to federal officials. Conway wrote to Calhoun that his territory could not accept the boundary proposed by the Choctaws and argued in favor of Congress's 1824 line. The Arkansas legislature submitted its own compromise proposal to Congress, suggesting a series of geometric divisions linking the southwestern corner of Missouri, the falls of the Verdigris River, and the confluence of the Kiamichi and Red Rivers. Such a boundary surrendered the state's new northwestern corner but retained most of the Choctaws' disputed claim for Arkansas. Even the president entered the fray. Monroe suggested incorporating the Kiamichi and Poteau Rivers into a new western Arkansas line. It would give both Americans and Native cultures part of the disputed territory but neither would receive it all. The Choctaw delegation rejected each of these compromises. In response they demanded a boundary even farther east than their first proposal, one that would stand several dozen miles east of the current line.[52]

As the treaty-making dragged into 1825, the Monroe administration, soon to leave office, decided to settle the matter once and for all. Calhoun declared that if Arkansans did not accept the Choctaws' first compromise line, the federal government would have no choice but to enforce the 1820 Doak's Stand boundary and evict an estimated five thousand trespassing settlers.[53] Begrudgingly the American settlers acquiesced. On January 20, 1825, the Choctaws approved a new treaty, with an eastern boundary of their reserve beginning "on the Arkansas [River], one hundred paces east of Fort Smith, and running thence, due south, to Red River."[54] The new Choctaw reserve line depended vaguely upon the measurement of unspecified footsteps from some random point at an abandoned outpost. It stood farther west than the 1820 Choctaw boundary but some forty miles east of the 1824 western boundary of Arkansas. Most importantly, the 1825 boundary demonstrated the ability of Native cultures as well as American settlers and politicians to shape the trans-Mississippi West. The line conformed more to what the Choctaws proposed than what settlers and politicians in Arkansas desired, much to the latter's chagrin.

News of the 1825 Choctaw treaty dumbfounded Arkansans. The *Arkansas Gazette* seethed that it "cuts off ... about 3,000 of our citizens, who ... the United States engage to *remove* to the east side of the line established by this treaty, and that this line is to be and remain the *permanent* boundary between the United States and the Choctaws!"[55] Pleading for mercy, Conway persuaded the federal government to allow American settlers on Choctaw land to remain long enough to bring in the year's crop.[56] Meanwhile the Arkansas legislature and new territorial governor George Izard tried without success to convince the Choctaws to push the line westward or at least not settle too close to the new boundary.[57] With little chance of either transpiring in the near future, the territorial delegate's brother James Conway led a survey team to mark the new Choctaw limit by the end of 1825. He reported to the *Gazette* that the American settlers he encountered during the work "complain much of our government, and say, that, to be driven from their farms, must inevitably cause their ruin."[58] That same year another survey established the western territorial line declared in 1824.

The two boundaries created an overlap of almost three and a half million acres in the roughly forty miles between the western line of Arkansas and the eastern line of the Choctaw reserve.[59] Resentment ran high among Arkansans who could not understand their failure to cajole the Choctaws.

In the hopes that they could eke out victory against at least one of their Native neighbors, Arkansans shifted their focus to the land contested with the Cherokees in the mid-1820s. The 1824 territorial boundary granted Arkansas most of Lovely's Purchase, which many viewed as a consolation prize to the tracts lost to the Choctaws. Delegate Conway wrote to his constituents in 1826 of the General Land Office's intention to demarcate twenty townships in the Purchase and open them to settlement.[60] Such a development would surround the Cherokees' diamond-shaped tract eventually and likely lead to their eviction from Arkansas as had happened to the Quapaws. Governor Izard expressed the attitudes of many American settlers in Arkansas when he described the occupation of Lovely's Purchase as a vital step toward the extinguishment of Cherokee claims in Arkansas.[61] As the *Gazette* argued haughtily: "To every friend of Arkansas, this change from the rude and uncultivated to the civilized and improved state, must afford real and sincere gratification."[62] Having lost their struggle with the Choctaws for the time being, Arkansans took solace in the hope that they might bully out the Cherokees.

Both the Choctaws and Cherokees struggled to protect their reserves against the designing Arkansans in the late 1820s. The Choctaws insisted that interloping American settlers leave their land and used the 1825 boundary survey to identify squatters for eviction.[63] Congress took up the matter in early 1827 but expressed little concern for the Native cultures. Conway told his colleagues that his territory did not want to upset its neighbors but that Arkansans "occupied the country before it belonged to the Indians."[64] By contrast Senator Thomas B. Reed of Mississippi, eager to clear the remaining Choctaws from his own state, portrayed them as wild children of the forest for whom life in Arkansas would be a much better fit. He described Native groups in the Deep South as "seeking a new abode on our Western borders, where the aboriginal character of the red man may be preserved from entire annihilation."[65] Such a vision contrasted with the "civilizing" intent of the western reserves, however. In any case, according to the *Arkansas Gazette* only about a dozen more Choctaws emigrated to their vast trans-Mississippi reserve in the summer of 1827.[66] Apparently the Choctaws did not share Reed's vision of their untamed future. The lack of tribal interest in the reserve infuriated Arkansans unhappy about its loss to them but inspired hope that they might yet push the Native newcomers west beyond their territorial limit.

The Cherokees nursed grievances of their own in the mid-1820s. The thought of new American settlements surrounding their 1817 reserve and perhaps forcing them to abandon it incensed the Cherokees. To that end, those living in Arkansas approved a tribal death penalty for anyone who supported the idea of selling or exchanging more land.[67] In 1826 the territorial legislature exacerbated the situation when it memorialized President John Quincy Adams about the Cherokee's perceived failure to "civilize," the ultimate goal of the trans-Mississippi cessions. Politicians in Little Rock described the Cherokees as "a restless, dissatisfied, insolent, and ambitious tribe, engaged in constant intrigues with neighboring tribes to foment difficulties, produce discord, and defeat the great object of the Government, in promoting the civilization of the Indians, and preserving peace among them." The legislature considered surrounding the Cherokees with American settlements the only way to "paralyze their wicked efforts" and bring prosperity to the region.[68] This portrayal contrasted with the Cherokees' years of successes as agriculturalists and merchants in Arkansas, however.[69] As one Cherokee leader wrote to the *Arkansas Gazette*, their manner of life, from houses to clothing to food, compared favorably with the Americans. He argued that literacy rates ranked higher among Cherokees, and warned not to "tell about hemming us in, by your settlements, to civilize us. But if civilization consists in pitched battles, to murder one another, or in shooting our neighbors and brothers, in streets and places of public resort, then we are in a woeful state of barbarism." Regardless, in October 1827 the legislature organized Lovely County to manage the Americans moving into the Purchase and elsewhere in the northwestern corner of the territory.[70] The Cherokees felt the noose tightening around their Arkansas holdings.

Feeling hemmed in on all sides, several Cherokee leaders traveled to Washington, D.C., in early 1828 to protect their interests. Tom Graves of 1823 murder trial fame joined the small delegation, all members of whom disregarded the tribal death threat to seek a new land deal. After several months of closed-door negotiations, the Cherokees and the Adams administration signed a new treaty on May 6, 1828. The delegation surrendered their 1817 reserve north of the Arkansas River. In exchange they received an estimated seven million acres farther to the west. A diagonal line running from the southwestern corner of Missouri to the northeastern corner of the Choctaw reserve near the abandoned Fort Smith formed the eastern boundary of this new Cherokee reserve. News of the treaty reached Little Rock about a month later and met with support there. Even though the new Cherokee reserve included much of Lovely's Purchase, many residents applauded gaining control over the old cession in the north-central part of the territory, nearer to most of its American settlements.[71]

Arkansans' jubilation waned once the treaty's full ramifications grew clear, however. The Cherokee accord contained a provision that took the territory completely by surprise. The preamble stated in part that both the Cherokees and Choctaws worried about further Arkansas designs on their overlapping land claims. After all, both the 1825 Choctaw treaty and the 1828 Cherokee agreement created reserves with land that overlapped the territory's 1824 western boundary. To that end the treaty's first article focused on the limit of Arkansas itself: "The western boundary of Arkansas shall be, and the same is, hereby defined, viz.: A line shall be run, commencing on Red river, at the point where the eastern Choctaw line strikes said river, and run due north with said line to the river Arkansas, thence in a direct line to the southwest corner of Missouri."[72] Thus the federal government used tribal treaties rather than legislation to drag the western boundary of Arkansas more than forty miles eastward. In doing so it made the polity correspond to both of the recently established Native reserves and dispatched the hassling matter of overlapping jurisdictions. The Cherokee delegation showed as much sagacity in boundary affairs as had their Choctaw counterparts. The latter struck the first blow against the 1824 Arkansas line and the former finished it off.

Astonished Arkansans decried the change to their western boundary. The new territorial delegate, Ambrose H. Sevier, expressed the sentiments of many. He wrote to Secretary of War James Barbour pleading against moving the boundary "one inch further to the east."[73] Barbour acknowledged that Arkansas had lost some land as a result of the treaty but also pointed out that the territory had purged from its boundaries practically all Native groups. From Barbour's perspective, Arkansans would come to view the agreement as mutually beneficial for American settlers and American Indians in the region.[74] Eventually Sevier agreed with the secretary. Ten months after the Cherokee treaty's negotiation, he admitted that the boundary shift "would have been made sooner or later" and that "it ought to give universal satisfaction, because the personal sacrifice has been comparatively nothing, and the remuneration munificent."[75] Having secured a wholly American territory without bloodshed, Arkansans grew to embrace their western boundary once the initial shock of its creation wore off. One of its few supposedly negative aspects involved the role that Native reserves played as dueling grounds. By removing the boundary farther to the west, men who sought to defend their honor without running afoul of territorial laws banning the practice found it necessary to travel greater distances to cross the line.[76]

In the fall of 1828 a survey team marked the new line between southwestern Missouri and the Arkansas River, and it remains the northwestern boundary of Arkansas today.[77] The federal government sought to mitigate the treaty's impact on American settlers in Lovely's Purchase as well. Congress provided liberally for those now required to move east of the new territorial boundary. It approved a bill to provide all heads of household, widows, and single men affected with up to a half section of land—three hundred twenty acres—anywhere on the public domain in Arkansas.[78] The relief program benefited only legal settlers with proof of land ownership, however. The *Arkansas Gazette* worried that squatters who had established farms on public land throughout the territory with the intention of legalizing their holdings later might lose their property to refugees from the west.[79] Local officials expressed little concern about the dispossession of such settlers, however. To manage the transition, Governor Izard called a special session of the territorial legislature and issued a proclamation to settlers in Lovely's Purchase to prepare for removal.[80]

The myriad reactions expressed by Arkansans echoed disputes among the Cherokees about the 1828 treaty. Some considered carrying out the death penalty on the delegation that handed over land in northern Arkansas that Cherokees had occupied and improved for more than a decade. Yet, according to the *Gazette*, attitudes among the Cherokees also changed as the months passed. The notion of escaping the insatiable appetite of land-hungry Americans struck many as appealing.[81] One issue of great concern to Cherokees and American settlers alike remained—the fate of their soon-to-be-abandoned farms and plantations. In accordance with the 1828 agreement, the federal government assessed the value of land surrendered by the Cherokees and compensated them for the changes made to it.[82] Arkansans looked forward to receiving "much valuable, improved, and desirable property" from the Cherokees. Inheriting farms already in production, they believed, would help them civilize the region more rapidly than their supposedly savage predecessors did.[83] Such attitudes reflected the countless paradoxes that emerged when Native and Euro-American lifestyles collided in the trans-Mississippi West.

The compensation provisions of the 1828 treaty reminded some Arkansans of the plight of those displaced by the 1825 Choctaw agreement. The *Gazette* asked why only those settlers evicted by the Cherokees should receive aid when many had lost land to the Choctaws as well. "Those people were as much entitled to the sympathies of the government, as those residing in the country recently ceded to the Cherokees," the newspaper opined, but "the government has turned a deaf ear to their prayer."[84] Sevier raised the matter in Congress in the waning days of 1828 by introducing legislation to provide for the affected settlers.[85] Attorney General William Wirt dismissed the effort, however, arguing that the federal government had agreed to relieve Americans leaving the new Cherokee territory only.[86] Sevier hoped a change of administrations in March 1829 would give a better result. President Andrew Jackson's attorney general concurred with his predecessor much to Sevier's chagrin.[87] For several years the territorial legislature requested land reparations for settlers displaced by the Choctaw but to no avail.[88] Americans south of the Arkansas River did not receive the generous assistance bestowed upon their counterparts on the north bank.

By 1830 the political divides of western Arkansas seemed stable at long last. Clearly marked boundaries separated the various groups that had competed for it over the previous decade. But the actions of the federal government that year complicated matters in the area again. The Indian Removal Act, supported by the Jackson administration as the best way to clear out Native peoples living east of the Mississippi River for American settlement, passed Congress that summer. It gave Jackson the power to designate lands in the Louisiana Purchase's residue—specifically west of Arkansas and north and west of Missouri—into reserves for Native cultures from any state or territory. The law did not mandate their forced removal but provided the legal backing for such a policy.[89] Responses to the Indian Removal Act ranged from enthusiasm to derision. Many resolutions against removal flowed to Washington, D.C., in late 1830 and early 1831, for example, written primarily by citizens from northeastern states.[90] This decreed exodus met with scorn in Arkansas Territory as well, albeit for reasons of security rather than ideology. American settlers there feared removal would lead to an influx of embittered, hostile Native cultures that would pass through the heart of Arkansas on their way toward western exile.

Arkansans proposed civil and military suggestions to the federal government to protect themselves as enforced relocation proceeded throughout the 1830s. Their legislature memorialized Congress to provide a free quarter section of land to anyone who moved to the western edge of the territory. Encouraging American migration and settlement along the boundary, it believed, would create a first line of defense to buffer the rest of the territory. It might also inspire newcomers from other western states with experience in dealing with potentially hostile Natives. Such a forbidding human barrier could obviate the need for expensive military patrols along the boundaries between Arkansas and the Choctaws and Cherokees as well.[91] The desire to build up an American presence to compete with Native neighbors intensified after Congress designated unorganized land west of Missouri and Arkansas as a supposedly permanent "Indian country" in 1834.[92] In response the legislature asked the U.S. Army to abandon Fort Gibson, located forty miles west of the territorial line after the treaty of 1828, and send those soldiers to the reoccupied, closer Fort Smith.[93] From the Arkansans' point of view, with the support of the military they and their plucky frontier compatriots could defend themselves against any threat from the west.

Native populations west of Arkansas Territory expressed concerns about the implications of removal as well. Many feared intertribal conflict between voluntary and forced migrants. Potential competition with other cultures in this new "Indian country" also inspired concern. In addition, they worried about the long-term fate of their reserves, believing that demarcated territories represented the only viable insurance for their societies' survival. As removed Choctaws arrived in the reserve south of the Arkansas River, they sought to reassert their cession's integrity. In the spring of 1836 more than seventy Choctaw leaders begged Congress to survey and mark their boundary with Arkansas Territory again to help the Choctaws protect their property. As the memorial noted nervously: "We have ever been the friends, in peace and in war, of the United States, and having from time to time ceded away our country east of the Mississippi, and removed to our present abode, where we wish to remain in peace and friendship with our American brothers, knowing that there is now no resting place, should we be forced from where we now are, this side of the

Rocky mountains, we submit it to your honorable body with the greatest confidence, that this question shall be put at rest."[94]

Well skilled in boundary-making, the Choctaws wanted to protect their future by using the tools of the American system. Such behavior contrasted with the popular image of a people who refused "civilization" as Americans perceived the term. Indeed, the Choctaws and Cherokees acted just as doggedly as their American counterparts when it came to boundary affairs.

As the federal government carried out the removal policy, the population living near the Arkansas western boundary grew ever more diverse. Yet those living on either side of the line bore remarkable similarities. While traveling through the region, Major Ethan Allen Hitchcock noted that boundary residents carried on the same lifestyle regardless of their cultural background. He stated of the Cherokees: "There was nothing to distinguish appearances from those of many of our border people except the complexion . . . and superior neatness."[95] Nonetheless, the worries of western Arkansans remained acute after their territory joined the Union in June 1836. Many settlers living near the boundary demanded a show of force to defend them from the imagined threat of their Native neighbors, an argument they had not made since 1820. Residents of Washington County, in the northwestern corner of the state just east of the Cherokee reserve, appealed to their elected officials for protection in 1838. They argued that "by our local situation in the far west, on the extreme frontier of the United States, in the neighborhood of the untamed savage of the prairies, we are entitled, on principles of humanity, to the efficient protection of our Government."[96] While American settlers believed in the power of invisible lines to stabilize the region, they wanted a military reinforcement of those divisions.

Members of Congress and other federal officials considered the defense of the "border states" as well.[97] In a proposal to the secretary of war in 1837, acting quartermaster general Trueman Cross suggested two military barriers paralleling the westernmost state boundaries.[98] Cross's barrier would stretch from Fort Snelling on the upper Mississippi River at present-day Minneapolis to Fort Towson on the north bank of the Red River, just north of the Republic of Texas. Roads would link a line of forts including Gibson and Smith to deploy troops as needed. Cross's proposal represented the most substantial attempt to impose a separation between Euro-American settlers and American Indians since the 1763 Royal Proclamation Line. It planned to contain over three hundred thousand relocated Native peoples with a two-thousand-mile-long line guarded by seven thousand American soldiers.[99] Cross's vision came into partial effect over the next few years. Yet territorial expansion to the Pacific Ocean in the 1840s meant that "Indian country" found itself in the middle of the country rather than on the western edge. Native territory shrank dramatically with the creation of Kansas and Nebraska Territories in 1854, squeezing ever more cultures into the present-day state of Oklahoma.

While the nation's focus turned elsewhere after the 1830s, the boundary between Arkansas and Indian Territory remained controversial for decades for those living on both sides of it. The southern portion between Arkansas and Choctaw lands proved the most problematic. Later surveys indicated that James Conway's 1825 line did not run due south from the vague starting point near Fort Smith but instead veered slightly west of south. Whether accidental or intentional this error gave Arkansas approximately 136,000 more acres than the treaty allowed. The Choctaws and Chickasaws—who combined their western reserves in 1837—appealed before and after the Civil War for a resurveyed boundary that followed the letter of the law.[100] In 1870 they asked the federal government "to protect its loyal and obedient people against the rapacity and craft of those who desire and hunger, by crooked means and seemingly just legislation, to possess themselves of [Choctaw and Chickasaw] lands, to invade their country like locusts, and to devour their substance."[101] An 1877 survey remarked Conway's line with an eye toward compensation rather than restoring the sliver of land to its legal owners, however. After paying legal fees and the bills of delegations to Washington, D.C., the Choctaws and Chickasaws netted less than thirty-five thousand dollars for the thousands of acres lost.[102]

The Arkansas-Cherokee boundary provided grist for the mill of controversy in the late nineteenth century as well. As late as 1892 the Arkansas legislature pondered asserting the state's claim to the part of Lovely's Purchase that remained outside the state boundary after 1828. Roughly half of the parcel lay beyond that line and solons of Little Rock sought to reunite the purchase. Since the land to the state's west remained Indian Territory and therefore the domain of the federal government, they needed only to negotiate with Congress rather than the Cherokees for it. This reasserted land claim included the town of Tahlequah, the tribal capital since the mid-nineteenth century. A Cherokee newspaper in Tahlequah dismissed the Arkansas claim and bemoaned its long buffeting by the federal government: "We have sold the United States more than half our lands, and she ought to be satisfied, but [is] not content with a half loaf."[103] In the end, the notion of reuniting Lovely's Purchase came to naught. The boundary remains intact to this day, perhaps best recognized by the massive Cherokee Casino just across State Line Road from Siloam Springs, Arkansas.

Not until 1905 did the line between Arkansas and the Native nations to its west reach its final alignment. For years the municipal authorities of Fort Smith, Arkansas, complained about a small strip of land between the political boundary and the Poteau River created by the 1825 line. The Choctaws and Chickasaws exercised little control over this territory and it developed into a haven for illegal activities. To resolve the matter Congress approved a slight alteration to the boundary, shifting it westward to the Poteau River for about seven miles. The action assigned to Arkansas an additional 130 acres. The land sold for over twenty-three thousand dollars, and the state gave those profits to the Choctaws and Chickasaws.[104]

The final significant proposal to alter this contested boundary emerged in 1905 as well. For years debate had raged locally and in Washington, D.C., over the fate of Indian and Oklahoma Territories. Congress established the former, present-day eastern and south-central Oklahoma, in 1889 by amalgamating the reserves of the Choctaws, Chickasaws, Cherokees, and several other Native cultures. It organized the latter, the remainder of the modern state, in 1890. Inspired by the federal government's recent willingness to hand over land near Fort Smith, the Arkansas legislature sensed its chance to reclaim territory lost via the 1828 Cherokee treaty. In the spring of 1905 the assembly resolved to annex the Cherokee, Choctaw, and Chickasaw reserves. Legislators claimed support for the notion among the Native groups affected, who "by right of ancient treaty belong in large part to this State."[105] Apparently the passage of more than eighty years had changed Arkansans' attitudes about the presence of American Indians in their state. Federal authorities did not support the idea, however, and united the two territories to join the Union as Oklahoma in 1907.

Evolving throughout the nineteenth century, the western Arkansas boundary reflected many years of intense competition and compromise. Numerous groups including American settlers and disparate Indian populations played essential roles in this transformation. Each faction used the legal tools available to it, whether local or national legislation or sweeping treaties, to protect its immediate and long-term interests. Native cultures proved adept at manipulating the boundary-making process to their own ends, much to the frustration of those who hoped to contain such cultures through the same means. Yet the defined and defended boundary helped keep the peace between Indians and settlers. In few parts of the trans-Mississippi West did the importance of these invisible lines shine through as clearly as in the area now divided into Arkansas and Oklahoma, where diverse groups struggled to exert control "on the extreme frontier."

Markers at the pre-1836 northwestern corner of Missouri, with Iowa to the left and Missouri to the right of the road

CHAPTER 4

Blood Will Be Shed
The Missouri-Iowa Boundary

Fans of college football may recognize the phrase "Border War," which has been applied to longstanding gridiron rivalries between numerous universities in the West. In the early nineteenth century a far more serious border war took shape in the trans-Mississippi West. Over several decades the divide between Missouri and Iowa evolved through a process of legal wrangling, political posturing, and even the threat of armed combat. The situation grew so intense that one of the governors involved was worried "that blood will be shed; and if blood begins to flow, it is impossible to foretell where the matter will end."[1] While modern scholars often view this "war" as little more than an amusing anecdote, the struggle over the Missouri-Iowa line represents something far more important. It illustrates the intrinsic value Americans placed upon state boundaries and the determination of people living within them to protect both the property they defined and the political identity they created.

The land eventually split into Missouri and Iowa came to the United States through the Louisiana Purchase in 1803. The line separating these two polities originated as a boundary between settlers and American Indians created five years later. By 1821 it found a new purpose as the northern line of the new state of Missouri. Several contradictory surveys of this line fueled uncertainty about authority along it, a problem exacerbated in the late 1830s as more settlers established homes and communities in the contested zone. For many, it echoed the antebellum debate between a state's rights and federal power. By 1839 the struggle between Missouri and Iowa erupted into a conflict marked by two distinct phases. The first, consisting of bombastic statements from politicians on both sides, was met with amusement by those living in the disputed region. But when this spat threatened to turn violent, the conflict entered its second phase, and residents of the contested region eventually took matters into their own hands and negotiated a settlement. Another decade passed, however, before the dispute saw a final resolution. Through it all, Missourians and Iowans demonstrated a sincere dedication to enforcing their boundary claims and providing an invisible barrier against their avaricious competitors on the other side.

Federal activity in the Louisiana Purchase commenced shortly after the ink dried on the 1803 treaty. To learn more about the lands it had just bought, the Jefferson administration authorized several expeditions in the years that followed. Although Meriwether Lewis and William Clark have kidnapped the textbook version of the story, several other expeditions scouted their own parts of the Purchase.[2] Captain Zebulon M. Pike, for example, led an often-overlooked trek in late 1805 to determine the source of the Mississippi River. He also sought to establish American influence with native cultures there to supplant British influence from Canada. In August 1805 Pike led almost two dozen soldiers and naturalists up the river from St. Louis. Shortly thereafter, Pike noted a feature in the Mississippi called "the rapids De Moyen," which featured prominently in the later boundary dispute between Missouri and Iowa. The eleven-mile-long stretch of rapids, located above the confluence of the Mississippi and Des Moines Rivers, complicated riverine travel through not only its rapids but also its plethora of sandbars and shoals.[3] The Des Moines received its name in 1673 from French explorer Jacques Marquette, who corrupted the name of an Illinois Indian settlement near there.[4] For Pike's company more than a hundred and thirty years later the rapids marked the start of the unknown, for "although no soul on board had passed them, we commenced ascending them, immediately."[5]

Three years after Pike and his crew navigated the Des Moines Rapids, and only five years after the United States took possession of the Louisiana Purchase, the first precedent for dividing the future states of Missouri and Iowa emerged. As with several other lines in the early trans-Mississippi West, Native societies played a large part in its creation. The Osages represented the most powerful American Indian group of the lower Missouri River valley, wooed for decades by the French, Spanish, and British. Their reaction to the Louisiana Purchase reflected this long-maintained independence: Osage leaders threw a note from St. Louis merchants bearing news of the deal into a fire, rejecting the new American claim to their homeland.[6]

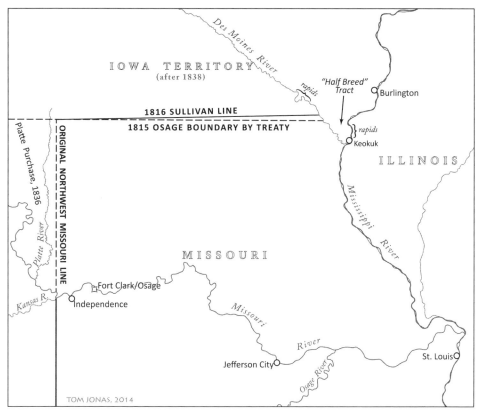

Treaty boundaries in northern Missouri Territory, 1808–24

Nonetheless, in 1808, Missouri's territorial governor, Meriwether Lewis, treated with the Osages with the help of his old colleague William Clark. Lewis hoped to open more land to American settlement up the Missouri River by delineating the limits of Osage territory. He also sought to punish the Osages for raids on regional American settlements that they viewed as a threat to their survival. "The establishment of a boundary has long been desirable," the governor observed, "and the want of one, settled by treaty, has never ceased to create *doubts*, and sometimes *embarrassments*, of the most serious nature, in our courts of justice."[7] The treaty conference took place at Fort Clark, a post on the Missouri River about forty miles east of present-day Kansas City. Lewis and Clark convinced several prominent Osages to remove themselves and all their people between the Missouri and Arkansas Rivers, and between the Mississippi River and a north–south line anchored at Fort Clark (later renamed Fort Osage). The vaguest part of the treaty mentioned a surrender of their claims north of the Missouri River as well.[8] Yet the regional tensions that culminated in the War of 1812 prevented either side from enforcing the treaty provisions for the time being.

Following the war, the lower Missouri River valley stood poised to experience rapid growth. By removing British interference in western affairs, attacking Native allies of the British, and promising land pensions to veterans, the federal government reinforced its claims to the region. Although some wanted Clark, the territorial governor since 1813, to continue the campaign against American Indians in the style of Andrew Jackson versus the Creeks in Alabama, he used the olive branch instead. Clark called a massive conference at Portage des Sioux, just north of St. Louis, in the summer of 1815. With few options since the departure of the British, more than a dozen Native groups—including the Osages—pledged their friendship to the United States.[9] In this new spirit of cooperation, the Osages made formal their land cession north of the Missouri River, an area over which they had at best a tenuous grasp at the time. With Osage consent, in 1816 a survey team led by John C. Sullivan defined the line between Osage and American territory, one that soon constrained other Native cultures in the region as well. Sullivan's team started at the confluence of the Missouri and Kansas Rivers, above Fort Osage at present-day Kansas City, and proceeded due north for a hundred miles. From there, the surveyors turned eastward and drew another straight line to the Des Moines River a few miles upstream from its confluence with the Mississippi River.[10] By following magnetic rather than true north, though, this second line tended slightly northward the farther east Sullivan traveled, an error that came to haunt later generations.

Sullivan's line took on additional significance as the 1810s came to a close. The increasing population of Missouri entitled the region to pursue statehood, and, as the only surveyed boundary north of the Missouri River, the 1816 line struck many as a sensible northern limit for the new state.[11] In 1820 Missouri voters adopted a constitution that included a series of boundaries for the new state. The document identified—in a seemingly thorough fashion—the northwestern and northern lines respectively as

> a meridian line passing through the middle of the mouth of the Kansas river, where the same empties into the Missouri river; thence from the point aforesaid north, along the said meridian line, to the

intersection of the parallel of latitude which passes through the rapids of the river Des Moines, making the said line correspond with the Indian boundary line; thence east, from the point of intersection last aforesaid, along the said parallel of latitude, to the middle of the channel of the main fork of the said river Des Moines; thence down and along the middle of the main channel of the said river Des Moines to the mouth of the same, where it empties into the Mississippi River.[12]

Missourians thus specified the Sullivan survey—"the Indian boundary line"—to avoid the hassle and expense of reinventing the wheel. Yet the 1816 line did not intersect any rapids at its eastern terminus, in contrast to the wording of the constitution. With a dearth of American settlement anywhere near most of the line, though, it mattered little at the time. When President James Monroe signed the proclamation admitting Missouri to the Union in August 1821, Sullivan's line performed double duty as an Indian boundary and Missouri's limit north of its namesake river.

Another three years passed after statehood with little American interest in the northern line of Missouri until a new agreement between the federal government and Native peoples of the Mississippi River valley affected the area significantly. The Ioway and the Sac and Fox nations, which dominated the Mississippi near its confluence with the Rock, Des Moines, and Iowa Rivers, surrendered their claim to land within Missouri in 1824. Agreements between all the parties referenced the Sullivan line, reinforcing its longstanding service as both a Native and American boundary. The Sac and Fox treaty also set aside a small parcel between the Mississippi and Des Moines Rivers as a sanctuary for "half-breeds" of that group.[13] The 1816 survey's importance resonated throughout the early nineteenth century. No fewer than fifteen pacts between the federal government and American Indians referred to the Sullivan line, reinforcing its value for many people in the region.[14]

Within a decade of statehood, Missourians contemplated the first of many proposed alterations to their northern frontier. Settlers pushing westward called for the state to annex a parcel between the straight northwestern line and the Missouri River to the west, a region they called the Platte Purchase. Others working their way up the Mississippi River wanted to alter the northeastern corner of the state. To that end, Missouri's legislature petitioned Congress in 1829 to let them annex the small "half-breed" tract created by the Sac and Fox treaty, yet their request fell on deaf ears.[15]

Two years later Missouri politicians offered an aesthetic reasoning for the northeastern addition: "It is a wedge in the corner of the State, disfiguring the form, and destroying the compactness, of our territory."[16] Apparently few Missourians cared that they had made their own wedge into Arkansas with the "Bootheel" in the early 1820s. But Congress saw no pressing need to add the mixed-ethnicity population to Missouri. Perhaps in compensation it acquiesced to the state's northwestern expansion by ceding to Missouri the Platte Purchase in 1836, extending a line due west from the state's northwestern corner to its namesake river (see chapter 2).[17]

With one of their hoped-for alterations to the northern boundary approved, many Missourians called for a new survey of it. The press in particular fantasized about how much land their state might now govern. One St. Louis newspaper hoped that the line intersected the Missouri River near the Council Bluffs, named after a meeting there between Lewis and Clark's Corps of Discovery and local American Indians in 1804.[18] In fact the bluffs stood some fifty miles north of Missouri's line. To clarify the situation, Missouri's legislature authorized a new survey of the boundary in 1837 and invited the federal government, as the ultimate authority over western territories, to take part.[19] When it declined, Missouri carried out the effort alone, building upon Sullivan's 1816 precedent.

Unlike Sullivan's trek, which labored from west to east, the 1837 commission led by Joseph C. Brown proceeded westward from a series of rapids at the Great Bend of the Des Moines River. According to Missouri's constitution, "the parallel of latitude which passes through the rapids of the river Des Moines" served as the northern boundary. To Missourians that meant a series of breaks in that river. Sullivan's line, however, did not intersect any rapids along the Des Moines. Brown's work exposed the contradiction in Missouri's constitutionally declared limits. His team therefore surveyed due west from a random point he selected along the rapids at the Great Bend. Therefore, although they traversed the same general area, Brown's and Sullivan's lines did not correspond. Instead, the 1837 Brown boundary stood anywhere from ten to fifteen miles north of Sullivan's line, the discrepancy a result of the latter's survey error caused by following magnetic rather than true north.[20]

Had the region between north of the Missouri River remained vacant of American settlement—as the first makers of Louisiana Purchase boundaries expected—the controversy might have remained an academic one. An influx of Americans into the region in the 1830s, however, complicated matters. While most Sacs and Foxes had surrendered their ancestral lands in present-day Illinois and Wisconsin by the 1830s, a holdout band led by Black Hawk resisted removal. Contemporaneous with the forced removal of remaining Indian people from the Deep South, the Andrew Jackson administration authorized troops to punish Black Hawk's followers and open more land for American settlement. As one scholar quips: "Black Hawk went to war to keep the white man out of the country; the result of the war was to bring the white man in."[21] In September 1832 the various Sac and Fox factions, including those peaceable to and hostile toward the federal government, accepted a demand to cede "a portion of their superfluous territory" west of the Mississippi River, embracing most of present-day eastern Iowa.[22]

The "half-breed" tract constituted an exception to this 1832 cession, however. In eight years a small group of settlers, many of whom worked for the American Fur Company, had built several small towns along the Mississippi River. One of these took its name Puck-a-she-tuck ("foot of the rapids") from that river's Des Moines Rapids, the same navigated by Pike in 1805; the town now calls itself Keokuk, after one of the conciliatory Sac and Fox leaders of the 1830s. In May 1834 the U.S. Army established Fort Des Moines in the tract to prevent trespassing by American settlers. Less than a month later, an act of Congress allowed tract residents to sell their land, and Americans flocked in eagerly.[23] So many came northward from nearby Missouri that its legislature approved a constitutional amendment annexing the area in 1835, passing both houses by a total vote of ninety-one to two, although the federal government refused the expansion.[24] Following the Sac and Fox cession, "a torrent of immigration . . . poured into this Western Paradise," in the words of contemporary booster John Plumbe, Jr.[25] To govern this blossoming population, Congress attached the region to Michigan Territory in 1834. Shortly before Michigan achieved statehood in 1837, the distant western lands came under the authority of the new Wisconsin Territory, which straddled the Mississippi River and extended onto the Great Plains.[26]

Almost immediately, this new territory found itself embroiled in controversy with Missouri over their boundary west of the Mississippi. In 1837 Wisconsin's territorial legislature met in Burlington, a small Mississippi River town in present-day Iowa just north of the contested line, while its permanent meeting-place rose far to the northeast in the future capital city of Madison. The body disagreed with Missouri's recently completed survey of the line that stood not too many miles south of their temporary digs. Contrasting the 1837 boundary with Sullivan's 1816 one, Wisconsinites claimed that Missourians had carved a large slice of fertile land from their southwestern flank. To that end, the territorial legislature petitioned Congress to reaffirm the 1816 line. Legislators suggested that, since few Americans knew anything about the Des Moines River at the time of the 1808 Osage treaty, the rapids referenced in it must have been located elsewhere. The Wisconsinites declared "the lower rapids of the Mississippi, known, from the time of their first discovery by civilized man, as the Des Moines rapids, or rapids of the Des Moines river" between Burlington and Puck-a-she-tuck as the proper termination point for the line. Additionally, the Wisconsin authorities worried about their free territory's loss of land to the slaveholding state south of the boundary.[27]

The dispute over which rapids—those in the Mississippi navigated by Pike in 1805 or in the Des Moines at its Great Bend identified by Brown in 1837—anchored boundaries referenced in treaties and legislation intensified by the end of the decade.[28] Missouri's general assembly made specific mention of the problem in its 1837 legislation when calling for another survey, directing its commissioners to explore the Des Moines River "to ascertain the true location of the rapids" upon which the line hinged.[29] Before the debate entangled its territory too much, Wisconsin's legislature persuaded Congress to organize a new polity from its land west of the Mississippi River in 1838. The body described Wisconsin Territory as being "too large and unwieldy for the perfect administration of prompt justice."[30] Federal officials agreed and created Iowa Territory that summer, extending from Missouri to Canada and from the Mississippi to the Missouri Rivers. Thus the boundary argument with Missouri now fell to the brand-new polity with its capital also at Burlington, the political center of two different territories in as many years.

To clarify their newly inherited contest, Iowans convinced the federal government to survey the political divide again in 1838. Missouri declined to participate since it had completed its own study the year before.[31] Conducted by experienced military engineer Albert Miller Lea, the federal expedition produced a detailed report in early 1839. It focused on an essential question regarding the boundary's eastern end: "Where are those rapids?"[32] Lea identified four possible lines in response, all of which offered legitimacy as well as faults. Sullivan's 1816 line possessed the benefit of seniority over all others, but its inaccurate survey tending northeastward did not meet the mandated criterion of "a parallel of latitude." The second candidate, a straight line drawn from the old northwestern corner of Missouri as located by Sullivan, did not intersect any rapids on its eastern end. The other two boundaries suggested by Lea crossed through rapids in the Mississippi and Des Moines Rivers, respectively, either of which he considered legal considering the vagueness of the phrase "the rapids of the Des Moines river." One of these last two corresponded to Missouri's 1837 survey, standing roughly a dozen miles north of the other limits he defined. In his report Lea also included letters from lawyers, surveyors, and even explorer, politician, and Indian agent William Clark offering contradictory opinions about which rapids should anchor the line according to laws and treaties. As a result, instead of clearing up the confusion, Lea's report added to it.[33]

Iowa's territorial population in the late 1830s grew faster than a bumper crop. The effervescent booster John Plumbe, Jr., rhapsodized about the prospects of the upper Mississippi valley, comparing it to "a beauteous and fascinating female, whose transcendant [sic] attractions must be *seen*, to be appreciated."[34] The federal government encouraged even more growth by surveying southern Iowa Territory for farms, timber works, and stone quarries.[35] For the era, the region's demographics offered a study in diversity. A Missouri newspaper described Iowa's population in 1839 as consisting of "the staid and phlegmatic German—the enterprising and industrious New Englander—the ardent and chivalrous Kentuckian—the hospitable and accomplished Virginian, the persevering and energetic Ohion [sic] and Hoosier, all congregating upon our shores and each furnishing his *quota* of the future character as well as prosperity of [the] Territory."[36] Estimates of Iowa Territory's population that year ranged as high as thirty thousand.[37]

Much of this settlement was along the Mississippi River, but by the late 1830s American farmers also favored the lower Des Moines River in the southeastern part of the territory. To provide them with local government, in 1836 Wisconsin's territorial legislature had created the immense Van Buren County, named for Martin Van Buren (then vice president), but Iowa politicians pared down the county in the years to come. Within a few years a dozen small towns flourished along the banks of the Des Moines, which bisected the county from northwest to southeast.[38] Migration proceeded at such a pace that soon territorial officials asked for federal money to build a road connecting many of Van Buren County's towns.[39] These towns stood on the front line of the impending clash with Missouri over

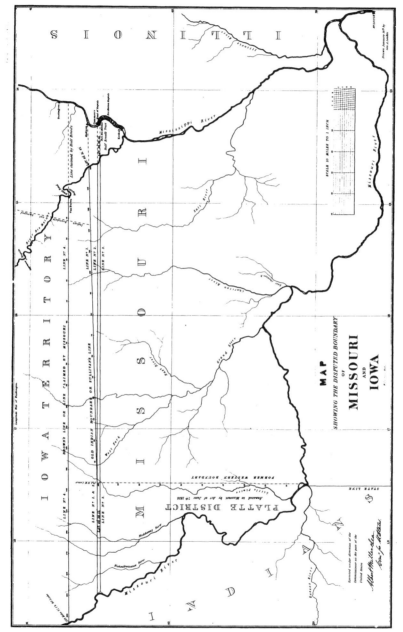

Four potential Missouri–Iowa Territory boundaries as identified by Albert Miller Lea in an 1839 federal report

the vague boundary. The first of many skirmishes involving the polities' overlapping claims arose in the spring of 1837. A grand jury meeting in the county seat of Farmington indicted one N. Doose "for exercising the office of Constable in the county, by authority of the State of Missouri."[40] Competition between local law enforcement over authority in this contested region inspired legal threats and even more serious tension in the years to come. Indeed, the more settlers moved into southeastern Iowa Territory, the greater the need to clarify its boundary with Missouri grew.

Van Buren County town-builders of the late 1830s encountered a unique group of Americans already living there, a culture that demonstrated well the complexity of life in such a contested region. A group of squatters had come to the "half-breed" tract shortly after the Black Hawk War searching for "freedom from the restraints imposed by the morality, the religion, the industrious habits and the taxing propensities of the old States." Nicknamed the "Hairy Nation" for their unkempt appearance, they cared little for the boundary dispute. Instead, laden with ballots and booze, members of this hirsute society exercised "their undoubted and undisputed right of sovereignty" in Missouri and Wisconsin and Iowa Territories at the same time. Dwellers in the "Hairy Nation" dined in their humble homes with politicians from both sides simultaneously. They avoided each side's tax collectors by claiming to acknowledge the opposite jurisdiction, depending on which revenuer appeared at their doorstep.[41] Decades later Alfred Hebard looked back on these settlers he had known personally. Hebard considered them predecessors to the "border ruffians" of Missouri and Kansas in the 1850s, "smart enough to make trouble" for political leaders and private citizens on both sides of the contested zone.[42] Yet few could imagine the trouble on the horizon at the time.

While some saw the uncertain authority along the boundary as an opportunity, others demanded action to clear up the confusion. Newspapers on both sides covered the contentious issue in the summer of 1839. Asserting its rights as an independent republic within the larger Union, the *Missouri Republican* of St. Louis argued that neither Congress nor Iowa's leadership possessed the authority to alter Missouri's boundaries.[43] Indeed, in 1839 the Missouri legislature approved two laws that further complicated matters along its northern limit. First, it created Clark County in the northeastern corner of the state, at the confluence of the Des Moines and Mississippi rivers. By law, Clark County extended north to the boundary surveyed in 1837, well beyond the 1816 line. The assembly also approved legislation that reasserted the state's northern limit as the one surveyed in 1837. Meanwhile, the editor of Iowa's *Territorial Gazette and Burlington Advertiser* pledged to stand firm against any challenge, regardless of Iowa's territorial youth.[44] Still, the newspaper worried that "the controversy between this Territory and the State of Missouri is beginning to wear a serious aspect."[45]

As the press fanned the flames, pressure increased on politicians in Burlington and Jefferson City, the governors in particular. The chief executives knew confrontational tactics well, and both had recently demonstrated a willingness to use force to get their way. To the south, Governor Lilburn W. Boggs threatened a war of extermination against Mormons living in Missouri in 1838. Facing state-sponsored religious disfranchisement, a radical faction within the church had attacked several non-Mormon settlements. In response, Governor Boggs ordered the state militia to kill or run off every Latter-Day Saint in the state. His campaign forced the Mormons across the Mississippi River into Illinois, where they reorganized at Nauvoo.[46] The next year, Boggs proved just as willing to defend his state from "foreign" invasion as from religious insurrection.

Boggs's counterpart to the north, Robert Lucas, had won an appointment from President Van Buren in 1838 as the first governor of Iowa Territory. No stranger to such office, Governor Lucas had served as the chief executive of Ohio from 1832 to 1836. During that tenure Lucas had ordered the state militia to Ohio's northwestern frontier to protect the state's claim over that of Michigan Territory to land along the Maumee River. Several hundred troops squared off near the present city of Toledo, Ohio, in 1835, until the federal government intervened. With the help of a fellow Democrat, President Andrew Jackson, Lucas retained the narrow region called the "Toledo Strip" for his state. Four years later, when faced with a similar situation in Iowa, "Headstrong Bob" Lucas acted in much the same way. As the governor of a young territory challenging a powerful state, however, he found himself in a drastically different political position.[47]

The most serious tensions between Missouri and Iowa took place in two distinct phases throughout the second half of 1839. The first consisted primarily of a heated exchange between Governors Boggs and Lucas in regional newspapers. The two traded threats over land they both considered theirs, in what one historian aptly calls "the merry war of proclamations."[48] As the man with the most experience in boundary matters, Lucas fired the opening salvo. He drew inspiration from a plea sent by Van Buren County commissioners in July 1839 reporting on efforts by

Clark County, Missouri, officials to collect taxes on farms the Iowa commissioners considered part of Iowa Territory, not Missouri. The commissioners knew of Lucas's defense of Ohio's limits several years earlier and begged his assistance.[49] Shortly thereafter newspapers printed a statement issued from Lucas challenging what he considered Missouri's attempt to "obtain a surreptitious jurisdiction" over a sliver of his territory. Lucas insisted upon judicial action to reach a suitable settlement, and he admonished any territorial resident against fraternizing with or even speaking to Missouri officials. The governor also noted that, if necessary, he would appeal to President Van Buren for assistance against the creeping authority of Missouri.[50]

Lucas thus issued a public declaration that Boggs could not ignore. Several weeks after Lucas's proclamation appeared in print, Boggs responded with one of his own. The Missouri executive declared his intention to defend his state's integrity as defined by the 1837 boundary survey. Boggs also pledged militia support to Clark County officials to compel payment of taxes in the disputed zone. Emphasizing his state's rights within the federal system, Boggs identified the United States as the real second party to the dispute since the national government held Iowa Territory in trust, as it would until Iowa joined the Union through statehood. In doing so, Boggs sought to reduce Lucas to a nonentity and go over his head to officials in Washington, D.C. Boggs also expressed regret that "a people whose language, habits, pursuits and principles are the same, and whose mutual interest prompts them to be neighbors in sentiment as well as locality" found themselves in such a resentful situation.[51] Missouri newspapers including the influential *Republican* of St. Louis expected the state to rally behind the firm stance of its governor.[52]

A month later, Lucas fired back with another lengthy missive, referencing several points made by Boggs. Lucas expressed dismay that Boggs had taken a militaristic stance, yet he adopted just such an attitude himself. Using the Boggs's logic, Lucas warned that by traipsing into Iowa Territory the Missouri militia essentially would declare war on the federal government. The territorial governor also addressed directly those living in the disputed part of Van Buren County, asking them "to be calm and discreet in all your acts." He went on:

> Look up to the civil authorities of the United States for your protection. Should you even be threatened with extermination by the all powerful arms of Missouri, be not dismayed. You are neither

slaves that you should pay tribute to a foreign government, nor passive members of a defenceless [sic] community, that you should be taxed without your consent. You occupy the exalted station of free and independent citizens of the United States. . . . [Y]ou may rest assured, that should the President of the United States authorize us to repel *force by force*, should our territory be invaded, it will be promptly done, regardless of the boasted prowess and superior numbers of the Missouri militia.[53]

By hinting at the free soil of Iowa Territory, in contrast with slaveholding Missouri, as well as Boggs's brutality against the Mormons several years earlier, Lucas added fuel to the fire between the two politicians.[54]

While the governors traded insults in the popular press, most residents of Missouri and Iowa apparently viewed the dispute as a comic one. The *Missouri Republican* thought that it "would be well if Governor[s] Boggs and Lucas can arrange this matter by a newspaper war."[55] The most famous incident of the entire episode took place during this exchange, when a Missourian cut down several hollow trees used for housing beehives by settlers who believed themselves to live in Iowa Territory. Tried in absentia, the vandal received a fine of $1.50.[56] This story provided a moniker for the entire conflict, the "Honey War." It also inspired a Missouri wag to pen a satirical poem about the situation, poking fun at the governors in particular, suggesting that they settle things themselves with a good-natured brawl (see appendix).[57] Around the time the poem hit the press, however, an incident in the boundary region transformed the conflict into something much more disconcerting.

The second phase of the boundary dispute began in October 1839, when Clark County sheriff Uriah S. Gregory crossed into the disputed zone to levy taxes. Gregory had attempted to do just that in August and September of 1839 but met with little success. In mid-October Gregory tried to collect from four residents. Two refused to pay, including John W. Davidson, the Van Buren County prosecutor, whose property Gregory reportedly "molested." Gregory then returned to the safety of the Clark County seat at Waterloo, in undisputed Missouri land. In response Lucas wrote a letter of support to Sheriff Henry Heffleman of Van Buren County, whom he expected to "be as prompt and vigilant in enforcing the laws and protecting the citizens of the United States within this Territory, as those of Missouri possibly can be, in their intrusions upon our neighbors."[58] Gregory

The Honey War theater, 1839

went back to the disputed zone on October 24, and he met with officials from Van Buren County, including Heffleman, to insist upon his authority to collect taxes from those living south of the 1837 boundary. The sheriffs' discussion escalated, with "several warm speeches on both sides; amounting almost to a declaration of war" according to a Missouri newspaper.[59] When the news reached Jefferson City several days later, Boggs ordered the state militia to support Gregory in carrying out his lawful duties.[60] While the *Missouri Argus* cautioned all involved to "consider the advantages of peace over discord," the stances of both executives suggested the use of weapons rather than words.[61]

Two days after the sheriffs' tense exchange, residents near the controversial boundary attempted to solve the problem themselves. Delegates from Clark and Van Buren Counties—including private citizens and local militia officers—met in Monticello, Missouri, about forty miles south of the disputed territory. After pledging friendship to their neighbors, the Missourians proposed to share jurisdiction until the federal government could sort out the matter. In response, the Iowans stated that they could not accept concurrent authority but would agree to both sides suspending tax collection for the time being. Upset at the rejection of their proposal, the Missourians drafted a resolution calling for Sheriff Gregory to "proceed to a more energetic discharge of his civil duties" and dismissed their counterparts to the north. Shortly thereafter, the adjutant general of Iowa's territorial militia reported to Governor Lucas that the conference had only aroused passions on both sides.[62] Rumors also swirled about Missourians attacking property in Iowa, including the burning of a house that had killed two children.[63] These vague stories did little to dampen emotions. For the next month, however, the situation calmed somewhat as both sides considered their next move.[64]

On November 19, 1839, Gregory returned to northern Clark County—or southern Van Buren County, depending on one's perspective—to tax the marginal region. Heffleman tracked him down quickly and arrested the interloper. Governor Lucas received the news happily and bestowed upon Heffleman "the approbation of every citizen of Iowa." A local court put Gregory on trial for exercising an illegal jurisdiction and jailed him at the county seat in Farmington.[65] When word of the arrest reached Missouri's militia officers they dispatched a brigade to prepare state defenses immediately, and mustered four additional divisions in the northeastern counties.[66] Shortly thereafter forty troops headed north, intending to break Gregory out of jail. Rumors of a large force under Heffleman's command awaiting them worried the Missourians, though, and the militia officers proceeded to Farmington alone to attempt a negotiated settlement. The Iowans rebuffed them, and the officers returned to Waterloo to await reinforcements. In the meantime Heffleman transferred Gregory to Burlington, where Heffleman met with Lucas in a failed attempt by the Iowa governor to resolve the situation on his terms.[67]

The potential for bloodshed increased with each passing day, much to the dismay of many people on both sides of the indeterminate line. As the *Missouri Republican* observed: "It is every way probable that a collision will ensue, as the excitement is becoming very intense and gradually extending over a much larger portion of the people of the state and territory."[68] Lucas wrote to Secretary of State John Forsyth expressing his concern in a more melodramatic fashion: "I am apprehensive that blood will be shed; and if blood begins to flow, it is impossible to foretell where the matter will end." Iowa's territorial legislature also memorialized Congress for help, asking protection for "that which our stronger sister is attempting to wrest from us by force."[69] Neither side wanted to back down first, but both shied away from advocating a settlement through carnage. Regardless, both Lucas and Boggs ordered their militias to rally to the defense of land they both claimed as their polity's own.

The seriousness of the situation contrasted markedly with the appearance of the soldiers marching toward the front lines. Both the Missouri and Iowa militiamen struck many observers as humble, perhaps even laughable. About eight hundred northeast Missourians mustered in the camp near Waterloo wearing uniforms that were anything but uniform. They resented the state's unexpected declaration that the militia must supply itself, and they looted a store for food and blankets. The Missourians captured supplies heading upriver toward Iowa Territory as well, and they even blocked the mail from reaching their enemy.[70] With overland transport still a rudimentary process, Missouri's long riverfront enabled it to affect the ability of Iowa to make war. Iowa's troops, numbering upwards of six hundred, found conditions much the same as those in Missouri. Farmington struck one observer as "a military camp, and the streets a place for military parade."[71] Armed with everything from muskets and pitchforks to hoes and spears, from scythes and clubs to "an old fashioned sausage stuffer," the Iowans also had to provide their own food and uniforms, which they appreciated as much as the Missourians.[72] One militiaman remembered sarcastically: "We were willing to shed our blood for our beloved Territory and, if necessary, to kill a few hundred Missourians, but we were not going to do that and board ourselves besides."[73] Nonetheless, they raised the rallying cry "Death to the Invading Pukes!" in reference to Missouri's nineteenth-century nickname of convoluted origin.[74] But the cold and snow of the coming winter represented the worst enemy for the ill-supplied forces of both sides encamped less than twenty miles from each other.

The campaign began on December 3, 1839, when General David Willock led a few dozen Missouri militiamen in a tax-collecting sortie into the disputed region. They gathered a significant amount of money before skedaddling back to Waterloo before Iowa's force could catch them. A Burlington newspaper responded by accusing the Missourians of high treason for having invaded Iowa Territory and by extension attacking the federal government.[75] Yet many residents of northern Missouri and southern Iowa concluded that the pseudo-war had to stop. Speaking about the plight of both militias, Missourian Thomas L. Anderson expressed the sentiments of many when he exclaimed: "in the name of the God of Mercy and Justice, gentlemen, let this monumental piece of absurdity, this phenomenal but cruel blundering have an end."[76] To that end, while Willock engaged in clandestine maneuvers, the Clark County court, Iowa's territorial legislature, and Lucas entered into peace talks of their own. Additionally, on December 9, a citizens' committee in Marion County, Missouri, seated at Palmyra about sixty miles down the Mississippi River from the boundary, drafted a resolution demanding an immediate end to hostilities and called for a federal solution to the crisis.[77] In poignant words the Missouri legislature resolved unanimously on December 14 that "if that much to be deplored time should come when we shall be required to shed the blood of each other, we here pledge ourselves collectively and individually to endeavor by every means in our power to allay the horrors and calamities of the civil war."[78] As quickly as it started, the "Honey War" drew to a close.

A diverse reaction to this sudden peace came just as swiftly. A committee of concerned Missourians from Lewis County, located between Clark and Marion Counties, spoke out against the perceived surrender to Iowa.[79] Meanwhile, families criticized both governors in the press for having sent their fathers, brothers, and sons out into the cold for a week to fulfill the bombastic pledges they had made months earlier.[80] The *Iowa Sun* of Davenport bristled at the expense and inconvenience caused by both governors. The newspaper noted the legacy of their past belligerence: "It is therefore perhaps not to be wondered at, that these two renowned chieftains, being placed in command so near each other, should, like two mighty ram goats, feel a desire to knock horns together, and make a noise in the world."[81] Sheriff Heffleman released the incarcerated Sheriff Gregory, and Iowa's territorial courts eventually dropped the charges against him.[82] The reaction nationwide to this peculiar tale did little to encourage its participants to remain in the field; even the *Mercury* of Charleston, South Carolina, a newspaper that would praise the forceful defense of a state's rights twenty years later, lampooned the conflict.[83]

The most graphic reaction to the "Honey War" came from the frostbitten militiamen themselves. One wrote:

> About the time we got our fires burning, we received information that we would be turned home. . . . However, being determined to have our sport, we retired a short distance outside of the old Colonel's blazed encampment, taking with us a quarter of venison that we had the good luck to kill on the way, which we severed in two pieces, and hung up, in representation of the two Governors, and fired a few rounds at them, until we considered them dead! dead!! They were then taken down, and borne off by two men to each Governor, enclosed in a hollow square, with the muffled drum, and marched to the place of interment, where they were interred by the honors of war. We fired over their graves, and then returned to the encampment.[84]

No one appreciated the end of the farcical conflict more than those expected to fight it. Neither Boggs nor Lucas approved of the extralegal decision, and they released resolutions to that effect, but the governors no longer enjoyed enough support to maintain their border patrols, and the militiamen trickled away from the encampments.[85] While volunteer troops on both sides of the line might have believed in the justness of their cause, the lack of logistical support dampened their initial enthusiasm and led them to feel like pawns for their whimsical governors.

With the dawning of the 1840s, peace at last returned to the overlapping Missouri–Iowa Territory boundary region. Yet the debate over which line marked their division still demanded settlement. Federal politicians, awakened to the matter through reports trickling eastward, took it up in early

1840. Senator Lewis F. Linn of Missouri spoke in Congress with sympathy for both sides, and he asked his colleagues to help resolve the argument.[86] To that end, the House of Representatives debated a bill to reaffirm the more southerly 1816 line as the official northern boundary of Missouri. Some Missouri newspapers scorned the notion when it reached their state at the end of winter, however, refusing to give up their state's land claim.[87] Whig-allied newspapers in particular used the bill as an example of the incompetency of their Democratic elected officials, including Senators Linn and Thomas Hart Benton.[88] Although he also disapproved of Linn's proposal, fellow Democrat Boggs reminded his constituents that the federal government should take the lead in arbitration.[89] He had little choice, especially considering the mutinous attitude of his state militia by that time.

Another issue of concern on both sides involved the almost-defenders of Missouri and Iowa. Members of both militias requested reimbursement for expenses incurred during the several weeks of war preparation. Missouri's state legislature approved appropriations in early 1841 for the troops sent to the northern frontier, as well as those who took part in the earlier anti-Mormon campaign.[90] As leaders of a territory, however, Iowa officials expected the federal government to pay their bills. Secretary of War Joel R. Poinsett stated that since he had not approved the use of the territorial militia he saw no reason why Congress should reimburse.[91] The House of Representatives requested Iowa's muster rolls nonetheless, which militia officers provided in the late summer of 1840.[92] Eventually Congress decided not to pay Iowa's militia, although for the next fifteen years the territory (and state after 1846) petitioned federal politicians to change their minds.[93]

In the winter of 1840–41, both Missouri and Iowa Territory changed leaders. Thomas Reynolds won an election to succeed Governor Boggs in the waning months of 1840, and in the spring of 1841 a new Whig administration in the White House replaced Democratic Governor Lucas with Whig John Chambers in Iowa. With both polities under new management, their approach to the boundary dispute changed.[94] For the first time in four years, Missouri proposed cooperation with federal authorities to "finally and peaceably" mark the boundary. In a plan drawn up by the legislature in late 1840, surveyors approved by the state senate and U.S. Congress would work jointly. In the meantime, county officials would desist from levying taxes on inhabitants of the disputed territory.[95] Iowa demanded its own voice in the project to defend local interests, but Missouri argued that the federal surveyor would look after the territory's needs. In the end, even in an atmosphere of renewed cooperation, the idea fizzled.[96]

Congress resumed consideration of this festering issue a year later. Once again the perennial debate about which rapids anchored the boundary—*in* or *of* the Des Moines River—proved controversial. U.S. Army surveys of the Des Moines River in 1840 and 1841 provided reports of numerous obstacles that one might consider rapids in that flow.[97] Missouri's secretary of state also forwarded more than a dozen letters from individuals affiliated with the 1816 survey and the 1820 constitutional convention, all of which insisted upon rapids in the Des Moines River.[98] But members of a House of Representatives committee assigned to investigate the matter disagreed. They believed that the better-known Des Moines Rapids in the Mississippi River as the ones intended to shape the line instead of one drawn through "an unknown and unbroken forest, inhabited by and belong to the savage, and on some one of twelve *ripples*."[99] Congress thus found itself enmeshed in a decades-old dispute with little sign of concluding.

When the House of Representatives debated a bill to clarify the boundary on July 20, 1842, member John C. Edwards of Missouri objected on the grounds of state's rights. He feared that the federal government would try to use its authority unconstitutionally to alter the shape of his home state.[100] Edwards suggested that Iowa did not need the disputed region since it already had "ample territory, and enough to spare for two more states." Other congressmen reminded him that Missouri ranked as the largest member of the Union by size already. Undeterred, Edwards continued: "Iowa is encroaching upon us, and grasping part of our territory; and the United States, like all tender mothers, is taking sides with her infant child against the older one, in sustaining her groundless pretensions."[101] In response, Iowa's delegate, Augustus C. Dodge pointed an accusatory finger at "gigantic, avaricious, grasping Missouri."[102]

The population of Iowa Territory remained on the rapid increase in the early 1840s, regardless of the pseudo-conflict with Missouri. As growth made Iowa's statehood ever more likely, the boundary issue remained in need of resolution. A constitutional convention met in the new capital of Iowa City in the fall of 1844 and pondered that question among many others. Delegates first proposed defining the southern line as "up the Des Moines to the old Indian Boundary line or North line of Missouri," flexible and vague wording that could be interpreted as meaning either the 1816 or 1837 limit.[103] Eventually they settled on a boundary up the Des

Moines River "to a point where it is intersected by the old Indian boundary line, or line run by John C. Sullivan in 1816." Camaraderie across the mutual yet ill-defined line thus remained elusive even five years after the near war. Indeed, when Missouri politicians heard of the proposal, they complained to the federal government that Iowa was trying to cleave off a part of their state south of the 1837 line.[104]

As Iowa worked its way toward equal membership in the Union, the contested zone between it and Missouri remained controversial. Congress proposed yet another survey in 1844, but Missouri refused to participate, marking the end of congressional efforts to resolve the dispute. In the meantime, the farther west settlement pushed from the Mississippi River, the wider the conflict spread. For example, Adair County, Missouri, two counties west of Lewis, stretched toward Iowa in 1845. Iowa prosecutors indicted Adair County's Sheriff Preston Mulnix early that year "for usurpation of office" but territorial Governor Chambers pardoned Mulnix before a trial commenced.[105] In an address to Iowa's territorial legislature in December 1845, the new territorial governor James Clarke said that the region remained at peace.[106] Missouri's legislature remained unconvinced, memorializing to Congress shortly thereafter that "feelings of the people bordering upon the line have become excited, until a civil war is at any moment liable to be kindled."[107]

Similar challenges developed after Missouri's legislature carved Schuyler County out of the northern reaches of Adair County. As had become the tradition of the region, Schuyler County's sheriff feuded immediately with his counterpart in Davis County, Iowa Territory, to the north. As a result, by the end of 1846 the neighboring counties had indicted each other's sheriffs for exercising authority beyond their jurisdiction.[108] As a story that never seemed to die, one filled with drama and comedy, the ongoing travails of the Missouri-Iowa boundary drew attention nationwide. A Philadelphia newspaper, for example, informed its readers after the Schuyler-Davis rigmarole: "Iowa has repeatedly captured and imprisoned the Sheriffs of Missouri and sentenced them to the penitentiary for the faithful discharge of their official duties."[109] One struggles to blame the Pennsylvania press for its slight exaggeration, considering the hyperbole that often saturated the boundary region itself.

In the meantime, Iowa politicians negotiated their territory's path toward statehood, including debates with the federal government about its proposed boundaries, mostly the northern and western limits (see chapter 2). But its long-contested southern line received due attention as well. The territorial legislature asked congressional permission to seek a settlement with Missouri in the U.S. Supreme Court. While the proposal made sense, the legislative branch ignored the request for the time being. Until both polities could appear in court on the same footing, Iowa remained an extension of the federal government, and most congressmen preferred to rid themselves of the interminable headache.[110] Therefore, Iowa's constitution as approved by the federal government in the summer of 1846 identified its southern boundary as the northern line of Missouri created by its 1820 constitution, with no reference to the surveys of 1816 or the 1830s.[111] By the end of 1846, Iowa earned its promotion to statehood, and the two members of the Union could finally settle their long-standing dispute as equals.

After years of bitter words and near bloodshed, Missourians and Iowans met in the courtroom rather than on the battlefield in late 1847. The states filed cross suits with the Supreme Court and sent lawyers to voice their arguments in its stately room in the U.S. Capitol.[112] Missouri's representative touted the 1837 Brown survey that showed evidence of rapids in the Des Moines River and thus justified the most northerly limit. By contrast the Iowans focused their attention on defining the long-contested rapids and insisted that only those in the Mississippi River could have been known to Americans in 1820 when Missouri's constitution referenced them. Without identifying which line should divide Missouri from Iowa, the latter's lawyer disputed the premise on which his counterpart rested his case.[113]

The Supreme Court waited more than a year to render its verdict in *Missouri v. Iowa*, apparently considering the matter far less pressing than many others associated with it. Associate Justice John Catron delivered the unanimous opinion of his colleagues on February 13, 1849. The justices noted that Missouri's state boundaries as initially described in its constitution depended on the 1816 Sullivan boundary. It had represented a separation between state power and American Indian land in no fewer than fifteen treaties. The federal claim to such land acquired from Native cultures changed only semantically when Native title gave way to territorial status. In addition, all public land surveys in the area started at Sullivan's line. The court also mentioned the rapids upon which the line depended for an eastern terminus. It considered the Mississippi River's Des Moines Rapids as most likely those referenced in treaty and law. Since the rapids extended fourteen miles up the Mississippi, the line could intersect them anywhere and still be considered legal.[114] Catron's opinion also addressed the innumerable controversies over the previous dozen years that had resulted from the vague political division and issued a stern reminder to both parties: "And it is further adjudged and decreed, that the State of Missouri be, and she is hereby, perpetually enjoined and restrained from exercising jurisdiction north of the boundary aforesaid dividing the States; and that the State of Iowa be, and she hereby is, also perpetually enjoined and restrained from exercising jurisdiction south of the dividing boundary established by this decree."[115]

The final arbitration of the Missouri-Iowa boundary, therefore, marked a victory for the new state to the north. Demonstrating again the nationwide interest in this story, the *New York Herald* reprinted a story from an Iowa newspaper praising the decision for preserving Iowa as "unshorn and unmutilated."[116] Albeit somewhat disappointed in the verdict, Missourians seemed relieved to put the affair behind them.

Sullivan's 1816 survey line remains the foundation of the official division between the two states, extending due east from the Missouri River to the original northwest corner of Missouri, and from there slightly north of east to the Des Moines River, creating an extremely obtuse angle. By 1851 the Sullivan line was surveyed again, allowing sheriffs to assess property and collect taxes in peace.[117] Undoubtedly many settlers appreciated the calm after so many years of tension. Given the transient nature of western American settlement, though, many who had lived through the 1839 conflict had moved on in the dozen years since. One newspaper noted the demise of the "Hairy Nation." Shortly after the Supreme Court decision, "the disputed territory soon became thickly settled by industrious and thriving citizens, in the former places of the Hairy Nation, who gradually left for regions where there is more freedom and less labor, more whisky and less tax-paying than the State of Iowa was about to impose upon them."[118] For those few who saw the lack of authority as an opportunity, the boundary's ultimate clarification spoiled their fun.

Several surveys and court cases conducted after the mid-nineteenth century clarified the line further. In 1895 Missouri and Iowa asked the high court to arbitrate on almost fifty-two acres of disputed land between them. Instead of overlapping jurisdictions, a lack of law enforcement over this region created a serious problem. Criminal activity within the small, unmanaged parcel had flourished as a result of the line's uncertainty. With the Supreme Court's help the two states resurveyed their common boundary peaceably and brought the recalcitrant acreage to heel.[119] Two decades later, a dam built on the Mississippi River at Keokuk, just above its confluence with the Des Moines, inundated the rapids that had caused so much controversy and obstructed riverine traffic above and below them.[120] Yet the line intended to intersect those rapids remains important to those on both sides. As recently as 2005, the Missouri Department of Natural Resources restored the surviving cast-iron boundary markers installed every ten miles along the Sullivan line in 1851.[121] The bucolic rolling hills bisected by the Missouri-Iowa boundary today contrast dramatically with the tension that crackled along that divide. Over the years, that hostility eventually morphed into good-natured amusement. A joke about the Supreme Court's 1849 decision, for example, lingered more than a century later: "An old farmer, so it goes, was delighted to hear that the decision put him in Iowa, not Missouri, 'cause I heard tell the climate "n" soil of Missouri ain't fitten ter raise decent crops.'"[122]

For more than forty years, politicians and settlers struggled to establish authority along the Missouri-Iowa line. This occasionally ludicrous contest reflected the lengths to which trans-Mississippi dwellers went to protect what they considered theirs, whether their land or their identity. Historians have generally interpreted the "Honey War"—when they have interpreted it at all—as little more than an amusing episode in the region's heritage.[123] In reality, the struggle represented something far more important. The line separating the two polities helped make one group Missourians and the other Iowans, and even their brief willingness to defend that identity with force illustrated the value they ascribed to their common boundary. It made them not only Pukes and Hawkeyes, but also a part of the broader Union, shaped by lines creating similar identities all across the United States.

Railroad sign along the Columbia River at the Oregon-Washington line

CHAPTER 5

Nature Has Marked out the Boundaries
The Oregon Country Boundaries

As U.S. territorial claims extended to the Pacific Ocean, the challenge of political organization followed, adapting to new circumstances and landscapes. Raging rivers, towering mountains, and vast basins in the Pacific Northwest all played an important role in determining its political future as the nationwide debate between boundaries of geography or geometry resonated loudly in this vast Oregon Country. More often than not, local settlers urged the adoption of topographical limits for its political communities. As one resident suggested in the mid-1840s, "Nature has marked out the boundaries."[1] But the federal government did not always depend on the landscape to transform its holdings into territories and states. Over a decade in the mid-nineteenth century, three future states—Oregon, Washington, and Idaho—emerged in the region, with boundaries that incorporated both natural features and straight lines. Nonetheless, the Oregon Country identity of residents in that vast landscape often transcended whatever kinds of political lines the federal government saw fit to establish there.

Although the United States did not possess sole claim to the Pacific Northwest until 1846, published accounts by numerous visitors familiarized Americans with it in the early part of the nineteenth century. With an eye toward potential U.S. control over the region, writers identified geographic features useful for transforming it into future territories and states. The creation of Washington Territory in 1853 brought the first division to the Oregon Country, blending geography and geometry to split it from the Pacific Ocean to the Rocky Mountains. This line reflected the competition between Americans living at Puget Sound and in the Willamette River valley, but it never severed the links between those regions. Throughout the 1850s, frequent cross-boundary contact of an economic, social, and even military nature reinforced the former unity. Overlapping jurisdictions created by Oregon's statehood in 1859 brought further controversy, particularly to the upper Columbia River basin. In the end, the haphazard manner in which both local and federal officials enforced it offers the most striking historical feature about the Oregon-Washington line.

European interest in the Oregon Country, a vast swath of the Pacific coast and interior between California and Alaska, had flourished in the latter half of the eighteenth century. British, Spanish, Russian, and American expeditions, both private and official, charted much of the coast by 1800. In 1792 the U.S.-based ship *Columbia* entered the perilous estuary of the river that would take its name, giving the United States its first claim to the area. By the early nineteenth century, the British and Americans emerged as the dominant parties contesting Oregon. The Lewis and Clark expedition of 1804–1806, along with *Columbia*'s voyage, gave Americans a stake in the far-flung territory, while the British created a more stable economic presence via the Hudson's Bay Company. Within a few years of the War of 1812, the booming otter, seal, and beaver fur trade demanded a reassessment of the Oregon Country. In 1818 the United States and Great Britain agreed to occupy the region jointly for ten years.[2] The next year the Spanish limited their claims to the Pacific Coast to land south of the forty-second parallel. Another agreement, with the Russians, affirmed the Americans and British as the sole imperial competitors for Oregon. After nearly a decade of joint occupation the two parties extended their agreement indefinitely in 1827.[3] Itinerant trappers contracted with several American firms to supply pelts, while Natives and Euro-American employees brought skins to the well-stocked British outposts, most notably Fort Vancouver about 120 miles up the Columbia River from the coast.[4] With such a vast territory so far from Anglo-American population centers to the east and a seemingly inexhaustible supply of critters to skin, sharing Oregon seemed a sensible and effective solution.

The nature of Euro-American life in the Oregon Country had changed by the late 1830s, however. Agricultural emigrants from the United States, lured by reports of excellent land along the Columbia River's tributaries, established farms in the heart of the region. Missionaries living among Native groups in the interior also added to the American presence. Several times over the next decade, the settlers congregating along the Willamette

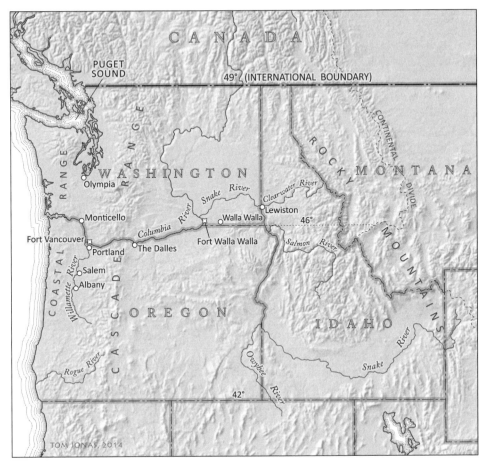

The Oregon Country

River, which flowed into the Columbia opposite Fort Vancouver, petitioned Congress to provide them a territorial government. In the spring of 1838 more than three dozen people signed a request for protection of their "germ of a great state," with political organization and a military presence.[5] Concerned more about maintaining the country's relationship with Great Britain than the pleas of a few distant farmers, Congress rejected the request.[6]

Regardless of this early defeat, American settlers kept up the pressure to establish some form of government as their ranks increased in the early 1840s. Petitions reached Washington, D.C., complete with warnings about British power in the Oregon Country as a threat to the joint occupation.

Dr. John McLoughlin, in charge of the Hudson's Bay Company post at Fort Vancouver, had long been credited with generosity toward the American newcomers. Now the farmers portrayed him as a tyrant threatening to expel settlers who wanted only to earn a virtuous living off the land.[7] In the spring of 1844 the House of Representatives Committee on the Territories offered a compromise solution. Instead of organizing a new territory along the Columbia River, the committee suggested placing American citizens there under the authority of the nearest organized territory, Iowa, seated at Burlington some fifteen hundred miles away.[8] This solution would provide a political voice for far-flung American settlers without upsetting the diplomatic balance with the British. This proposal soon died, however, due in large part to the problems of coordinating activity between the Mississippi and Willamette Rivers, even with the recently blazed Oregon Trail.

By the middle of the 1840s some five thousand emigrants had crossed the continent to the Willamette River valley and other destinations in the Oregon Country.[9] Their occasional requests for territorial organization sometimes resorted to wild exaggeration as they insisted upon the physical as well as patriotic connections between the Atlantic and Pacific coasts. One petition, for example, described the Rocky Mountains as "little more obstruction than a cornfield," in an effort to downplay that geographic barrier.[10] Yet Congress remained silent on the matter of political organization. For want of federal action, a group of settlers south of the Columbia River declared their own polity in the summer of 1845, a fairly common practice exercised by western settlers seeking law and order in distant regions. The provisional Oregon Territory created legislative, executive, and judicial offices, and gave special attention to property rights as one might expect in a self-organized farming community.[11]

The presence of ever more American settlers in the Oregon Country offered only one of several threats to the stability of the region and Anglo-American relations. Adding to the problem was the new president, James K. Polk, who had won election in 1844 on an expansionist platform including a desire for sole control of Oregon. At the same time, officials in London expressed willingness to compromise over the land that grew less valuable to them daily with the waning of the fur trade as global fashion trends changed to accommodate new tastes. Negotiations between the two governments led to an agreement to partition Oregon, but diplomats debated hotly just where the line should fall. British officials proposed using the Snake and Columbia Rivers west from the Rocky Mountains, or extending the existing national line along the forty-ninth parallel to the Columbia River and then down the Columbia to the ocean.[12] Major newspapers like *The Times* of London entered the fray, recommending an easy and sensible solution—extending the forty-ninth parallel all the way to the ocean. Such a cleavage would give "a sovereign but barren dominion" to what *The Times* considered a handful of rowdy Americans on the banks of the Willamette River, land that the British could well afford to surrender.[13]

When British and American emissaries concluded their agreement in the summer of 1846, the suggestion supported by *The Times* won the day. A simple extension of the forty-ninth parallel split the Oregon Country to the Straits of Juan de Fuca, leaving the British with Vancouver Island and the United States with the Olympic Peninsula. The Hudson's Bay Company retained navigation rights to the Columbia River south of the border, although the fur trade's decline made that allowance hardly worthwhile, and it pulled out of the region by the mid-1850s.[14] With the Oregon Country south of forty-nine degrees now in American hands legally, all that remained was to figure out what to do with it. Few expected the immense territory, roughly comparable in size with that other northwest—the Northwest Territory of the 1780s—to remain whole forever. Enough Americans had traveled to Oregon programmed with the mental image of their country's political institutions that they had considered potential divisions of the region for years. Many mid-nineteenth-century sources looked to its future as not one but several prosperous states divided by significant natural features well known to those first Americans living in the region.

Practically every American author to write about the Oregon Country in the first half of the nineteenth century gave attention to its geographical units. Some went so far as to foresee each becoming its own territory or state. As early as 1830 the most famous booster of Oregon, Hall Jackson Kelley, suggested four distinct parcels within the region. From west to east, Kelley identified the narrow strip between the Pacific Ocean and the Coastal Range, another linking the Willamette River valley and the Puget Sound, farther eastward to the confluence of the Snake and Clearwater Rivers (on the present boundary between Washington and Idaho), and finally east to the Rockies.[15] Nathaniel J. Wyeth, a New England entrepreneur and friend of Kelley, made several trips to Oregon in the early 1830s, and wrote an oft-cited report about the region. He identified three distinct segments of Oregon, each marked by its own unique climate and divided from the others by a range of mountains. Wyeth's first two corresponded to Kelley's, but he condensed the rest into an immense tract between the Cascades and the Rockies.[16] Additionally, overland guide Lansford W. Hastings produced his own image of the region in 1845. He combined the first two regions noted by Kelley and Wyeth, uniting everything from the Cascades west to the ocean, and divided Wyeth's last section along a small range in present-day northeastern Oregon.[17] All these sources utilized mountain ranges as physical divisions, not the rivers that ended up playing a more important role.

Not all writers of the Oregon Country ignored the divisive possibilities of the Columbia River, however. It served both as an essential corridor for trade and a daunting cleft through the landscape, particularly as it neared the ocean through its immense gorge. For some even after 1846 it represented a cultural divide between the British presence at Fort Vancouver, still occupied by a few Hudson's Bay Company trappers, and the Americans on the Willamette River. Regional promoters sought to wipe out this barrier and encourage American settlement on both sides of the Columbia. Alexander Ross, who thirty-five years earlier had accompanied John Jacob Astor to found a trading post near the river's mouth, wrote an extensive narrative of the region in 1849. Ross touted the agricultural potential of both the Willamette River valley and Puget Sound to farmers looking for a fresh start.[18] Similarly, Joel Palmer promoted settlement on both sides of the Columbia, regardless of the continued presence of the suspect British at Fort Vancouver. Palmer went so far as to envision how Americans could redraw the map of Oregon to create their preferred style of political communities:

> The country will, without doubt, be divided into at least three states. One state will include all the country north of the Columbia river. Nature has marked out the boundaries. Another state will include all that country south of the Columbia river to the California line, and west of the Cascade range of mountains. This country, however, is large enough to form two states. The country east of the Cascade range, extending to the Rocky mountains . . . would cover all that vast barren region of country which can never be inhabited by the white man. The western portion of this section is fertile. The line doubtless would be established between, leaving the eastern portion as Oregon territory, for future generations to dispose of.[19]

While Palmer's first two proposed states would have been significantly smaller than present-day Washington and Oregon, and the third a much larger Idaho, one nonetheless sees the nascent future of the region in his prediction.

Annexation to the United States did not end the uncertain authority of settlers living in the Oregon Country. For two years after the 1846 treaty they sent more requests to Congress to provide law enforcement and legal structures to manage their land and property. One petition noted that "although three thousand miles—nearly two-thirds of which is a howling wild—lie between us and the federal capital . . . we are *Americans* still." As such they expected the rights they had known in eastern states and territories.[20] The Whitman Massacre, the killing of thirteen American missionaries by neighboring Cayuses and Umatillas near the Columbia and Snake River confluence in November 1847, prompted even more pleas for organization and federal support.[21] After extensive debate, in late summer of 1848 Congress enacted and Polk approved the creation of Oregon Territory, which embraced the entire parcel from the Pacific Ocean to the Rocky Mountains between the forty-second and forty-ninth parallels.[22] Subdividing the massive parcel remained a challenge for another time.

The next four years passed with relative stability in Oregon Territory as the new government there grew more effective at managing its growing population. By 1850 almost eighteen thousand Americans had traveled to the country's new Northwest.[23] Of these, just over one thousand settled north of the Columbia River, in contrast to the more established towns and farms on the Willamette River and other southern tributaries.[24] A gradual resentment developed between the two factions, with the lower Columbia representing a dividing rather than uniting line. To address this, in August 1851 residents north of the river convened at a farm along the Cowlitz River near present-day Toledo, Washington, to propose a new territory split from Oregon by the Columbia. As one proponent noted: "Does it not seem that nature destined the Columbia river to be the dividing line between two . . . powerful States?" The meeting also proposed a slew of new counties to provide local organization.[25] The next spring Congress received a memorial from Oregon's territorial legislature expressing various concerns, including the question of its political future. Although they recognized the Columbia as an essential route in the territory, the legislators noted that "with wild and unconquerable fury, [it] has burst asunder the Cascade and coast ranges of mountains, and shattered into fragments the basaltic formations."²⁶ Such dramatic language hinted at a shifting attitude among territorial denizens toward the regional role of that vital river. Within a year the mighty Columbia shattered something else—the political unity of Oregon Territory.

The pressure to divide Oregon Territory along the Columbia River demonstrated the tension between the two centers of American population within it, along the Willamette River and at Puget Sound. The most prominent newspaper in the latter region turned a split of the territory into a full-blown crusade in the waning months of 1852. At Olympia, on the south end of Puget Sound, the *Columbian* issued a series of editorials exhorting "Northern Oregon" to reject its perceived political neglect from the other settlement center.²⁷ The newspaper demanded "*a legal divorce from the south,*" and by November 1852 enough support developed to call another meeting of activists in the north.²⁸ Conscious of their town's image as the leader of the uprising, many Olympians insisted that the meeting take place elsewhere to demonstrate their widespread grievances. To that end the assembly convened in Monticello, a town on the Columbia roughly equidistant from Portland and Olympia, then the population centers of the Willamette valley and Puget Sound, respectively, and near the Cowlitz River meeting site of the previous year.²⁹

The Monticello convention in late November 1852 memorialized Congress for a new territorial government bounded to the south and east by the Columbia River. Portraying the two sides of the river as incapable of cooperation, the memorial argued that they "must . . . always rival each other in commercial advantages, and their respective citizens must, as they now are and always have been, be actuated by a spirit of opposition."³⁰ The *Columbian* cheered the convention's work, asking "what stronger natural boundaries are necessary to mark a just claim for independent territorial existence, and future state sovereignty?"³¹ Not wanting their identity to depend on the remainder of the territory, the convention offered "Territory of Columbia" rather than "Northern Oregon" as a name for their new polity.³² Over the winter the proposal wended its way to Washington, D.C., where federal officials considered the need for a new territory on the Pacific shore.

In February 1853 the House of Representatives took up the matter of dividing the Pacific Northwest. Oregon's territorial delegate, Joseph Lane, embraced the notion warmly and lobbied his colleagues for its approval. When Congress drew up a bill to split the Territory of Columbia away from Oregon, however, it proposed an alteration to the boundaries proposed at Monticello. Instead of following the Columbia River from its mouth to its intersection with the forty-ninth parallel, the bill included only the lower portion of that flow, then followed the forty-sixth parallel due east to the Rocky Mountains. Lane made no opposition to the change as it affected land far upriver from either population center. Indeed, the House spent more time arguing over the name of the territory. The suggestion of "Columbia" seemed unwise—the nation already had a District of Columbia, after all, and Congress wished to avoid confusion. Instead they changed it to the patriotic and much-applauded "Washington." Somehow, this selection seemed less problematic even though the capital city bore the same name, hence the awkward clarification of "Washington State" ever since. In any case, two days before he left office in March 1853, President Millard Fillmore approved the law reorganizing the Pacific Northwest.[33]

The birth of Washington Territory attracted attention from media on both sides of the country. The *Daily National Intelligencer* of Washington, D.C., described the law as one of the few productive things accomplished during the recent legislative session, adding "another step to that ladder of empire on which this growing nation is mounting to the loftiest heights of political greatness."[34] Word reached Puget Sound by late April 1853, and the *Columbian* beamed with pride at the fruition of the movement it had encouraged.[35] The Olympia paper expressed no concern with the name change or use of geometry rather than geography on the eastern part of the boundary, preferring instead to focus on the mere fact of its culmination. By contrast, the *Weekly Oregonian* of Portland worried about the region's potential future subdivision. In the years immediately following Washington Territory's inception, its southern neighbor seethed with resentment at the thought of withering away, losing more land to the east and south until it ranked in size and significance with other tiny members of the Union like Delaware and Rhode Island.[36] Yet the line for which northern settlers pleaded proved less divisive than expected. Within two years of Washington Territory's organization, the two political communities split by the Columbia River and a straight line through the sagebrush found it necessary to cooperate in the face of a serious threat on their shared frontier.

Increasing American settlement in the Pacific Northwest worried the first inhabitants of the region, American Indians. Out of desperation numerous Native groups banded together to defend their homeland in the mid-1850s, leading to a series of chaotic battles on the fringes of both territories. A trading post on the Columbia River near its confluence with the Walla Walla River offered one place of particular concern. Congress even referred to this spot, called originally Fort Nez Perces but known commonly as Fort Walla Walla, in the 1853 law dividing Oregon Territory. Congressmen identified it as the closest point of reference for where the boundary shifted from the Columbia to the forty-sixth parallel. A few Hudson's Bay Company trappers remained there per the 1846 treaty, but American missionaries and settlers in the region trumped the lingering British presence.[37] With only a token U.S. Army presence in the region, both Oregonians and Washingtonians needed to join forces along their new divide to preserve their land claims. Although no indication exists that Native groups used the vague boundary to their advantage consciously, their resistance along the forty-sixth parallel made an already chaotic situation even more complicated.

Residents of the Columbia River region long recognized the potential for conflict with nearby American Indians. Oregon Territory's superintendent of Indian affairs, Joel Palmer—he of the three-state vision for the Northwest—warned Washington, D.C., in 1854 that it needed to act fast to conclude agreements with various local Native groups. His superiors in the Department of the Interior noted the increasing pressure in Oregon and Washington Territories, and delayed formalizing settlers' claims of ownership to Native lands.[38] To combat this inaction, Superintendent Palmer and Washington's first territorial governor, Isaac I. Stevens, treated with several native groups in 1854 and 1855 to create reservations in the far eastern portions of both territories.[39] Stevens's ambition to link his new territory to the eastern states with a railroad fueled his desire to settle land claims, which worried American Indians, especially along the plains of the upper Columbia River. These negotiations only delayed the inevitable, and by 1855 Natives fought back against the newcomers. In the months that followed, skirmishes flared from the Rogue River in southwestern Oregon through the Columbia River valley. This extensive conflict in the mid-1850s, sometimes called the Yakima War, was not the first nor would it be the last such struggle for control of the Pacific Northwest. Yet more than any other it demonstrated the way a political boundary's presence complicated frontier warfare.

Trouble between Native groups and American settlers developed in earnest in the early fall of 1855. Andrew J. Bolon, agent for the Yakimas east of the Cascades, set out to confer with his charges after receiving reports of migrants harassed on their way to the coast. The meeting escalated into violence, leading to the deaths of Bolon and several of his companions. Their murders at Yakima hands sent shockwaves of panic through the settlements to the west and forced American Indians to choose sides as well. Some native groups like the Yakimas, Palouses, Walla Wallas, and Klickitats joined forces to repel the invaders, while others like the Nez Perces, Colvilles, and Spokanes proclaimed their solidarity with the Americans.[40] A correspondent for the *Weekly Oregonian* reported nervously: "The times make a man feel every morning for the top of his head. Thank God I am bald."[41] With Governor Stevens east of the Rockies on official business, Governor George C. Curry of Oregon Territory took charge of responding to the threat. He called for volunteers to bolster the thin ranks of Army regulars for the move into Washington Territory. Thus Curry interpreted broadly a clause in Washington's 1853 organic act that established their concurrent jurisdiction over the Columbia River.[42] From his point of view, although the fight began north of the line, an invisible political barrier need not prevent his protecting Oregon Territory.

In early December 1855 three hundred Oregon volunteers, a few dozen militiamen from Washington, and a smattering of regular soldiers moved on a large American Indian force holed up at Fort Walla Walla. Its defenders preferred guerrilla tactics to a siege, and over the course of several days the two sides sparred along the Columbia River. Although the Natives possessed plenty of rations and ammunition captured from nearby agency stores, their forces splintered and retreated northward beyond the Snake River, where the conflict sputtered along for several more bloody months.[43] The cooperation of both territories on their shared frontier, an invisible line slicing through the heart of the battleground, aided their effort. Commanders from both sides directed each other's troops, and temporary fortifications arose on either side of the forty-sixth parallel, where Oregonians and Washingtonians alike massed to reinforce their claim to the Pacific Northwest.[44] Although the *Weekly Oregonian* reported battle news from "our little army in northern Oregon," most of the fighting took place on Washington's side of the boundary, which both territories ignored for the sake of their mutual interest.[45]

Fighting of an entirely different character broke out in early 1856, this time a war of words between the territorial governments in Salem and Olympia and the U.S. Army. The heretofore insignificant line dividing Oregon and Washington territories figured prominently in this new disagreement, which demonstrated the consequences of treating their shared boundary so cavalierly. The trouble began in January 1856 when Major General John E. Wool, commander of forces along the Pacific coast, criticized in California newspapers the militias fighting on the upper Columbia River. In response, the *Weekly Oregonian* insisted that General Wool had abandoned his duty to protect Americans, who had taken matters into their own hands. It also noted that his subordinates had requested volunteers from Governor Curry, but that Wool had rescinded those requests.[46] Oregon's territorial legislature memorialized President Franklin Pierce in support of Curry's call for volunteers and chastised Wool for his perceived aloofness.[47] When word of this spat reached Washington, D.C., Congress ordered an investigation into the conduct of every prominent figure involved in the battles on the upper Columbia.

The confused political and military authority along the shared boundary is evident in a series of letters between the territorial governors and federal officials. On February 12, 1856, as sporadic incidents of violence rattled nerves in the Pacific Northwest, Wool wrote Governor Stevens about the handling of the previous fall's conflict, criticizing Curry for sending volunteers to fight outside his jurisdiction. Wool argued that no real danger from Indians existed along either side of the Columbia River. Stevens refuted Wool's allegations a week later. The Washington governor expressed his gratitude to Oregon's defenders, who had escorted him safely through the contested region. Stevens asked Secretary of War Jefferson Davis to investigate Wool's actions, accusing the general of abandoning Americans to the mercy of unchecked savagery.[48]

Stevens resumed his tirade against Wool a month later, demanding that Davis remove the commander from his position. Without the troops from Oregon to protect Washington's eastern flank, Stevens warned, "a hurricane of war" would erupt "between the Cascade and Bitter Root" Mountains. In defense of his fellow executive, Stevens insisted that "the Oregon volunteers fought the Indians mainly of Oregon, and that, near the confines of the two Territories."[49] In a sense, the Cascades contained the authority he exercised from Olympia while economic and social ties from the upper Columbia River flowed toward Portland and other places in Oregon, which sought naturally to protect them.[50] Curry's efforts on behalf of eastern Washington thus struck Stevens as geographically logical. Political boundaries ranked second to the need to defend the territories' common frontier, at least according to Washington's aggravated territorial governor.

Having endured several weeks of criticism from both Stevens and Curry as well as an insulting resolution from the Oregon territorial legislature, Wool lashed out in the early spring of 1856. Defending himself in a letter to officers in the East, Wool declared his astonishment "to find that the legislature does not know the boundaries of its own Territory. Walla-Walla is in Washington and not in Oregon Territory."[51] The general portrayed the militias rather than the Natives as the real threat to peace in the Walla Walla region, which he considered "not worth the expense" of bloodshed.[52] Further, Wool accused Curry of inflating the threat posed by American Indians on the upper Columbia River in order to excuse an expedition: "It could not have been projected for the defense of the inhabitants of Oregon, nor for the protection of Oregonians in Washington Territory, for none resided there. What then could have been the object? Nothing but a crusade against the Indians, and a long war to enrich the country."[53] In another letter he again railed against Curry for "making war . . . beyond his Territorial jurisdiction against Indians not at war with the whites," while ignoring a threat of violence in southern Oregon. Apparently attacks on the Rogue River, near the territory's southern boundary, did not infect the governor with the same martial ardor. To send armed settlers north of the forty-sixth parallel struck the general as "wholly inexcusable, when the inhabitants of his own Territory were suffering under savage barbarities."[54]

In the months and years that followed, the army and the federal government attempted to sort out what had happened along the forty-sixth parallel. When Congress debated paying the bills racked up by volunteer forces, several members noted the complicating role of the political line. Senator John J. Crittenden of Kentucky described the region as "one Territory that had been made into two. Its being two territories does not vary the matter much."[55] Others supported Wool's portrayal of the bloody reprisals pushed by Curry and Stevens as wholly unnecessary, Representative Abram B. Olin of New York for example: "A great portion of the moneys expended in these disturbances, was spent in the expedition to the Walla-Walla country. It was waged by the Governor of Oregon out of his own Territory, and beyond his jurisdiction. . . . [He] fitted out an armed expedition, sent them into the Territory of Washington, and made war upon a people believed to be desirous of maintaining peaceful and friendly relations with us. There was no occasion for this expedition, and there is no justification for it; there was apparently no motive for it, except plunder and bloodshed."[56]

Similarly, historians tend to side with Wool over Curry and Stevens. D. W. Meinig, for example, argues that the U.S. Army provided far more stability to the region than local militia, who tended to want to clear it of Indians who might compete for land and resources. The Yakima War was just what Wool had claimed, an excuse for American settlers to contain their competitors to reservations and open the land for themselves.[57] Regardless of its controversy, Congress appropriated almost three million dollars—reflecting the inflated prices of goods on the Pacific coast—to pay the extensive bills for the frontier conflict in early March 1861.[58]

In war and in peace, Oregonians and Washingtonians worked closely across their boundary in the 1850s. The vaguely enforced political barrier along the Columbia River and the forty-sixth parallel came increasingly into question as Oregon Territory labored toward statehood. The *Weekly Oregonian*, a Republican organ, accused Democrats in Salem of plotting to erase the Columbia River as a divide and reunite all of the land west of the Cascades into one state. In the meantime, Congress considered an enabling act to bring Oregon into the Union in the opening days of 1857. They proposed to limit the state on the east at the 120th Greenwich meridian, a northward extension of the then-boundary between California and Utah Territory, between the Columbia and the forty-second parallel. Such a divide included the Coastal and Cascade Ranges and would have meant an area about half as large as the present state of Oregon.[59]

Wrapped up in debates over the future of territories nearer to Washington, D.C.—"Bleeding Kansas," in particular—Congress failed to pass an enabling act for Oregon. That did not stop the people of the Willamette River valley from organizing themselves, however. Fifty-two delegates assembled in Salem in the late summer of 1857, arguing issues like the impact of the recent *Dred Scott v. Sanford* decision on their territory and how to form an effective political authority. The proposed boundaries for Oregon also demanded attention. A subcommittee charged with deciding proper limits reported to the convention in mid-August with a series of lines depending heavily on natural features. The Columbia and Snake Rivers formed the northern and northeastern boundaries of Oregon, with a straight line drawn south from the confluence of the Snake and Owyhee

Rivers to the forty-second parallel for an eastern limit. This divide included a portion of Washington Territory between the forty-sixth parallel and the Snake, including the region around Fort Walla Walla in which Oregon's volunteers had recently fought.[60] The entire constitutional convention turned its attention to the matter of their proposed state's boundaries on August 24, 1857. In their deliberations two contrasting images of the state's shape emerged—one a north–south division along the Cascade Range, and the other an east–west split using the Columbia River. Both relied upon essential geographic features of the Pacific Northwest but with different visions for the area's political future.

Some members of the convention, with images of great peaks like Mount Hood and Mount St. Helens in mind, sought to shape their political future accordingly. In reaction to the committee's proposal, Charles R. Meigs of The Dalles, an important commercial town up the Columbia River from Portland, suggested ending the state's jurisdiction at the summit of the Cascades. Meigs called the Willamette River valley the heart of "Oregon proper" and described the mountains to its east as a barrier well suited to a political divide. Meigs told his colleagues that his fellow residents east of the crest "prefer being in vassalage to the government of the United States . . . [than] to being in vassalage to the country west of the Cascade mountains." He found support from Delazon Smith of Albany, a former speaker of the territorial House of Representatives. Smith wanted to make Oregon a long coastal state like California by reclaiming Puget Sound in a polity that spanned the Columbia, leaving the land east of the Cascades "naturally fit to be organized into a great inland state." Some delegates accused Meigs of trying to reserve the new territory for himself and his friends, one centered on his hometown. Others feared isolation and neglect from the eastern states if they did not include enough natural resources in their limits to guarantee future prosperity. Delegate Thomas J. Dryer, the editor of the *Weekly Oregonian*, remarked: "If we are going to have a state let us have a large state with land and room enough."[61]

The federal government tacitly, if not expressly, endorsed the north–south political division along the Cascade Range. A few months after the constitutional convention, for example, a report on the condition of American Indians proposed new boundaries for agencies in the Pacific Northwest. Two stood on the leeward side of the mountains, divided by the Columbia River, while another embraced the entire windward side with no regard to preexisting territorial lines. As the official suggested, "the proper boundaries have been distinctly marked by nature."[62] Another contemporary act declared that property surveys in both Oregon and Washington be "extended and made applicable also to the lands lying east of said mountains within said Territories."[63] Other such land acts involving the West referred almost invariably to only one polity at a time. Officials both near and far from the Oregon Country in the mid-nineteenth century paid little heed to the first lines drawn through it.

A north–south division along the Cascades garnered little support in the 1857 constitutional convention. In the end most delegates advocated the official divide along the Columbia River as established in 1853. Still, the inclusion of a portion of Washington's jurisdiction between the Snake River and the forty-sixth parallel as Oregon's would-be northeastern corner inspired some debate among the solons of Salem. Delegate Smith feared that encroaching on their neighbor would doom the statehood proposal in the halls of Congress, while Lafayette F. Grover of Salem urged the annexation of the Walla Walla region. He told the convention of a conversation with a resident from that place, who said that "all their business relations would be with the state of Oregon." Grover emphasized that including all land south of the Columbia and Snake Rivers as proposed by the boundary committee "is not only natural in itself, but it is natural in uniting the political destinies and social feelings of a people residing in the same valley; the 46th parallel dividing the Walla-Walla valley in two parts. It is of more importance than is generally supposed."[64]

Less than a month after the day spent debating Oregon's boundaries, the convention presented the lengthy document to their constituents, who ratified it in early November 1857. Buried near the end of the document appeared Article XVI, which defined the limits of the political community. The northern line passed "easterly to and up the middle channel of [the Columbia] river . . . and in like manner up the middle of the main channel of the Snake river, to the mouth of the Owyhee river" and from there due south to the forty-second parallel. Recognizing the controversy of taking a bite from their northern neighbor, though, the article closed with a provision that "the Congress of the United States, in providing for the admission of this State into the Union, may make the said northern boundary conform to the act creating the Territory of Washington."[65] In response, Delegate Dryer expressed his displeasure in his *Weekly Oregonian*. The editor complained about the hesitancy of the boundary section, which might allow Congress to constrain the new state.[66] Washington's territorial legislature rankled at the notion as well, albeit for different reasons. It sent a bitter message to Washington, D.C. grumbling that "the seeking thus to appropriate a valuable portion of the domain of an adjacent Territory without its knowledge or consent, is an act of gross injustice, wanting in courtesy and right; and that our delegate in Congress be, and he is hereby instructed to exert his influence to confine Oregon, so far as this Territory is concerned, to her present boundaries."[67]

When statehood threatened to solidify the heretofore-vague line between Oregon and Washington Territories, both sides scrambled to protect land they viewed as theirs.

Congress received Oregon's proposed constitution in early 1858, but waited until May of that year to debate it. Senators drew up a bill to admit the state, but exercised the option given to them by the convention to change the state's boundaries. Replacing the southern banks of the Columbia and Snake Rivers as Oregon's northeastern limit, the Senate drew a line up the former "to a point near Fort Walla Walla, where the forty-sixth parallel of north latitude crosses said river; thence east on said parallel, to the middle of the main channel of the Shoshones or Snake river," and from there followed the eastern division prescribed by the Oregonians. Officials in Washington, D.C., also granted "the residue of the Territory of Oregon" east from the Snake and Owyhee River confluence to the summit of the Rockies to Washington Territory.[68] Not until February 14, 1859, did federal officials approve Oregon's statehood with limits conforming to the 1853 law that established its northern counterpart.[69] News of Oregon's promotion reached the new state about a month after President James Buchanan signed the measure. The rejection of Oregon's tenuous claim to the Walla Walla region, especially the fast-growing town of that name there, furrowed few brows amid the general euphoria of Oregonians' new role in the federal republic.[70]

Following Oregon's ascension to statehood, Washington Territory became an awkward unit extending from the Pacific shore to the Continental Divide, embracing all of two present-day states (Washington and Idaho) and portions of two more (Montana and Wyoming). Yet the unifying forces that spanned the Columbia River remained in place even with this unequal political relationship. In 1862, for example, the *Weekly Oregonian* supported a proposal for an insane asylum funded and managed jointly near Vancouver, north of the river in Washington.[71] Questions about authority over the Walla Walla region also remained prominent in the early 1860s, particularly after a series of nearby gold strikes. Several thousand argonauts, many fleeing the Civil War in the eastern states, headed to the Clearwater, Salmon, and Snake Rivers near the northeastern corner of Oregon.[72] As a result, the people between the Columbia and Snake Rivers and the forty-sixth parallel found themselves the focus of yet another feud over the political fate of the Pacific Northwest.

Many Washingtonians advocated reorganizing the territory to manage better this large population far from the seat of government. Newspapers played a central role in shaping and encouraging the debate. The *Washington Standard*, printed in Olympia, rejected a split along the Cascades and instead pushed for a geometric division extending from the northeastern corner of Oregon up to the international border. Looking to their counterpart to the south for inspiration, the *Standard* claimed: "We must imitate our neighbors in Oregon, who acted wisely in refusing to give up her immense pastures east of the mountains, which will become in a few years the most valuable portion of the state."[73] Reflecting the expansive interest in boundary affairs held by Beaver State denizens, the *Weekly Oregonian* considered the matter as well. It supported the old idea of splitting Washington Territory along the Cascades, setting the stage for a future state the size of Ohio on the shore of the Pacific.[74] In response, the *Standard* accused Oregonians of meddling in the hopes of economically and politically dominating a new territory established on the upper Columbia River.[75]

Attention to the Walla Walla region increased throughout 1862 as the mines to its east expanded and rumors swirled about its potential future as the heart of a new territory. The *Washington Standard* insisted that it "would prefer to remain in our present condition rather than to be separated from the Walla Walla valley."[76] Some residents of that area supported a cleavage from the West Coast, however, particularly considering their Democratic majority as contrasted with the Republican strength in Olympia. The *Washington Statesman* of Walla Walla, for example, observed the discussion with keen interest, but supported the *status quo* for the time being and opposed any division that might affect its ties to the mining regions to the east.[77]

Reports of the pressures in Washington Territory reached the nation's capital in the winter of 1862–63 as it reeled from the defeat at Fredericksburg and cheered the victory at Stone's River. Some members of Congress proposed resurrecting the forty-sixth parallel as a boundary, continuing from its then-terminus on the Snake River to the crest of the Rockies.

Olympia would once again manage all land to the north of that line, while to the south parts of other territories would be added to make a proposed Shoshone Territory. The *Washington Standard* decried Congress's idea: "It would still leave Washington six hundred miles in length . . . with two intervening ranges, and their valleys of varied adaptation; and our Eastern population, who have been the most clamorous for a Territorial government of their own, would remain an integral portion of Washington."[78] Nonetheless, the measure passed the House of Representatives. In the Senate debates, however, the new polity—named either Idaho or Montana at various points—spanned the Rockies, splitting from Washington along a line due north from confluence of the Snake and Clearwater Rivers. President Abraham Lincoln signed just such a measure in March 1863, creating Idaho Territory.[79]

Reaction to the latest division of the Pacific Northwest came faster than in previous instances, the result of the transcontinental telegraph, which sent news to the East Coast at lightning speed. A prominent California newspaper, the *Sacramento Daily Union*, noticed that Washington Territory had "the lawless, semi-secesh portion sloughed off" and expected prosperity from the newly condensed political community.[80] Olympia's *Washington Standard*, happy to have avoided a split down the Cascades, believed that "the prosperity of both sections is now a fixed fact" and looked optimistically toward the future cooperation of Puget Sound and the upper Columbia River.[81] The *Washington Statesman* of Walla Walla also approved of the separation, which blunted the threat of its town losing influence to nearby Lewiston, a mining camp near the confluence of the Snake and Clearwater Rivers, which became Idaho's first territorial capital.[82] Politically, the Walla Wallans preferred the devil they knew.

As the Civil War came to a close, the boundaries of the Pacific Northwest solidified in fact and in practice as settlers throughout the region adapted to life in their political communities. But the most contested part of the Oregon Country—the Walla Walla region between the forty-sixth parallel and the Columbia and Snake Rivers—did not remain quiet for long. Regretting the federal government's alteration of their northeastern frontier in 1859, Oregonians appealed to Congress to grant them their constitutionally described boundaries. In late 1865 Oregon's legislature memorialized Congress that "a large number of the citizens of the Walla-Walla county, in the Territory of Washington, have expressed to this assembly their earnest desire that such county be annexed to and included within the jurisdiction of this State."[83] They made the same request for a geographic rather than geometric northern boundary three more times in the 1870s.[84] Washington's territorial legislature instructed its delegate in Congress to oppose strenuously any loss of land to its southern counterpart.[85] Although they accepted the Columbia River as a political divide, Oregonians found it convenient to ignore the forty-sixth parallel. From military expeditions to political control, in the minds of many state officials the Walla Walla region belonged within the limits of Oregon.

In the summer of 1876, as the country celebrated its centennial, Congress pondered the matter of Oregon's request for an extension of its line to correspond to its constitution. A majority of the House Committee on the Territories supported the proposal, replacing "an *artificial* boundary" with "a *natural* boundary." Representative Greenbury L. Fort of Illinois, however, disagreed with his colleagues. He identified several potential problems that would follow such a boundary shift, including the question of whether settlers in the region would pay for relocating inmates from their towns serving sentences in the territorial prison or living in the insane asylum to appropriate facilities in Oregon.[86] Aside from the occasional committee debate, though, Congress ignored Oregon's petitions on the northeastern boundary matter, thus preserving the alliterative appeal of Walla Walla, Washington. Similar disappointment followed dozens of requests to add Idaho's northern panhandle to Washington or create an entirely new territory out of the upper Columbia River basin in the late nineteenth century.[87]

For a century, the boundaries of Oregon in its constitution and in federal law stood in opposition to one another. Not until 1957, one hundred years after the Salem convention, did the state legislature propose amending the constitution to correspond to the forty-sixth parallel. It sent to Oregon voters a proposed amendment that "the State of Oregon shall be bounded as provided by section 1 of the Act of Congress of February, 1859, admitting the State of Oregon into the Union of the United States," which appeared on the ballot in the fall of 1958.[88] With the electorate's approval, Oregon relinquished its claim to the southeastern corner of its neighbor.[89] Not all northwesterners gave up the thought of altering the region's political divides, however. Numerous suggestions for redrawing the boundaries along geographic features percolate into the early twenty-first century. In 2003, for example, one Washingtonian appealed unsuccessfully to the state legislature for "a mutually agreeable realignment of current state boundary lines ... to generally follow the crest of the Cascade mountains from the border with British Columbia, Canada, southward at least to the Columbia River, and preferably all the way to the Oregon/California border."[90] Some ideas never fade away completely.

The boundaries of the Oregon Country, a blend of geographic and geometric features, have never seemed quite real to residents of either political community. A shared identity traced back to the early nineteenth century proved difficult to replace. Economic and social interests crossed the lines and in serious circumstances even military forces ignored them. Settlers living along them, particularly in the Walla Walla region, both praised and railed against the lines when each suited their interests of the moment. Their occasional enforcement reflected the feeling that the states shared too much to bother with some invisible line cutting through rivers and basins. Many early sources expected mountains rather than rivers to split the Northwest into future polities. Although the towering physical barriers often cropped up in discussions about the region's future, they did not serve as such until Congress created the Idaho-Montana boundary in 1864. The debate between geographic and geometric boundaries and the effectiveness of either resonated throughout the lands drained by the Columbia River. As some of the most flexible state boundaries in the trans-Mississippi West, the Oregon Country lines illustrated the difficulty of establishing authority when social connections and natural regions challenged political barriers time and time again.

Swimming pool at the Cal-Neva Casino Resort on the north shore of Lake Tahoe

CHAPTER 6

A State Bordering upon Anarchy
The California-Nevada Boundary

Frank Sinatra, one of the twentieth century's most popular entertainers, personified the notion that people of many backgrounds find opportunity, intrigue, and even romance in boundaries. Consider his 1953 hit "South of the Border," for example. But Sinatra's interest in borders extended to state boundaries as well. For several years in the early 1960s the crooner owned the Cal-Neva Casino Resort on the north shore of Lake Tahoe. The resort's name originated from its location, straddling the line between California and Nevada. A pool was situated so that guests could swim back and forth from one state to another. In addition, a line of gold to the west and silver to the east was painted on the ballroom's immense stone fireplace. This seemingly innocuous backdrop for photographs and celebrations belies the controversial heritage of the boundary between the Golden State and the Silver State. As their numbers grew in the nineteenth century, American settlers on both sides of the Sierra Nevada sought stability for their lawless region, which one territorial governor described as being in "a state bordering upon anarchy."[1] Battles over the California-Nevada line demonstrate some of the myriad complications inherent in transforming northern Mexico into part of the American West.

Few residents of the region anticipated the boundary controversy between California and Nevada in the mid-nineteenth century. Ethnic tensions within the Mexican Cession, especially in gold rush era California, suggested instead that the region's most important political boundaries would be those that might split the Anglos to the north from the Hispano majority to the south. In his multivolume geographic history of the United States, D. W. Meinig considers the process of creating the California-Nevada boundary "controversial but much less consequential" than California's perceived ethnic polarization.² Yet years of competition and chaos along the state line suggest otherwise. In addition, the widespread debate on whether to use geographic or geometric boundaries resonated loudly on both sides of the Sierra Nevada. The question of using that range or a simpler series of lines as a boundary dominated arguments that lasted for more than two decades. An increasing population in the western Great Basin, disputes with American Indians and the Mormon Church, a rebellion within the legal limits of California, and other problems at the local and national level kept the region in tumult for many years.

The 1848 Treaty of Guadalupe Hidalgo, through which the United States acquired northern Mexico, presented many challenges to the U.S. government. The need to organize the large land cession represented one of the most daunting and immediate. Unfortunately, the lack of well-defined boundaries within the region under Spanish and Mexican control denied federal officials a useful precedent.³ In particular this complicated determining the size of California, as many scholars of the Golden State have observed. Writing in the 1880s, Hubert Howe Bancroft suggested that the Colorado River served as an informal eastern boundary for pre-1848 California, but that without a European population in the Great Basin, no official divide ever emerged there.⁴ In the 1910s, Cardinal Goodwin described a series of early-nineteenth-century maps that indicated several eastern limits for California, most stretching through present-day eastern Nevada and western Arizona. Goodwin noted that the "maps published at different periods all agree in making the 42d parallel the northern boundary of California,—the line established by United States and Spain in the treaty of 1819—but that is about all they have in common."⁵

More recent historians concur with Bancroft and Goodwin. Warren A. Beck and Ynez D. Haase identify several Spanish and Mexican expeditions into the Sierra Nevada in their *Historical Atlas of California*, but they indicate little interest beyond the coastal settlements.⁶ In his study of the northern Mexican frontier, David J. Weber includes a map showing the territory of Alta California in 1824 as stretching from present-day southwestern Wyoming to northern Baja California, but he does not discuss the origins of these limits.⁷ Malcolm L. Comeaux also notes how the northern Mexican territories "vaguely blended into each other."⁸ This uncertain divide between Alta California and Nuevo Mexico probably caused little concern in the first half of the nineteenth century. Aside from a few small settlements near missions in present-day southern Arizona, connected primarily

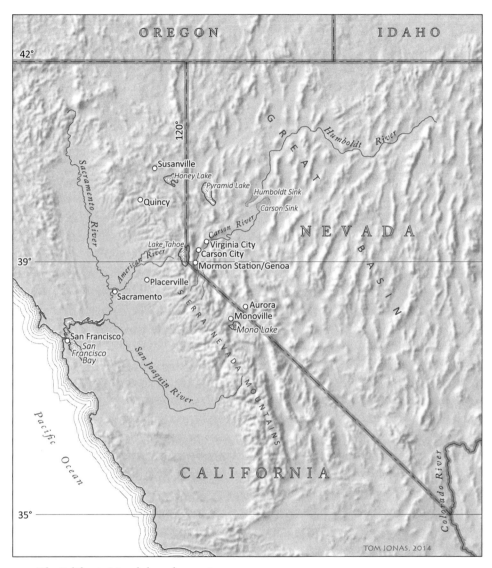

The California-Nevada boundary region

southward to Sonora, Hispanos boasted almost no presence between the Rio Grande valley and the Pacific coast.[9]

American observers in the 1840s also offered conflicting definitions of California's size. In his 1845 guide for overland emigrants, Lansford W. Hastings described Alta California as all the land between the forty-second and thirty-second parallels and west of the "Cordilleras mountains," a term then used for the western Rockies. This included all of the present states of California, Nevada, and Utah, as well as significant portions of Arizona, Colorado, New Mexico, and Wyoming, and slices of Baja California and Sonora. Hastings's California was "more than four times as large as Great Britain [and] twice as large as France" and occupied the majority of the Mexican Cession of 1848.[10] That year, an Army Corps of Engineers report by Colonel John C. Frémont included a map depicting the entire tract taken from Mexico. Frémont's map portrayed New Mexico contained to the Rio Grande valley and gave the rest of the cession to California, which further encouraged Americans to view the latter polity as immense.[11] The James K. Polk administration also embraced that proposed division of the new Southwest, which dispatched the problem of organizing the expansive territory by granting most of it to California.[12]

Members of Congress gathered at the U.S. Capitol in the summer of 1848 to debate how to manage the Mexican Cession, a concern made all the more difficult by sectional controversies in the eastern states. Senators outpaced their colleagues in the House of Representatives when it came to imposing order on the distant lands. An omnibus bill debated in late July called for three territories carved out of all land acquired recently: California, New Mexico, and Oregon. While the provisions for Oregon Territory included boundaries—the Continental Divide and the forty-second parallel—the Senate identified the other two polities only by name.[13] This lack of defined boundaries contrasted with the legislation Congress had crafted for other western territories to that time, suggesting that members did not comprehend fully the land that had come into the country's possession at the end of the war with Mexico.

The need to provide structure to the Pacific Coast grew more acute after federal acknowledgement of the gold discoveries in California in December 1848. Senator Stephen Douglas of Illinois proposed making one immense state out of the cession immediately, while reserving to Congress the right to create "new States out of any portion of said territory which lies east of the summit of the range known as the Sierra Nevada or California mountains."[14] Douglas's plan prompted strong debate in the Committee on the Judiciary, the majority of which opposed his plan. In a report to the Senate, the majority argued that Congress possessed the constitutional power "to *admit* new States—not to *create* them" out of preexisting states, as outlined in Article IV of the Constitution. No federal legislation could trump that document.[15] A minority of the committee offered a new proposal—a state bounded by the crest of the Sierra Nevada and territorial status for everything else.[16] Whichever plan they chose, Douglas insisted that his colleagues act soon, hoping "that Colt's pistols will not continue to be the common law of that land."[17]

Two additional proposals to define California emerged in Congress in early 1849. Representative Henry Hilliard of Alabama suggested a state bounded by the Pacific Ocean and the Sierra Nevada to the west and east, and forty-one degrees thirty minutes and thirty-four degrees thirty minutes to the north and south. About two-thirds the size of the state that eventually emerged, this California centered on San Francisco and left Los Angeles and San Diego within the limits of a radically enlarged Texas. It also marked the only time a proposal for California did not extend north to the forty-second parallel, a generally accepted precedent hearkening back to the 1819 treaty dividing Spanish lands from the Oregon Country.[18] The final proposal, also made by Douglas, flirted with the ridiculous. It vaguely bounded California by the ridge separating the Great Basin from the Colorado River drainage and by the desert trails that had been followed by Colonel Frémont on his recent survey.[19] What an honor for the military man, that his very footsteps might shape the political future of the West. Yet the debate dragged on, and the cession remained for a year and a half without any authority beyond a provisional military government.

The failure of Congress to provide California with either a territorial government or a plan for statehood did not bode well for its rapidly growing population. Gold-seekers flocking to California inflated its number of non-Native residents from fourteen thousand to about a hundred thousand in two years, more than sevenfold.[20] While the gold rush was international—people came from all over the world in the hopes of striking it rich—Americans were most interested in the region's legal and political fate. American fortune-seekers brought not only mules and pans but also knowledge of a system of government that they sought to re-create in the West. To that end, military governor Bennett C. Riley called for a constitutional convention in the summer of 1849. He had orchestrated local affairs in the manner of a territorial governor since the war's end but he realized that the burgeoning region needed a more stable authority.[21] In answer to Governor Riley's call, forty-eight delegates gathered in Monterey on September 1, 1849, to discuss a proper course of action. Most had lived in California since before the gold rush, with disparate origins: members hailed from sixteen states and several foreign countries. Nine Hispano members provided a voice for their recently absorbed community as well.[22] The convention reflected the ethnic diversity of California's 1849 population and the broad interest in their political fate.

Many issues demanded attention from the solons of Monterey, but California's size proved one of the most controversial. During the course of the convention, which lasted for approximately six weeks, delegates debated no fewer than nine different eastern boundaries for their state. Some included geographical features, while others depended upon various lines of latitude and longitude. The westernmost proposed eastern boundary corresponded to the summit of the Sierra Nevada, while the easternmost of the nine called for a boundary along the 105th meridian, which passes through present-day Denver, Colorado. A delegate who insisted upon a geographic division stated: "Nature, sir, has marked out for us the boundary line of California. God has designated her limits, and we ought not to go beyond the line traced by the Omnipotent hand. The snowy range of the Sierra Nevada separates two communities of this country; [and] they can have no connexion, either political or social." Most delegates advocated a smaller state, insisting that California take only what it could realistically control and let the federal government handle the rest. Supporters of a larger state, however, noted that their option would solve the problem of what to do with the rest of the cession, or at least postpone the matter until sectional tensions in eastern states eased.[23]

It was the ninth and final line considered by the convention that created the present California-Nevada boundary. This geometric barrier ran south along the 120th meridian from the 42nd parallel to the 39th, then angled southeast to the intersection of the Colorado River with the 35th parallel and continued down that river to the Mexican border. The delegates explained their choice of lines in a statement addressed to their fellow Californians: "In establishing a boundary for the State, the Convention conformed, as near as was deemed practicable and expedient, to great natural landmarks, so as to bring into a union all those who should be included by mutual interest, mutual wants, and mutual dependence. No portion of territory is included, the inhabitants of which were not or might not have been legitimately represented in the Convention."[24]

Yet their "great natural landmarks" included only the Colorado River, as the other boundaries consisted solely of geometric divisions along points of latitude and longitude. In all fairness to the delegates, the hinge of the eastern boundary—the intersection of the 120th meridian and the 39th parallel—corresponded to a natural feature, but only accidentally. By coincidence this essential point rested in the southern end of Lake Tahoe, although none of the delegates could have known that at the time.[25] The representatives in Monterey also emphasized the multicultural aspect of their convention by congratulating paternalistically the nine Hispano members who had adapted so quickly to American notions of territorial division.[26] But declaring the state boundary and enforcing it were two different things, as residents on both sides of the Sierra Nevada learned in the years to come.

American residents of California had some regional company in pondering the political fate of the Mexican Cession. Hundreds of miles away across the stony peaks and sagebrush flats, another group sought to carve out a place to call their own. Mormon settlers near the Great Salt Lake, who had arrived there in 1847, believed they had a right to control part of the vast province too. Like Californians, the Mormons thought big. Several months before the Monterey convention a meeting in Salt Lake City proposed an immense state of Deseret, a long-sought sanctuary for the Latter-Day Saints. Deseret encompassed parts of nine present states, with a boundary along the Sierra Nevada crest and coastal access to southern California. While federal officials resisted granting the group such an enormous domain, the dilemma of whether to provide them with their own polity or place the growing population under the jurisdiction of another territory remained. The Mormons continued to play an important role in affecting the political fate of the region, especially on California's newly declared eastern line.[27]

Before the end of 1849 four out of five California voters approved the constitution drawn up in Monterey, and state government commenced by the middle of December. The document then headed east to the nation's capital, along with senators- and representatives-elect from that state. California thus became the first state in over fifty years to request admission to the Union without a congressional enabling act (excepting the special case of Texas), an action that met with resentment in the capital. Senator Henry Clay of Kentucky objected that "the people of California, by their convention, have appropriated not only a Territory to themselves, but they have assigned boundaries, without ever having consulted Congress one way or another."[28] The House of Representatives even debated a resolution insisting that "California should not be admitted into the Union as a State without a restriction of her limits."[29] By early May 1850, however, a

Senate committee assigned to study the matter concluded that the boundaries drawn in Monterey represented the best division for the state.[30]

Beyond the question of California's limits, Congress considered other controversies in 1850. Through the efforts of Clay and others, a series of bills passed over several months that summer produced the Compromise of 1850, which attempted to address territorial and slavery concerns across the nation and keep the sectional peace. On September 9 Congress approved three acts of concern to the West—statehood for California, territorial status for New Mexico and Utah, and definition of the boundary between New Mexico and Texas.[31] Regarding the two new territories, Congress avoided mentioning specific lines or features when creating their limits. Instead, they defined Utah Territory as bounded "on the west by the State of California" and New Mexico Territory as everything left over after the establishment of California and Utah.[32] In one busy day the federal government thus imposed political organization on the Southwest.[33] Without an official survey, though, the eastern California boundary existed only on paper, a condition that created both opportunities and dangers for those who came to live near it.

California's statehood established an expansive political community with boundaries that encompassed more land than all but one other state, the newly redrawn Texas. The Golden State also contained diverse environments, some better known and more appealing than others. In particular, the quality of its land east of the Sierra Nevada did not seem encouraging. Countless overland emigrants passed through the area on their way to the goldfields and farm country of California, but few decided to remain there. Even the first American politicians from the Pacific Coast told Congress that their state's eastern boundary traversed an arid land with limited agrarian potential. Indeed, they considered the western Great Basin "of little or no value."[34] Although a handful of settlers made their homes east of the Sierra Nevada in the 1850s, most migratory Americans continued across the mountains. Those who staked a claim in the Great Basin discovered the challenges of such a rugged life exacerbated by their isolation from any effective political or legal authority.

The presence of several suspect factions drew California's attention to its eastern line in the early 1850s. The first significant American settlement in the area appeared in 1850 when a colony of Latter-Day Saints arrived in the Carson River valley, in the farthest western reach of the theocratic Utah Territory. They came to create an outpost to profit from overland migrants nearing the end of their journey. National media observed the establishment of Mormon Station, but the press doubted it would prosper in such a rugged region. Indeed, several years passed before the supply post boasted a year-round, self-sufficient population.[35] In 1851, California's legislature expressed concern about other groups living along the state's perimeter, as well. The body petitioned Congress that "our State is unprotected from the different tribes of Indians that live upon our borders" and demanded a chain of defensive forts along the boundaries. The War Department did not acquiesce, however, even after Californians insisted upon the need to defend those invisible lines "from the incursions of either internal or external enemies."[36] With most American Indians in the Great Basin considered non-threatening and a paltry number of American settlers along California's lines, such a military presence seemed unwarranted. But the presence of Indians and Mormons, both seen as obstreperous groups by most mid-nineteenth century Americans, soon inspired more controversy east of the Sierra Nevada.

Following the founding of Mormon Station, settlers of various faiths and ethnicities filtered gradually into the lands east of California's eastern boundary. As months passed, officials on both sides of the line believed the region needed political representation. California acted first, with an extralegal and overreaching gesture it considered necessary nonetheless. In 1852 the state legislature proposed a Pautah County outside California's eastern limit in an effort to organize the settlements just past the state's hinged boundary in western Utah Territory.[37] But Pautah County failed to find much support among the isolated residents, unsurprising considering its dubious legality. Seeking a more effective solution, some of the area's non-Mormon settlers wrote to Sacramento requesting annexation by California. In response to this threatened dismemberment, Utah's territorial legislature established Carson County in 1854. This local authority overlapped most of the extralegal Pautah County, and Utahans dispatched officials to provide law and order over the land that belonged legally to them.[38] These competing counties illustrate the complexity of controlling this isolated yet growing region.

Trouble increased as non-Mormons made further inroads east of the Sierra Nevada. In September 1853, a fifth-generation German American named Isaac Roop settled near Honey Lake, establishing the first American presence in northeastern California.[39] Others soon followed, building farms and ranches near Roop's home in the mid-1850s, which developed into the town of Susanville, named for his daughter. Believing erroneously that Susanville stood east of the boundary, they bemoaned their dismissal by officials in Fillmore, then the capital of Utah Territory.[40] Without a survey to clarify their location in California, however, the Honey Lake denizens received little attention from that state either. In 1854, the creation of Plumas County in northeastern California extended a new jurisdiction over Honey Lake, but officials at its seat at Quincy—separated from Susanville by several mountain ranges—exercised little authority there. On the rare occasions when tax collectors made the difficult journey to Susanville, the settlers declared that as residents of Utah Territory (as they thought they were), they owed no money to California.[41] These settlers used the uncertain boundary to their advantage throughout the 1850s.

The gauzy nature of this region's boundaries attracted national attention, especially in conjunction with the controversial population that dominated Utah Territory. That decade witnessed numerous proposals to end what many outsiders considered a theocracy under the Mormon president and territorial governor, Brigham Young. Newspapers in the nation's capital repeatedly demanded action. Editorials in the *Washington Evening Star* proposed altering the lines of not only California but also the territories of Kansas, Nebraska, New Mexico, and Oregon to balkanize Young's polity.[42] The *Star* argued that such a plan would provide the rights of citizenship to Mormons while ending the political control of the church, even if such extended territories would prove harder rather than easier to govern. As one *Star* writer suggested: "We must divide, in this case, to conquer."[43] Western boundaries developed religious as well as political overtones when it came to the thorny issue of Utah Territory.

Federal officials reconsidered the California–Utah Territory boundary in the mid-1850s. In 1856 California's legislature asked Congress to approve shifting the state's longitudinal eastern line from the 120th meridian to the 118th meridian, a little over a hundred miles to the east.[44] This would embrace all of "Pautah County" (and Carson County) and assure the residents of Honey Lake of their location in California. Non-Mormons living in the disputed territory petitioned Congress in support of the idea and bemoaned the Mormon officials sent by Governor Young. A House of Representatives committee discouraged the alteration, however, arguing that "the State of California is already too large" and that its officials could not adequately control the territory within its limits already.[45] Distracted by chaos on another boundary at the time—that between Missouri and Kansas Territory—Congress spared little concern for the more distant contested region.

Residents along the California-Utah boundary in the mid-1850s knew not to which government they owed allegiance, and they grew increasingly resentful of their perceived dismissal by both. As other westerners did in similar circumstances, they considered radical solutions. In the spring of 1856, for example, a group of settlers met at Isaac Roop's home to declare a temporary government for the Honey Lake valley. Roop and the others titled their new jurisdiction the Territory of Nataqua, a Paiute word for "woman" (chosen with the inference that the new territory lacked more of that vital resource than any other). The western line of Nataqua matched California's eastern boundary on the 120th meridian—meaning that the territory's founders lived outside of the political community they were trying to create. In any case, it served as little more than a local effort to legalize land claims.[46] Roop's biographer James Thomas Butler believes the Honey Lake residents did not consider Nataqua a long-term solution anyway but instead "a frontier land club or claim association, created to protect squatters who had no legal government."[47] Regional authority stumbled again when Utah Territory rescinded the law creating Carson County in early 1857, placing the distant settlers under the direct authority of officials in Salt Lake City.[48] Even while playing at territorial government, by the late 1850s residents of the divided region had lost more local power than they had gained.

In 1857 the region's settlers made another effort to shape their own destiny. For the first time, disaffected settlers in the Honey Lake and Carson River valleys came together to address their common concerns. In early July they met in Genoa, a non-Mormon town that had grown up around Mormon Station, and proposed a new territory embracing much of present-day Nevada and Arizona and a small portion of Utah. As with Nataqua, the western boundary of this territory named "Sierra Nevada" corresponded with California's geometric eastern line. Apparently, Honey Lake settlers had accepted their location within the Golden State by this point, however, and several Susanvillians considered it "folly, and worse than folly, to attach the people of this Valley to a State about which they know nothing, and care nothing."[49] They convinced their Carson valley colleagues to amend the western line of Nataqua to the summit of the Sierra Nevada, pending the approval of California to the cession, which the activists considered both logical and likely.[50]

Opinions about this latest extralegal western territory appeared in numerous newspapers. The *Sacramento Daily Union* made no objection to the mountain crest as a political divide, particularly as a way to cast off the troublesome population at Honey Lake.[51] But San Francisco's *Daily Alta California* adopted a more hostile tone: "We should never be led to suspect that they emanated from a 'Honey region,' as they are particularly vinegarish."[52] The nation's capital also received the news with skepticism. Washington, D.C.'s *Daily National Intelligencer* portrayed the territory's founding fathers as "desperadoes" who claimed far more land, particularly to the south, than they could realistically control, a bewildering stance especially for people who wanted their own polity because their rightful officials were too far away.[53]

Events in Salt Lake City also affected the boundary region in 1857. Governor Young, concerned about widespread hostility toward his church and his political authority, recalled the faithful from the far-flung corners of the Great Basin that fall. Although he explained this as a first step toward shifting Mormon colonization efforts toward the Columbia River, many people outside the church suspected an ulterior motive. Outsiders accused Young of preparing to fight the federal government for Utah and recalling his followers because he needed all the help he could get for this. The following "Mormon War," a nearly bloodless standoff settled by replacing Young with a non-Mormon governor, had far-ranging consequences. For those living along the eastern Sierra Nevada, this episode resulted in a significant depopulation. As many as a thousand Mormons living there packed "their household goods and gods, with their wives and their cattle" and headed for the Great Salt Lake, one eastern newspaper reported.[54] Without a significant Mormon presence in the western part of the territory, the remaining settlers pushed harder than ever for separation from Utah. With the encouragement of California's delegation, Congress considered new plans for a Great Basin territory identified as either "Sierra Nevada" or simply "Nevada." Ignoring the incessant pleas of the Honey Lake residents, federal officials wanted California's eastern line to serve as the new territory's western boundary.[55]

Relations between California and Utah Territory grew more unstable in the late 1850s. A vigilante execution of a man suspected of murder for hire in the summer of 1858 reminded everyone of the need for law and order. Regional newspapers hesitated to support mob justice but saw no realistic alternative without more effective political organization. As a correspondent for the Placerville, California, *Mountain Democrat* observed: "It is well known throughout California that the Carson Valley and the valleys East of the Sierra mountains, have been for a long time past infested with thieves, and been the great resort and lurking place for all fugitives from justice."[56] The new governor of Utah Territory, Alfred Cumming, appointed a federal judge to oversee these western reaches, which Cumming considered in "a state bordering upon anarchy." He also pressured the legislature to resurrect Carson County to restore some local control.[57]

After several ineffective efforts to transform western Utah Territory into a new polity, many boundary-area residents felt frustrated. But the events of 1859 exacerbated the preexisting tensions beyond anyone's expectations and forced influential people near and far to reconsider the region's future. After several years of limited placer mining near the Carson River, in early 1859 prospectors discovered traces of what eventually developed into the Comstock Lode, a rich source of gold and silver. A full-fledged rush began by the summer, especially when hordes of Californians headed for the new diggings just beyond their limits. These migrants imported the system of local mining laws they had developed during the rush to California a decade earlier, recognizing what one such district described as "the isolated position we occupy, far from all legal tribunals, and cut off from those fountains of justice which every American citizen should enjoy." But these tested forms of temporary authority did not meet all the needs of the western Great Basin.[58]

With an increasing American population of farmers, ranchers, and fortune-seekers, the need for local government grew more acute. Eastern newspaperman Horace Greeley observed the pressure on a cross-country journey in the summer of 1859. He witnessed a mass meeting at Carson City, several miles from Genoa, where assembled firebrands proposed yet another extralegal movement to establish "the embryo State of Nevada."[59] Those at the convention in Carson City emphasized the lawlessness of their surroundings, even with a restored county government and federal law enforcement, and stated their belief that separation offered their best hope for success.[60] Chipping into the debate, Greeley proposed using the mineral rushes to address the Mormon situation. He suggested altering Utah Territory "by cutting off Carson Valley on the one side, and making a Rocky Mountain territory on the other, and then let them go on their way rejoicing."[61] The federal judge appointed for Carson County opposed Greeley's idea for a nefarious reason of his own. He argued instead that by remaining in Utah Territory the burgeoning western population could eventually take control of its entire expanse for themselves, and thus overwhelm the Mormons to the east.[62]

This "embryo state" attracted interest beyond the Carson River valley. With another political rebellion in the offing, residents of Susanville again pressed for a polity east of the Sierra Nevada. Isaac Roop won election as the provisional governor of this latest extralegal Nevada Territory, even though he lived several miles outside of its boundaries.[63] The movement received support from various sources, including the California legislature. Although politicians in Sacramento ignored the repeated secession attempts of Honey Lake, in January 1860 they once again petitioned Congress to transform western Utah Territory into a separate polity.[64] The *Territorial Enterprise* of Carson City also trumpeted the idea in several editorials.[65] For the third time in four years, the region's American inhabitants sought to establish their own government.

While the Mormons had little reason to cheer this influx into land still part of the territory they dominated, another group also grew wary of the demographic shift. Relations between settlers and American Indians, which had been limited and generally peaceful until the mineral rushes, further complicated life in the area. The Northern Paiutes represented the largest Native population in the western Great Basin. As late as 1857 a regional newspaper portrayed the Paiutes as eager participants in the transformation of the area into an agrarian paradise: "All the other tribes are warlike, insincere, treacherous, and the most of them bloodthirsty. Should a Territory be organized, the Pah Yutes would promptly unite with the whites, and identify themselves with the peaceful progress of the country."[66] But the influx of Americans after 1859 changed the balance of power east of the Sierra Nevada dramatically, and in 1860 the Paiutes determined to fight for their homeland.

American migrants in the new Southwest expressed worry about a Native uprising as the Carson County mineral rush continued in 1859. Stories circulated that a band led by the influential Northern Paiute Winnemucca wanted compensation for the loss of territory to miners and ranchers—including sixteen thousand dollars for the Honey Lake valley—and the newcomers worried what might happen if he did not receive it. As winter came to a close in early 1860, factions of Paiutes detained migrants passing through the western Great Basin. Several settlers and Paiutes were murdered, actions that increased suspicion between the groups but went unpunished, a result of the lack of effective legal jurisdiction. Two settlers appealed to "Governor" Roop: "We believe that the Indians are determined to rob and murder as many of our citizens as they can, more especially those on the borders." Roop appealed to the U.S. Army for help, but commanders in San Francisco could not respond to the request of an extralegal executive. Thus spurned, the provisional governor warned "our citizens who are exposed on the border settlements" to prepare for a full-scale war. The *Territorial Enterprise*, meanwhile, cautioned its readers that the actions of a few Paiutes should not lead to an extermination campaign against the society as a whole.[67]

Three months of relative quiet followed, as Americans in western Utah Territory accepted the admonition of the *Enterprise*. In early May 1860, however, Paiutes living along the Carson River killed seven Americans at William's Ranch. Miners, farmers, and ranchers grabbed their guns and organized militias in Carson City and Virginia City, the two largest towns in the area.[68] Two days after the attack, more than a hundred volunteers led by Major William M. Ormsby moved north to a Paiute settlement near Pyramid Lake, between the Carson River and Honey Lake valleys. Ormsby's patchwork of troops proved no match for the Paiutes, however, who ambushed and killed nearly four-fifths of the invading force, including their commander.[69] With Utah Territory incapable of keeping the peace on its western boundary, nearby California came to the rescue. Governor John G. Downey sent wagonloads of supplies and weapons across the Sierra Nevada through snow upwards of twenty feet deep to protect settlements populated mostly by people who had until recently lived in his state.[70]

California's questionable jurisdiction hampered its ability to help, as did a lack of good information about the conflict. News reports crossed the mountains irregularly and they often conflicted or came from sources of dubious quality, as exemplified by this correspondent for the *Mountain Democrat*: "I did, on one occasion, seize my pen, with the view of giving you a history of events here, but being fired with martial ardor, I mistook that peaceful instrument for a bayonet, and charged upon a pine board by my table, thinking it was an *Injun*. My ink bottle being capsized in the effort, I doubted not that I beheld the blood of the savage, as I had heard that the Pah-Utahs are all black hearted rascals; so without more ado, I started down town to 'fight my battle o'er again,' at the gin cocktail manufactory, but I fainted on the way, and was carried home, in a state of insensibility."[71]

Ultimately a volunteer force five times that of Major Ormsby's ill-fated command, backed by more than two hundred regular troops sent from California at Governor Downey's request, dispersed the Paiutes and ended the uprising by the summer of 1860.[72] Although the Paiutes remained in the area, conflict with American settlers waned after their defeat and they adapted to life on nearby reservations, including one at the scene of their great victory at Pyramid Lake.[73]

The feud with the Paiutes demonstrated several realities about conditions in the western Great Basin. First, its new and growing American population faced serious competition for the region. Second, Utah Territory could not adequately protect the settlers on its western edge. Third, California could offer logistical support but not legal protection for people living beyond its limits. Fourth, the provisional government of Nevada Territory had no power even when it came to matters of life and death. Carson City's *Territorial Enterprise* thus feared forever languishing under "an organized mob or vigilance committee."[74] Also reflecting on recent events, a newspaper in Salt Lake City portrayed the Carson River valley as "a most unremunerating burthen upon Utah" and a "worthless, unaccountable scab" that should be picked off immediately.[75] San Francisco's *Daily Alta California* summed up the situation: "Never before has any civilized community approached so near without falling into utter anarchy, as this."[76] More than any other factor, the conflict with the Paiutes demonstrated the necessity to impose a new order on those living between the authority of California and Utah Territory.

In Washington, D.C., the question of how to stabilize the Great Basin offered just one of many worries in the winter of 1860–1861. The election of Abraham Lincoln as president the previous fall had inspired a cascade of secession declarations from seven southern states, the most serious threat to federal supremacy in the nation's history. Congress and outgoing president James Buchanan worked to stabilize the West in early 1861 to prevent the most distant Americans from also trying to leave the Union. In mid-January, for example, Senator Henry M. Rice of Minnesota proposed admitting the entire trans-Mississippi region as five new or altered states. Under his plan California would extend eastward to the 114th meridian, just east of the present Nevada-Utah line, and thus encompass the entire disputed region east of the Sierra Nevada.[77] Instead, several weeks later Congress approved a different idea, the creation of a new territory for the western Great Basin.

On March 2, 1861, President Buchanan signed a bill establishing Nevada Territory, part of a busy final week in office in which he also approved the creation of Colorado and Dakota Territories. To the north, east, and south, geometry provided Nevada's limits—the 42nd parallel, the 39th meridian west of Washington (within a few miles of the 116th Greenwich meridian), and the 37th parallel, respectively. The western limit, though, caused the most controversy. Congress declared as Nevada's western boundary "the dividing ridge separating the waters of the Carson Valley from those that flow into the Pacific; thence on said dividing ridge northwardly" to the summit's intersection with the 41st parallel, and from there due north to the 42nd parallel. This definition included a significant portion of northeastern California, to take effect only if politicians in Sacramento consented.[78] Such a revised line made some natural sense, considering the logistical problems posed by the Sierra Nevada. Yet the federal government had encroached upon the legal limits of California, an astonishing action considering the tense atmosphere of early 1861. As southern states questioned federal power over local sovereignty, one might have expected Congress and the president to treat the defined limits of states with more respect.

Although unconnected to the birth of Nevada Territory, California's government revisited its eastern boundary in early 1861. The state government approved a survey of the line to end the question and invited the federal government—the final voice on all territorial matters—to participate. It was hoped that surveying the line would silence the malcontents at Honey Lake and establish jurisdiction over several new mining districts, including the booming Esmeralda region centered at Monoville and Aurora, towns located near the vague line extending southeast from Lake Tahoe.[79] The residents of Monoville had appealed to legislators in Sacramento for a local government, and they received their own Mono County abutting the unmarked state limit shortly thereafter.[80] In the early 1860s, the conflicting boundary visions of California's state government and the federal government through the creation of Nevada Territory resulted in ever more unrest.

Word of the new Nevada Territory arrived in Carson City via the Pony Express a little more than a week after President Buchanan penned his signature.[81] A newspaper there believed that California's legislature would acquiesce promptly to the boundary shift, placing Honey Lake and other disputed lands in the new territory.[82] When the news crossed the mountains, however, Californians viewed the proposal with skepticism. The *Sacramento Daily Union* worried about the loss of valuable mining deposits.[83] San Francisco's *Daily Alta California* suggested two possible courses of action, "one to reject the proposition, the other to pass an act of conditional cession, to become valid if ratified by the people of the tramontane strip on popular vote."[84] To complicate things further, some Californians wondered whether the legislature could approve the shift at all, or whether the matter should appear before voters statewide since it would mean changing the boundaries as defined in California's constitution.

With Nevada Territory claiming jurisdiction westward to the summit of the Sierra Nevada, and California declaring its authority eastward to the geometric limits established in 1850, the federal government created a new disputed region in the already unstable area. In contrast to the pre-1861 situation in which no one exercised effective control, the new crisis was one of overlapping jurisdictions with immediate consequences. A man accused of murder at Honey Lake in the spring of 1861, for example, stood trial in Carson City rather than Quincy, the seat of California's Plumas County.[85] In another incident, the sheriff of Plumas County confiscated some livestock from a rancher at Honey Lake to settle an outstanding debt. The rancher demanded the return of his property, which he considered seized by an interloping law enforcer, and convinced a Carson City judge to affirm Nevada's control over his home.[86] Farther south, the residents of the new Mono County in California took advantage of their marginal position by electing delegates to the first Nevada territorial legislature in the fall of 1861.[87] At that session Nevadans created a series of counties, including one incorporating Honey Lake.[88] While making a new government for the western Great Basin solved some problems, it created plenty more.

When the California legislature convened for its annual session in early 1862, Nevada Territory mounted a campaign to complete the boundary shift suggested by Congress. James W. Nye, appointed by President Lincoln as the first (and only) governor of Nevada Territory, traveled to Sacramento with a delegation to plead the case. Several speakers appeared before the state legislature in mid-March, including Isaac Roop, the former provisional governor of Nevada Territory and now a member of its council (the equivalent of a senate). Roop told the assembled body that Honey Lake looked eastward rather than westward for protection and leadership and should be allowed to join the new polity. Governor Nye portrayed the land east of the summit as inaccessible and generally worthless to California yet at the same time essential for Nevada's future. He also downplayed the disputed region's agricultural and mineral potential, painting a picture of a desolate area the Golden State should be only too happy to abandon, offering contradictory opinions about its worth. "This line," Nye insisted in reference to the summit boundary, "had been fixed by mountains so high that they milked the clouds."[89] Deliberations lasted for weeks, but when the California legislature adjourned later that spring it retained the "slice of our glorious State" so coveted by Nevada.[90] Perhaps as a consolation, in the summer of 1862 Congress extended Nevada's limits some sixty miles to the east, placing more of Utah Territory under Carson City's jurisdiction.[91]

The political boundaries of the region remained frustratingly confused throughout 1862. When residents of the Honey Lake valley feared another Paiute uprising, for example, they appealed for help to Carson City rather than Sacramento. Nonetheless, they sought supplies from Plumas County businesses for their defense, once again straddling the vague divide.[92] Perhaps the most prominent example of boundary confusion emerged when Congress approved an act for a transcontinental railroad and telegraph line. In the act, Congress instructed the Union Pacific to build "westerly upon the most direct, central, and practicable route, through the Territories of the United States, to the western boundary of the Territory of Nevada." At the same time it told the Central Pacific to work "from the Pacific coast . . . to the eastern boundary of California."[93] If both railroads built to their mandated ending points, therefore, the lines would pass each other for miles, with the Central Pacific stopping on the sagebrush flats of the Great Basin and the Union Pacific halting at the crest of the Sierra Nevada. Congress hoped the eastern California and western Nevada lines would eventually become one and the same, but California's resistance thwarted that intention. The fact that the federal government failed to rectify its convoluted actions added to the region's woes.

As the Civil War began its third bloody year in early 1863, marked by the great clashes at Chancellorsville, Gettysburg, and Vicksburg, the ongoing battle over the California-Nevada line turned violent as well. The previous year politicians in Carson City renamed their most northwestern county after Isaac Roop and seated it at Susanville, the settlement built around his ranch near Honey Lake. After Governor Nye appointed officials to manage this new Roop County, authorities in Plumas County, California, sought to enforce their rightful jurisdiction there. In early February 1863, Plumas County Sheriff Elisha H. Pierce organized a posse and crossed the Sierra Nevada to arrest the territorial authorities infringing upon his county. Near Susanville they encountered a gang led by Roop himself. Roop's mob detained Pierce and his men, but the sheriff escaped and returned across the mountains for reinforcements. California newspapers reported that Nevadans had armed Roop's gang to reinforce their claimed authority, exacerbating the "Honey Lake imbroglio."[94]

A few days after his first attempt, Sheriff Pierce returned to Susanville with about a hundred men, intent on rescuing the captured members of his first posse. The settlers resisting his presence secured their captives in a fortified log cabin, and Pierce commandeered a nearby barn for his own headquarters. On February 15, 1863, as the Plumas County men barricaded the barn, the Honey Lake firebrands warned them three times to desist, after which they opened fire. For three hours the two camps exchanged sporadic shots, wounding three men in the barn. After the inconclusive skirmish, both sides agreed to negotiate their differences. They resolved that neither Plumas nor Roop Counties should exercise jurisdiction in the contested zone until the territorial, state, and federal governments resolved the boundary dispute.[95] Within a few days, representatives for both Governor Leland Stanford and acting territorial Governor Orion Clemens (brother of Mark Twain, who followed Orion to Nevada) began an investigation. A Nevada Territory report warned of "a bloody border warfare of the most desperate character" if the dispute did not end soon.[96] Gordon N. Mott, a judge and newly elected delegate to Congress from Nevada, reinforced his adopted home's position in the regional press. In a letter published on both sides of the Sierra Nevada, Mott argued that Nevada had every right to establish counties stretching to their summit. Without a formal survey to mark California's eastern limit, he found their claim to the disputed region tenuous. In response, a San Francisco newspaper proposed a quick, rough demarcation to assert its state's claim. Such a temporary boundary, even if proved inaccurate by later careful studies of the line established by law, might calm things down, the press hoped. In the meantime, both governors refused to surrender what each considered their rightful jurisdiction.[97]

The perennial problems at Honey Lake offered only one source of controversy. Acting Governor Clemens recognized another potential source of violence along the ill-defined boundary.[98] He wrote to Governor Stanford that "we know the same difficulties are liable to arise in reference to another portion of our border line—the Esmeralda section of the country" southeast of Lake Tahoe.[99] Correspondents from the Esmeralda district mining town of Aurora echoed Clemens's concern, noting that several incomplete surveys had failed to place the town in either polity definitively.[100] Attempting to secure the booming mines to Nevada, Clemens created Esmeralda County seated at Aurora. Coincidentally, the Golden State legislature approved shifting the seat of Mono County from Monoville to Aurora. As a result, in the spring of 1863 local officials and district judges owing allegiance to both sides vied for control of the region. Addressing the convoluted situation, Nevada's legislature declared that, in Aurora, California laws not in conflict with Nevada laws would be legally valid for the time being, until a formal survey. Thus, that marginal town served as the seat of two contentious counties, one in a state and one in a territory, at the same time.[101]

In the interest of ending this debacle for good, members of the California legislature reaffirmed their original 1850 boundary in the spring of 1863, regardless of Nevada Territory's organic act or the hopes of Honey Lake settlers. During the debates in Sacramento a strong resistance grew to using the summit of the Sierra Nevada as a boundary. Several politicians noted the challenges for survey teams scaling rugged peaks and sheer cliffs to mark the line. As one state senator quipped: "Adopt the summit of the Sierras as the dividing line and there would be no time within the next 6,000 years when we could quite agree where our jurisdiction terminated." Some officials also worried that using the ridgeline as the political divide would lead to court battles over lode mines that followed veins through the mountains and crossed into the neighboring authority.[102]

Congress could not avoid the mess it had made in regard to Nevada Territory's boundaries when it considered promoting the region to statehood in 1864. That spring the federal government approved an enabling act for the polity to draw up a state constitution. Hoping to end the uncertainty, Congress revised Nevada's 1861 boundaries but in doing so once again created a new problem while resolving another. For the line extending southeastward from Lake Tahoe, the act specified that "the eastern boundary line of the State of California" would serve as Nevada's limit.[103] For the boundary extending north from the lake, however, Congress referenced the 43rd meridian west of Washington, which stood several miles west of California's line on the 120th Greenwich meridian. Again the federal government created an overlapping jurisdiction, this time one of some six hundred thousand acres.[104] One marvels at the bewildering inability of the executive and legislative branches to craft a common boundary between the two polities.

Following a brief constitutional convention and federal approval of its charter, Nevada entered the Union on October 31, 1864, just in time to cast three electoral votes for Abraham Lincoln and help secure his second term. Over the next several years Nevada grew twice, to the east another meridian to the present Nevada-Utah line, and to the Colorado River on the southeast, annexing the area now dominated by Las Vegas. It achieved its present shape by 1866, having experienced some of the most dramatic post-statehood boundary changes of any state in the Union.[105] In 1871, Nevada's legislature made one more attempt to expand to the summit of the Sierra Nevada, but by then improvements in transportation and communication helped the two states exercise a more effective authority over those who lived near their boundaries.[106] Nonetheless, situations common in boundary regions lingered. For example, federal surveyors informed a hotelier in 1876 that his property, on the south shore of Lake Tahoe, stood astride the boundary. He thus advertised that patrons could dine at the same table and yet in different states at the same time. The owner, apparently able to make the choice for himself, decided to pay taxes in California rather than Nevada since the former demanded lower rates.[107] To this day, though, the Nevada constitution retains that long-cherished hope of carving off a slice of California. In an article last amended by popular referendum in 1982, Nevadans identified their northwest limit as both the 1864 Nevada line and California's 1850 boundary. The article concludes: "All territory lying West of and adjoining the boundary line herein prescribed, which the State of California may relinquish to the Territory or State of Nevada, shall thereupon be embraced within and constitute a part of this State."[108] Absent such consent, the recognized barrier between the two remains the 120th west Greenwich meridian.

Credit for the establishment and maintenance of the California-Nevada boundary rests primarily with Pacific Coast politicians. Once the leaders of the Golden State determined their limits in 1849, they defended those lines even when confronted with extraordinary pressure. Most of the regional tension developed because, for well over a decade, no government could manage the western Great Basin adequately. For modern Americans, the loss of a few dollars at the Cal-Neva's slot machines represents the greatest threat along this boundary. But the ghosts of Isaac Roop, James Nye, and Elisha Pierce join Frank Sinatra's as reminders of the countless gambles that have taken place along the geometric divide between California and Nevada.

Fence line in the San Luis Valley, with New Mexico to the left and Colorado to the right

CHAPTER 7

Two Distinct Civilizations
The New Mexico–Colorado Boundary

Railroads flourished in the late nineteenth century West, from transcontinental routes to short lines built through mining districts and farmland. High in the mountains that straddle the boundary between New Mexico and Colorado, the Cumbres and Toltec Scenic Railroad today offers tourists a chance to experience the cinders and smoke of bygone days. Funded by both states, the excursion train weaves across the political divide eleven times as narrow-gauge aficionados snap photographs and struggle to find a signal for their cell phones. This innocuous ride through the Rockies belies the repercussions of that state boundary. When first established, it cleaved a Hispano population unified by history, religion, and culture, marooning several thousand in the Anglo-dominated territory to the north. The boundary affected the opportunities of Hispanos who found themselves suddenly living in southern Colorado, as well as Anglos who moved into the region in the years that followed. Frustrated by the sense of marginalization that came with living near the line, Anglo newcomers went so far as to propose secession from both Colorado and New Mexico to create a new polity. Ultimately the divide's ethnic implications resonated loudest, however, representing what one observer of the time described as an invisible line between "two distinct civilizations."[1]

The U.S. claim to the region evolved throughout the early nineteenth century. The Louisiana Purchase of 1803 included the Mississippi River's western drainage, in this instance the Arkansas River valley but not that of the Rio Grande, which flows through the massive San Luis Valley and southward into Spanish New Mexico. The Adams-Onís Treaty of 1819 clarified the area's U.S.-Spain border as the Arkansas River, with everything north and east of it claimed by the United States and everything south and west of it claimed by Spain. Mexico inherited the latter territory after winning its independence in 1821. The Treaty of Guadalupe Hidalgo in 1848 transferred to the United States almost all of the modern Southwest, including the states of California, Nevada, and Utah, and large parts of Arizona, New Mexico, and Colorado, as well as southwestern Wyoming.

To provide U.S.-style political organization, the state of California and the territories of Utah and New Mexico appeared by late 1850. The latter two embraced far more land than the modern states bearing their names. New Mexico's northern boundary reflected long-standing interest among its Hispano population to settle north of their historic heartland around Santa Fe. The Mexican territory's government granted vast tracts of land to local promoters to encourage settlement in the region during the 1830s and 1840s. Two of these—Conejos and Sangre de Cristo—shared 3.5 million acres of the San Luis Valley and surrounding mountains, the claims divided by the Rio Grande. A grant of 1.7 million acres to Guadalupe Miranda and Carlos Beaubien, associates of New Mexico's political aristocracy, extended from the Sangre de Cristo Mountains eastward onto the high plains. All three tracts overlapped the present New Mexico–Colorado boundary.[2] By the time U.S. forces conquered New Mexico in 1846 the northernmost grants remained in force on paper but their owners had done little to occupy them. Strong resistance from several Native cultures, the Utes in particular, retarded Hispano occupation of these lands.[3]

Congress acknowledged the land grants nonetheless when it crafted New Mexico's northern limit in 1850. The line stretched eastward along the thirty-seventh parallel from near Death Valley to the Continental Divide in the San Juan Mountains. From there it turned north to follow the Rockies' crest to the thirty-eighth parallel and then extended eastward through the Sangre de Cristo and Wet Mountains and the high plains south of the Arkansas River.[4] Arguing in favor of such a boundary in 1850, Missouri's Senator Thomas Hart Benton stated: "This is not merely a question of territory, but of people. We give government to people—not to woods and prairies."[5] Such a limit would embrace in a single polity a distinct ethnic community as it established new settlements in its northern reaches.

Throughout the 1850s northern New Mexico Territory experienced dramatic change as groups of subsistence farmers clustered in new villages in the San Luis Valley to occupy the land grants. Starting in 1851 with San Luis on Culebra Creek, just north of the present-day state boundary,

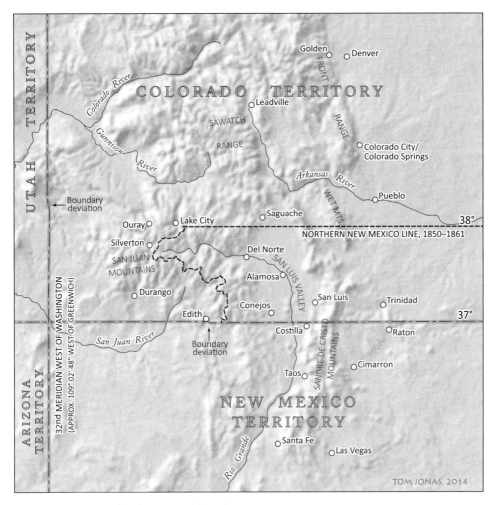

New Mexico–Colorado territorial boundary region

six villages supported by irrigation systems and defended by stout adobe structures appeared in the upper valley.[6] Ute bands and other Native groups that hunted in the area resisted this presence and made the Hispano settlement effort "like driving a salient into an enemy's lines," in the words of one historian.[7] Native resistance proved powerful enough to inspire the U.S Army to establish Fort Massachusetts in the northeastern part of the valley in 1852. From there it orchestrated several campaigns in the middle of the decade to protect the Hispano settlements, which in turn inspired more emigration from the south.[8]

The fort and the Hispano settlers perched on the far northern fringe of the immense New Mexico Territory, which covered by 1853 all of present-day New Mexico and Arizona and parts of Colorado and Nevada. Even with an influx of entrepreneurs and settlers from eastern states, New Mexico remained a predominantly Hispano region. A memorial to the federal government in 1851 from newly arrived Anglos bemoaned that "there is no hope for the improvement of our Territory unless Americans rule it."[9] Yet biculturalism remained the fact of the day. The legislature, consisting of a house of representatives and council (the territorial version of a senate), reflected New Mexico's predominant population. Debates took place in both Spanish and English, with translators assisting those who knew only one language.[10] Yet the body adopted Anglo-American practices as well. New Mexico's legislature furnished its rooms in Santa Fe's adobe Palace of the Governors, then nearly 250 years old, with pine desks, muslin draperies, and other examples of "republican simplicity." Within a few years some Anglos saw potential in the political future of the territory: "These are a new people, speaking a strange language, and whose whole method of thinking and acting, in all things political, had been widely different from that of the American people. . . . Although the republican system of the United States is beautiful in its simplicity, and easily understood by those who have breathed its atmosphere and been trained under it from early youth, yet a strange people, who have been reared in ignorance of its precepts, and deprived of all political training, must necessarily require time before they can work with ease and facility in the new harness."[11] By the late 1850s, the assembly looked and operated like any other state or territorial legislature of the mid-nineteenth century.

One member of New Mexico's legislature reflected its eclectic makeup as well as the multiethnic character of the territory's northern reaches. Midwesterner Lafayette Head had come to the region in 1846 as part of the invasion during the war with Mexico. In 1847 he married a widow, Maria Juana Martinez, and adopted her son as his own. Head served as a county sheriff, federal marshal, and militia officer during the eight years following his arrival in the territory. He won election to the council and rose to its presidency in the late 1850s. During that time Head moved to the new town of Guadalupe in the Conejos land grant along the upper Rio Grande, where he established himself as a rancher and miller as well as a politician popular with Anglos and Hispanos alike.[12] Head remained a major player in territorial politics in both New Mexico and Colorado for over a quarter century.

The fate of the predominantly Hispano settlements in the northern San Luis Valley changed through a seemingly unrelated development, the discovery of gold along the upper tributaries of the South Platte River in the late 1850s. This led to an influx of American newcomers, who lobbied for local political organization to re-create the institutions they had known back east. As one inhabitant of the new settlement of Denver insisted, the gold-seekers could choose "the government of the knife and the revolver, or shall we unite in forming here in our golden country, among the gulches and ravines of the Rocky Mountains, and the fertile valleys of the Arkansas and the Plattes, a new and independent state?"[13] To the latter end, in 1859 a group of miners declared their own Jefferson Territory, with boundaries along the 37th and 43rd parallels and the 102nd and 110th Greenwich meridians.[14] These lines encompassed all of modern Colorado, substantial parts of Nebraska, Wyoming, and Utah, and a thin slice off Kansas's western flank. They also incorporated the Hispano settlements of the northern San Luis Valley. The fortune-seekers made no effort to exercise authority over their self-proclaimed southern reaches, however.[15] For their part the Hispanos ignored the extralegal Jefferson Territory and continued to look to Santa Fe for leadership.[16]

As Kansas received its long-debated promotion to statehood, and as states in the Deep South seceded from the Union in early 1861, Congress took up the matter of reorganizing western territories to ensure their loyalty to the federal cause. Farmers on the northern plains and miners east of the Sierra Nevada earned Dakota and Nevada Territories respectively. In addition, after two years of Jefferson Territory's largely ineffective existence, Congress gave the miners flocking to the central Rockies a real local polity.[17] Like the founders of Jefferson, the federal government settled quickly upon a rectangular territory for simplicity's sake. The Senate instigated the process in early February 1861 via James Green of Missouri, chairman of its Committee on the Territories. He offered legislation to create an Idaho Territory (later renamed Colorado) between the thirty-seventh and forty-first parallels and the twenty-fifth and thirty-second meridians west of Washington. It encompassed land from Nebraska, Utah, and New Mexico Territories as well as tracts left over after Kansas's statehood the previous month.[18]

Although all the territories neighboring this new Colorado lost land for it, only New Mexico's contribution contained a non-Indian population worthy of the Senate's attention. Yet the Senate expressed little concern for the Hispanos affected by this transfer. Stephen Douglas of Illinois, for example, worried instead about the property implications of the boundary. Residents of New Mexico could own slaves according to territorial law, Douglas noted, and he fretted that the new territory might remove that right from the Hispanos transferred to it via the proposed boundaries. Senator Green dismissed the concern, responding that the line "does not cut off five inhabitants, according to my opinion, and not a single 'nigger,'" a remark that inspired laughter from his colleagues. He argued further that American miners preferred the thirty-seventh parallel, as evidenced by their choice of lines for Jefferson Territory.[19]

The contemporaneous debate in the House of Representatives contradicted at least one of Green's opinions, that regarding the number of Hispanos affected in northern New Mexico Territory. House members estimated that a political division along the thirty-seventh parallel would cleave off some three thousand Hispanos. Thomas Bocock of Virginia noted that a purely geometric line might "exhibit more beauty and grace" than the one then existing. Regardless, he thought it would prove unjust to the people of the San Luis Valley and encouraged his colleagues to reject it.[20] New Mexico's territorial delegate Miguel Otero expressed concern that Congress considered itself "at liberty to cut it up in any shape they please." Otero portrayed Hispanos in the upper San Luis Valley as content with their affiliation to Santa Fe. By contrast, Galusha Grow of Pennsylvania insisted that the rule of one territory differed little from another. The settlers in the San Luis Valley, he stated, "are not removed from [the] jurisdiction [of the United States], whether they are placed under the territorial government of Colorado, or remain under the territorial government of New Mexico. None of their rights have changed, and they cannot be changed by being transferred from one territorial government to another. Both the territorial governments are creatures of the United States, and are within the jurisdiction of the United States."[21] Ultimately cleanliness of territorial division proved the winning argument for both senators and representatives. By the end of February both houses authorized Colorado Territory with a straight southern boundary along the thirty-seventh parallel. President James Buchanan signed it into law a few days before he left office.[22]

Despite potential ethnic tension in the brand-new Colorado Territory, the Anglo majority of its first legislature demonstrated collegiality with their few Hispano colleagues. Two members from the southern region, Victor Garcia and Jesus Barela, won election to the territorial House of Representatives in August 1861.[23] Living farther from Denver than any other members, they arrived five days after the session commenced. Only moments after they took their seats the House voted on a resolution to appoint an interpreter for Garcia and Barela, recognizing that they did not speak the language of the body's deliberations. This measure failed due to the cost of a translator. Immediately afterward another member proposed that the territorial secretary, a presidentially appointed office, "furnish the gentlemen a competent Spanish interpreter." The House accepted this suggestion and shifted the expense of a translator from the territory to the federal government.[24] Anglos in the legislature saw no objection to the participation of Hispanos as long as they did not have to foot the bill to accommodate their needs.

Colorado's first territorial legislature created three counties with predominantly Hispano denizens—Costilla, centered at the town of San Luis; Guadalupe (later Conejos), stretching from the Rio Grande to Utah; and Huerfano, east of the Sangre de Cristo Mountains and south of the Arkansas River.[25] The opportunity to seek county offices provided Hispanos the potential to exercise their political voice within Colorado. In addition, they found an advocate in territorial Governor William Gilpin, a fan of the San Luis Valley since passing through the region in the 1840s, several years before Hispano settlement had pushed toward its northern end. He arranged to translate his opening address to the 1861 legislature into Spanish and sent complimentary copies to the new southern counties, labeled courtesy of Governor "Guillermo Guilpin." After his tenure as chief executive, Gilpin purchased a large ranch in the San Luis Valley from the Sangre de Cristo grant holders.[26] Another example of this accommodation between Anglos and Hispanos in early Colorado appeared in the pages of the *Rocky Mountain News*, Denver's first and most widely circulated newspaper. Starting on September 25, 1861, and occasionally thereafter the *News* printed the legislature's proceedings in both English and Spanish.[27] Editor William N. Byers made no mention of why he chose to include the occasional bilingual record. The use of both languages in several editions suggests either the presence of subscribers in southern Colorado or that Byers hoped to increase his readership there. In any case, the paper's bilingual efforts fell victim to a lack of space, supplanted by reports of Union defeats in the opening months of the Civil War. Regardless, Byers's willingness to engage the territory's Hispano population suggests a certain amount of ethnic cooperation, in conjunction with the official gestures made by Gilpin and the legislature.

Such efforts to blunt the Anglo-Hispano divide in Colorado did not impress New Mexico politicians, who nursed a grudge against the federal government for having sliced off several thousand of their residents. The topic resurfaced in Congress in May 1862 during a House of Representatives debate over dividing New Mexico in half to create Arizona Territory. New Mexico's delegate John S. Watts reminded his colleagues of the injustice done to the San Luis Valley's residents "merely for the purpose of beautifying the lines of the new Territory of Colorado." Colorado's delegate Hiram P. Bennet retorted: "We have certainly taken better care of that people than the Territory of New Mexico ever did," having provided its inhabitants with roads and post offices.[28] In January 1863 New Mexico's legislature memorialized Congress about another problem, that of the unsurveyed boundary between the two territories. It feared that Colorado authorities took advantage of the uncertainty to exercise "jurisdiction on a considerable territory south of their true boundary, and within the limits of this Territory."[29] New Mexico's resentment at the loss of its territory and citizens to the Anglo-dominated polity to the north simmered long after Congress decreed the line splitting them.

Problems of a far more serious nature developed in Colorado's half of the San Luis Valley in 1863. That spring two Hispano brothers living near Lafayette Head's ranch instigated terror among the region's Anglo population. Felipe and Vivian Espinosa, emigrants from New Mexico in 1858, made a home on the Conejos River in the village of San Rafael. They resented the American conquerors of their homeland and turned to horse theft and robbery of wagon trains. In March 1863, after soldiers ransacked the Espinosa home looking for stolen goods, the brothers departed in search of vengeance. The two ranged about a hundred miles north into the heart of Colorado, shooting and stealing from Anglos to avenge the wrongs done by Americans on their homeland. At one point they even considered assassinating Colorado's territorial governor, John Evans, during an official visit to Head's ranch to negotiate with the Utes. Army scouts and soldiers tracked down the brothers, along with a nephew who had joined their crusade halfway through, and ended their rampage by mid-October.

Although accounts vary, most sources attribute more than thirty killings to the Espinosas.[30] The terror they incited demonstrated that at least some in southern Colorado were scarcely impressed by efforts to ameliorate ethnic divisions there.

Nonetheless, Colorado's leaders maintained their efforts to reconcile the territory's two ethnic populations. To that end they lobbied for bilingual official documents for the benefit of all residents. Governor Evans noted in early 1864 that the federal government had turned down the territory's request for funds to print in both English and Spanish laws passed during legislative sessions. He called Colorado's Hispanos "a worthy portion of our people, whose disposition to perform the part of good citizens cannot be doubted." Evans asked the legislature to take up the expense itself to avoid "leaving those citizens uninformed as to the nature and requirements of the laws they are expected both to enforce and to obey."[31] The next year the body appropriated two hundred dollars to translate and print the territorial secretary's annual message in Spanish.[32] In 1867 the legislature authorized three hundred translated copies of three sessions' worth of laws. It also set aside money to print the governor's annual message and six hundred copies of the complete territorial statutes in Spanish.[33] Elected and appointed officials thus made good-faith gestures to promote ethnic harmony in early Colorado Territory.

Yet Colorado's predominantly Anglo population did not always try as hard as they could to include their Hispano neighbors in political affairs. The territory's early flirtations with statehood provide a case in point. In 1864, in an attempt to secure more electoral votes for President Abraham Lincoln, the Republican-dominated Congress offered statehood to Colorado, Nebraska, and Nevada Territories. That summer Colorado voters dismissed a proposed constitution as a ploy of Denver politicians to dominate the entire polity.[34] Undeterred, statehood advocates regrouped and called for a new constitutional convention in the summer of 1865. Fifty-five delegates assembled in Denver, none of whom were Hispano. Without input into the constitution's creation, the southern counties rejected the document overwhelmingly when it was put up for a territorial vote in October. Costilla voters rejected the constitution by a vote of 563–26, Huerfano refused it 257–98, and Conejos returned an astounding 465–1 vote against it.[35] Colorado voters overall approved the constitution, however, and forwarded it to Washington, D.C. Federal officials refused the document in the end, the enabling act of 1864 having expired.[36] Colorado's judicial branch also clashed with Hispanos. In 1868, for example, the territorial chief justice, Moses Hallett, overturned a lower court's decision when he learned that its arguments had taken place in Spanish. He declared that "the Spanish language . . . is not to be tolerated in this country."[37] Small wonder that Hispanos referred to Hallett as *el juez severo*, "the strict judge."[38] Cultural cooperation extended only so far.

While Coloradans lobbied for statehood, their neighbors to the south sought to amend their common line and return the wayward Hispanos to the New Mexican fold. In early 1865 New Mexico delegate Francisco Perea appealed to the U.S. House Committee on the Territories for the restoration of the San Luis Valley settlements. Perea dismissed "evenness and symmetry" in making Colorado's southern line as a "really unimportant object." He pleaded, "The land thus severed from New Mexico was a fair and fertile part of her domain, and its inhabitants were her own people, bound to her by every tie of ancestry, nativity, and association."[39] The *Santa Fe Weekly Gazette*, a staunch advocate of Perea's, applauded his request. It noted that New Mexico had never minded the lack of a straight northern line before 1861 and doubted that Colorado cared much about a straight southern line. The newspaper dismissed this particular use of geometry: "It is pleasant to the eye to see State boundaries running with parallels of Latitude and Longitude, and intersecting each other at right angles, but to secure this symmetry of appearance the Government is scarcely justifiable in doing a great wrong to a large population."[40] Whether out of affection for the tidier straight line or a lack of interest in amending it, however, Congress ignored Perea's entreaty.

Two years later a violent incident demonstrated the lingering ethnic tensions along the New Mexico–Colorado line. In the 1860s, Trinidad, a Colorado town east of the Sangre de Cristo Mountains on several major trading routes, boasted a diverse but fractured population including Anglo-Americans, Hispanos, and Indians. On Christmas Day 1867 a wrestling match between an Anglo stage driver and a local Hispano strongman degenerated into mob violence. After more than a week of chaos, ten Hispanos lay dead and local political and ethnic frustrations stood exposed. Historian William J. Convery points out that this so-called Trinidad War instigated "a debate over the very existence of the territorial boundary that separated Colorado from a contiguous zone of settlement in New Mexico."[41] Like the Espinosa rampage a few years earlier, the Trinidad War reflected a potential for regional violence inspired in part by the political divide opposed by Hispanos and defended by Anglos.

The Trinidad War reflected economic machinations as well. It took place in the wake of a gold rush in the Sangre de Cristos just south of the boundary, near the town of Cimarron. Coloradans eager to reinvigorate their stagnant economy sought to redraw the line and annex these new mines in the Moreño River valley.[42] Charles F. Holly, a former judge and territorial legislator, led the charge. A Cimarron correspondent to the *Rocky Mountain News* dismissed the notion and "his obese highness" for promoting it, however. The reporter rejected the notion that New Mexicans sought the transfer or that Coloradans needed yet another mining district. Echoing the 1861 boundary dispute, he stated that "New Mexicans love their territory, and never have, and never will consent to giving away any portion of their territory."[43] New Mexico's legislature responded by asking Congress not to permit the transfer and if possible to restore the land it had lost seven years earlier.[44] Anglos living in the San Luis Valley sent remonstrations of their own, calling it "a manifest injustice of our rights to attempt to force us back with a race for whom we can entertain no common sympathy whatever."[45] New Mexicans and Coloradans viewed their common boundary as malleable long after the line's establishment. The simmering ethnic hostilities proved obvious to visitors as well, including Samuel Bowles, editor of a respected Massachusetts newspaper. He toured Colorado in 1868 and recognized the territory's cultural split. Bowles praised the majority Anglos as virtuous and capable but dismissed Colorado's "Mexican" residents as "ignorant and debased to a shameful degree."[46] These divergent populations remained locked in a very public contention, regardless of political efforts to bridge the divide.

The lack of a survey to demarcate jurisdiction and property exacerbated the situation. Newspapers as far away as Wisconsin printed stories discussing the territorial tension and the need for a formal boundary investigation.[47] To that end the General Land Office appointed Ehud N. Darling to lead a team along the thirty-seventh parallel in the summer of 1868. Darling and his crew marked the shared line with stone piers and completed their work late in the year.[48] The surveyor praised the San Luis Valley for its potential both in farming and ranching. In addition Darling reported one community more affected than any other by the boundary, Costilla, a San Luis Valley town in the Sangre de Cristo grant. He found that the line bifurcated it but placed most of its estimated sixteen hundred residents in New Mexico.[49] The act of surveying the territorial boundary reinforced its complicating presence as well as the overriding authority of American institutions. The Hispano population living along the boundary now knew its precise location but admired it as little as ever. As one historian suggests, Darling's survey made the region "a white man's land."[50]

While Darling's team marked the political barrier in stone, the Catholic Church reinforced the fracture in its own way. In the 1850s its diocesan boundaries followed federally imposed territorial boundaries. The Hispano population of the San Luis Valley thus looked southward for religious as well as political leadership. Several times a year, priests trekked north to perform baptisms, weddings, and funerary rites for the faithful on the fringes.[51] In 1858 Bishop John B. Lamy of Santa Fe established the first parish in his territory's northern reaches at Conejos, a town on the river of the same name, a parish that included nearby Guadalupe and San Rafael.[52] Meanwhile, Catholic fortune-seekers heading to the central Rockies in the late 1850s fell under the jurisdiction of the Kansas diocese. Its bishop, John B. Miege, sent a request to Rome to transfer the distant region to Santa Fe. As a result the Holy See expanded Lamy's authority to include all or part of six modern states: New Mexico, Arizona, Colorado, Wyoming, Montana, and Utah. The revised 1861 boundary between New Mexico and Colorado Territories thus made little immediate impact on the Catholic Church. Lamy sent priests Joseph P. Machebeuf and John B. Raverdy to Denver to tend the flock in that burgeoning town, but they remained under Lamy's direction from Santa Fe.[53]

As Colorado Territory's population evolved, its religious institutions changed as well. By 1867 the increasing Catholic population in Colorado and beyond made operating the immense diocese of Santa Fe impossible. Lamy requested a division using the territorial boundaries for simple expediency. In February 1868 word arrived from the Vatican that Colorado and Utah Territories now embraced a separate vicariate apostolic, one also split along the boundary between those two polities three years later. Machebeuf won appointment as the first bishop based in Denver. After 1868, Conejos served as the parish center for Colorado's half of the San Luis Valley, embracing territory over 120 miles from north to south and over 30 miles from east to west. Not until the late 1880s did the diocese establish more convenient parishes for Catholics in the region by staffing churches in Alamosa and Del Norte. Thus the political break with New

Mexico evolved into a religious break as well. Whether they liked it or not, by the 1870s Hispanos in the San Luis Valley turned toward Denver for both political and spiritual guidance.[54]

Residents made statehood efforts on both sides of the thirty-seventh parallel in the 1870s, with mixed results. Since New Mexico's annexation to the United States in 1848, some living there had attempted to win a star on the national banner, but they faced ethnic resistance, especially in the eastern states. An 1874 article in *Harper's Weekly* suggested that "there is surely no reason for giving a hundred thousand people in New Mexico, who do not speak our language, and who can not read or write, an equality with the State of New York in the Senate of the United States."[55] Another *Harper's* article in 1876 made a similar claim, portraying the territory's population as "virtually an ignorant foreign community" undeserving of statehood, as opposed to Colorado's more American citizenry.[56]

North of the line Coloradans touted their cultural diversity, a claim made more palatable nationally since Anglos represented the majority population. Officials recognized the multilingual nature of the territory by printing laws and proclamations not only in Spanish and English but also German, reflecting new immigration.[57] Attempting statehood again in 1875, thirty-eight constitutional convention delegates meeting in Denver included three Hispanos. One of these, New Mexico native and rancher Casimiro Barela represented a wealthier faction of Hispano culture than most subsistence farmers in the San Luis Valley. The Trinidad-area resident would go on to speak for southern Colorado and Hispano interests in the territorial and state legislature for forty years.[58] Through the efforts of delegates Barela, Jesus M. Garcia, and Agapeta Vigil, as well as Lafayette Head, Colorado's proposed constitution included several provisions of interest to Hispanos. It decreed the printing of all laws and legislative journals in English, Spanish, and German until at least 1890, by which time the convention expected all state residents to homogenize as English literates. In a statement released with the constitution, some members of the convention assured "the Spanish-speaking population of the State an equal opportunity of being fully informed of the provisions of the fundamental law, as well as all laws passed in compliance therewith."[59] Colorado's solons of statehood continued their efforts to blunt the ethnic divisions within the territory.

Colorado's first state legislature following its promotion to the Union in August 1876 illustrated this ongoing struggle. Meeting in Denver the following November, the General Assembly included seven Hispano members in its house of representatives, a seventh of its total membership. Legislative business took place in English but the house provided a translator for their southern members' comprehension.[60] Not all observers looked favorably upon this system, however. A Kansas City newspaper remarked that the Hispano house members "are full blooded Mexicans, with full blooded, unpronounceable names. They employ an interpreter, and manage to keep the run of matters as well as though they could speak United States talk."[61] Regardless of the progress made toward blending Anglos and Hispanos in Colorado, the multiculturalism met with disdain in the national court of public opinion.

Shortly after Colorado achieved statehood, its boundary with New Mexico Territory developed a new complication. The 1870s witnessed an Anglo influx into the southwestern reaches of the state, especially the San Juan Mountains where the Rio Grande originates and flows eastward into the San Luis Valley. The newcomers settled in lands long occupied by the Utes, who surrendered their claims to portions of present-day Colorado in 1863, 1868, and 1873. While the Utes retained vast stretches of western Colorado, they handed over the San Juan Mountains to prevent conflict with American miners.[62] Yet the fortune-seekers resented the lack of political influence their peers enjoyed in other parts of the state, a condition linked to the infancy of the mining region. In addition, they lacked sufficient transportation and commercial links to the rest of the state and region. The miners felt a sense of marginality that came from living on the edge of the state, several major mountain ranges removed from the large eastern slope towns. Oblivious to the Hispanos' frustrations inspired by the same condition, the Anglo miners pondered using the thirty-seventh parallel to unite forces rather than a divide. Better schooled in U.S.-style state-making than their Hispano counterparts, these newcomers believed that they possessed the tools for their own salvation. To overcome their sense of abandonment, within a few months of Colorado's statehood some of these southwestern miners pondered secession from the new polity to create one of their own.

For these frustrated silver-grubbers, the inadequacies of Colorado's government grew apparent following the first post-statehood election. The Republican Party had dominated Colorado politics since the territory's birth.[63] Its candidates in late 1876 came primarily from the northern reaches of the state. Denver's John Routt, for example, served as the last territorial governor and won the Republicans' nomination to serve as the state's first governor.[64] To fill the post of lieutenant governor, Republicans reached out to southern Coloradans, Anglo and Hispano alike. They chose Lafayette Head with his long political career in both New Mexico and Colorado, having served in the latter's legislature in 1872 before helping to write its constitution. Head's appeal bridged southern Colorado's cultural divide and his party membership all but assured him a job for the next two years, the length of executive terms in the state at the time. The Republican ticket swept the elections in October 1876 and the two top-billed candidates received support from a majority of voters in the San Luis Valley and the San Juans.[65] But voters in the southern mountains soured on Colorado's new government quickly since Head represented the San Luis Valley more than he did the new mining regions. Residents of the latter, raised in the American political tradition, felt disfranchised in a way they considered far more of an insult than that foisted upon their Hispano neighbors, and they considered drastic measures to rectify the situation.

Residents of Del Norte, a mining supply center on the western edge of the San Luis Valley, heard the first rumblings of political divorce in southern Colorado in the spring of 1877. It served as one of three seats of the U.S. district court in Colorado, with Denver and Pueblo securing the other two permanent locales. Yet Del Norte residents believed they deserved more.[66] Feeling abandoned by northern politicians of both major parties, residents of that town pledged to "cause havoc" among Colorado officials who turned a blind eye to their section.[67] State Representative Alva Adams spoke to a crowd in Del Norte about this sense of alienation on April 13, 1877. He suggested that Del Norte would one day become the capital of a new and wealthy state encompassing the agricultural lands of the San Luis Valley and the mineral riches of the San Juans. In language similar to antebellum secessionists Adams intoned: "May the day soon dawn when . . . we shall cut off forever from the political magnates of the north, who, as they hold the balance of power, have compelled us to sue and beg for our political rights, and then have denied our rights." Adams decried southern Colorado's status as a "political province" and encouraged the audience to take charge of their own destiny.[68]

Adams's speech went over well with the crowd at Del Norte and provided a topic of regional conversation for months. Yet these firebrands expressed more concern with identifying what made them different from other territories—who and what they were *not*—rather than what unified their region—who and what they *were*. As a result, no clear boundaries for this proposed new territory emerged. Pueblo's *Colorado Chieftain* vaguely described a geographic area including the San Juan River basin, in the area of Four Corners, to provide local control for settlers hundreds of miles from the capitals of Colorado, Utah, Arizona, and New Mexico.[69] At first that newspaper insisted upon the loyalty of southern Colorado and called the proposal "an insane undertaking."[70] Yet the *Chieftain* also argued: "There perhaps [are] no two parts of any state in the Union so unnaturally linked together as the country north and south of the divide" between the South Platte and Arkansas Rivers.[71] Secession advocates agreed on one thing: a name for their proposed territory. They chose "San Juan" in reference to the profitable new mining district in those mountains.

Outside reaction to the talk of political divorce proved negative. Denver's influential *Rocky Mountain News* viewed the idea with surprise and amusement. When several weeks passed after Adams's speech without additional chatter about the idea, the *News* suggested that he had been "only jokin'."[72] More haughtily, the *Longmont Post* asked: "What does 'Jesus Maria,' attending his flock on the banks of the Purgatoire, care where the officers come from? What does John Jones, tramping over the mountains of San Juan, with pick and shovel, care where United States Senators come from, so [long as] he finds rich float?"[73] A newspaper in the northeastern Colorado farming town of Greeley dismissed the effort to create San Juan and another one in progress to split Dakota Territory in two, believing the country had enough polities within it already.[74] Even the *New York World* printed an article about the rumbles of secession. The *World* portrayed southern Colorado's Hispanos as frustrated with the higher taxes that came with statehood. It suggested that they looked to the thirty-seventh parallel now as a refuge rather than an obstacle. The *World* reported that many Hispanos had fled south of it to the less expensive New Mexico Territory. "A few years more and the 'greaser' will be a rare species in Colorado," the *World* forecasted. Along the same line it ridiculed the territory's Hispanos as resistant to statehood because they "are a slow set constitutionally, and if they can get away from schools, churches, newspapers and progress they are happy." The newspaper intimated that Anglos living in northern New Mexico wanted their own polity as well, in order to distance themselves from the supposedly less astute Hispanos.[75]

For the first few weeks of the San Juan movement it remained the purview of several southwestern Colorado mining towns, but the idea soon adapted to established intrastate tensions. Pueblo, removed by several mountain ranges from the San Juans and San Luis Valley, had long resented the strength of Denver. In the 1876 election most Pueblo voters had backed the defeated Democratic candidate for governor, unlike the Republican-leaning Denver area and San Juan and San Luis Valley counties.[76] In the secession talk, Puebloans saw an opportunity to dominate a vast, wealthy new territory liberated from Denver. By late May 1877, the city overlooked the political disparity of the previous year to usurp the movement that started in Del Norte, almost two hundred miles of rugged, winding roads away.[77] The *Chieftain*, Pueblo's most prominent newspaper, adopted the mantle of secession and became its most dedicated supporter. Fanning the flames for several months, the *Chieftain* touted political divorce and the obviousness of Pueblo to serve as the seat of San Juan's government. In this effort the *Chieftain* found an advocate in the *Colorado Transcript* of Golden, a mining supply town just west of Denver at the Rockies' base. Golden residents remained embittered over their loss of the capital to Denver a decade earlier, even though their claim to the seat of government proved tenuous at best. The *Transcript* did not expect the San Juan movement to succeed but it cheered any effort to embarrass Denver.[78]

The majority of secessionists, whether in Pueblo or the San Juans, expected to take some of New Mexico with them. While they advocated a watershed boundary to San Juan's north, they sought to gobble up two or three northern counties from that territory as well. Numerous vague references to the "Americanized" counties of New Mexico, "which are tired and sick of the misrule and despotism existing there and are anxious to join their fortunes with a more enlightened class of people," exposed ethnocentrism and vitriol aimed against the Hispano population of New Mexico.[79] Some Anglos in the territory, especially those in the Moreño mining district near Cimarron, supported the notion. That town's *News and Press*, for example, looked with favor on the idea: "The new territory would comprise the richest mineral and mining land in the world, the largest and finest grazing and agricultural lands in either state, and a population of white American citizens, of wealth, intelligence, and industry, whose interests are bound up in the advancement of their country."[80] Anglos intended to dominate San Juan just as they dominated Colorado, further squeezing out the Hispano voice.

Other New Mexican sources proved less enthusiastic about losing more of their northern reaches. The *Albuquerque Review* noted that the disaffected parts of Colorado had come from New Mexico in 1861, and it believed their territory had endured enough indignities and deserved to keep its northern counties.[81] Santa Fe's *Daily New Mexican* dismissed the Anglo frustration with an astute observation: "Some people can never be contented anywhere."[82] A Spanish-language periodical published in Las Vegas, an important commercial town perched along trade routes east of Santa Fe, used insulting comments about the territory in the *Chieftain* to defend New Mexico and its inhabitants. The *Revista Católica* printed a six-part series in the late summer of 1877, refuting negative impressions of the territory and extolling its virtues.[83] New Mexicans proved resistant to surrendering any more territory or enduring any more abuse.

Although most of the attention devoted to San Juan focused on Colorado and New Mexico, those in other nearby territories with similar political imaginations found the idea inspiring. The lack of defined boundaries encouraged those who sought the grandest transformation possible. Many reports in the national press included references to Utah Territory and how San Juan could provide a bulwark against the Mormon Church. In the minds of many activists, the San Juan River basin—the Four Corners area—provided the best framework for a respectable American territory. As Pueblo's *Chieftain* declared: "The San Juan Valley Americans want to be free from New Mexico, and our near neighbors in Utah have no sympathy with the Mormons, and desire to get away from under their rule."[84] The *Chicago Post* believed that the new territory would "effectively terminate the career of Mormonism and prove a cheap disposal of that mountain ulcer."[85] Both the Hispanos of New Mexico and the Mormons of Utah endured countless insults over the years as newcomers advocated what they envisioned as ethnically and religiously respectable territories.

The talk of political divorce built throughout the early summer. The *Chieftain* received dozens of letters in support of San Juan, an idea that "takes like wild fire among the people."[86] Secessionism reached a climax in Pueblo during the Independence Day festivities when local officials put on a public farce. They intended it to poke good-natured fun at the San Juan movement but also to demonstrate the seriousness of their grievances. The boisterous festivities included local Republican politician Eugene K. Stimson portraying San Juan's governor. Stimson shared the stage with Puebloans dressed as a Prussian bandleader named "Grasshoppersufferer," Russian emissaries "Pulldownyourvestsky" and "Wipeyourchinoff," "the Woman's Suffrage Strawberry Picnic Convention," and even the "Minister

Plenipotentiary and Envoy Extraordinary of the capitol of Colorado," who had come, it was said, to bid farewell to the southern part of the state.[87] Stimson gave an "inaugural address" reiterating the numerous complaints southern Coloradans had voiced against their northern brethren. He closed the sarcastic speech on a serious note, emphasizing the need for secession: "Tonight we are merely going through the form or pretence of a separate state government, but as we leave this place, how many of us are there but that will wish that it were the truth and that the separation had really been made."[88]

While the citizens of Pueblo guffawed, the joke seemed to run its course with the rest of the state. Interest in secession faded in the summer of 1877 as criticism of the idea grew more widespread. The *Rocky Mountain News* commented that the leaders of the San Juan movement "belong to that uneasy and discontented class of spirits who are never happy unless getting themselves and everybody about them in hot water."[89] Similar comments appeared in the pages of the *Denver Tribune*.[90] More distant newspapers viewed the effort with just as much scorn and amusement. A reporter for the *New York Tribune* observed the difference between the two parts of the state: "Northern Colorado is peopled from the eastern and northwestern states, and has the capital and intelligence of the state—in the south, Missourians, Texans and the plainsmen have mingled with Mexicans, greasers and Indians, constituting a population several degrees lower in intelligence and civilization." The New York press thought that, if established, San Juan would become just another territorial dumping ground for has-been eastern politicians.[91] In response to the criticism, the *Chieftain* churned out more articles pointing to supporters of San Juan around the region.[92] Nonetheless, the short-lived effort to reorganize southern Colorado and northern New Mexico fizzled in the weeks after Pueblo's Fourth of July spectacle.

Several factors help explain the failure of San Juan. First, many Americans of the post–Civil War era considered secession either from the nation or within it as one and the same, and found both distasteful.[93] Second, the area's sense of abandonment eased a few months later after the Denver and Rio Grande Railroad laid tracks into the San Luis Valley, facilitating transportation and communication.[94] Third, more candidates from southern Colorado appeared on the 1878 ballot. "Governor" Eugene Stimson ran successfully for state auditor, and Frederick Pitkin of Ouray, a mining town across the San Juans from Del Norte, won election as the state's second governor.⁹⁵ Finally, the lack of regional unity retarded San Juan's secession. The movement began in Del Norte, but Pueblo—which resented Denver more than it supported southwestern Colorado—usurped it as the center of the movement. Pueblo's physical separation from the San Juans and its arrogant assurance that it deserved to lead the cause angered the earliest advocates of political divorce. Additionally, a correspondent for Del Norte's newspaper found resistance to the notion in several nearby communities.⁹⁶ Lake City's *Silver World*, for example, remarked: "As to the silly twaddle about the organization of a new territory, it is but wasting space to discuss it. There is no probability of such movement being made, no possibility of accomplishing it if made, and no desire, in this portion of the State at least, for any change."⁹⁷ The ineffectiveness of the San Juan movement might lead some to dismiss it as an aberration, but it illustrates the tactics of people who perceived a threat to their rights as a result of their marginalized location.

When the threat of secession passed, the New Mexico–Colorado boundary's ethnic significance reasserted its primacy. Newspaper accounts reminded readers of the invisible line's legacy, in particular for the Hispanos sliced away from New Mexico. The *Daily New Mexican* of Santa Fe groused that the transfer contributed only "to the impoverishment of New Mexico and the enrichment of Colorado."⁹⁸ In late 1877 the *Denver Daily Tribune* sent a correspondent into the neighboring territory. His report from the San Luis Valley demonstrated the lingering tension between the two polities. The *Tribune*'s itinerant observed: "On the [southern] border of Colorado two distinct civilizations meet, and what may be termed two races of men come into contact, both sending out their influence among the other to some extent. . . . Within three hundred years of habitation, the Mexican has pushed north from Santa Fe a hundred and fifty miles. Within twenty years, the white man has gone south from Denver two hundred and fifty miles. Let the intelligent reader draw any conclusion he may."⁹⁹

This Anglo expansion facilitated by the railroad removed the last vestiges of Hispano control over southern Colorado in the late nineteenth century. Hispanos either moved south of the thirty-seventh parallel or resigned themselves to the new reality.¹⁰⁰ The political aberration of San Juan illustrated the contemporary shift in the region's power structure. It reinforced the steady loss of influence Hispanos living there had endured ever since their separation from New Mexico.

Other issues along the boundary refused to consign themselves to history, however. The turn of the twentieth century witnessed a new burst of controversy. This time it involved the community of Edith, a small town in the mountains west of the San Luis Valley. In the late nineteenth century a dispute arose as to which side of the line Edith stood on, a question made more problematic by the disappearance or destruction of many markers from the 1868 Darling survey. To clarify school and voting districts, Colorado's legislature asked the federal government for a new survey. Howard B. Carpenter led a team eastward from Four Corners between September 1902 and October 1903, installing iron posts along the thirty-seventh parallel. Carpenter's team discovered significant discrepancies between his line and that marked by Darling more than thirty years earlier. Darling's line stood about a half mile too far south in the San Juan Mountains and San Luis Valley, and weaved back and forth over the thirty-seventh parallel east of the Rockies. Carpenter placed most of Edith in New Mexico—in contrast to Darling's survey locating it in Colorado—with the main street bisected by the boundary. Several Hispano towns in the San Luis Valley experienced similar splits, with one placing a community in one state and its train station in the other. As the body responsible for New Mexico Territory, however, Congress failed to confirm Carpenter's work and left the boundary question unresolved.[101]

Lingering uncertainty simmered through the first decade of the twentieth century. In the summer of 1904 Colorado's secretary of state asked the General Land Office for information on Carpenter's survey to inform several residents near the line whether they lived in that polity or New Mexico. The office sent the information and told the Interior Department secretary, with jurisdiction over that agency, of the perils involved in adjudicating a boundary dispute between a state and a territory, something neither Congress nor the department could do. Instead it advocated waiting until New Mexico achieved statehood and letting the two states settle their dispute in court.[102] Four years later Congress declared Carpenter's line the federally recognized boundary, but Attorney General Charles J. Bonaparte expressed concern. He worried about the legal implications of slicing off a strip of Colorado by supplanting Darling's boundary with Carpenter's without the state's consent. Bonaparte also feared that individuals who committed crimes in the disputed strip might evade prosecution. President Theodore Roosevelt vetoed the measure upon the attorney general's recommendation in late 1908.[103] Resolution of the New Mexico–Colorado boundary dispute awaited equal status for the polities.

With its ethnically suspect population, at least from the perspective of many Americans in eastern states, New Mexico languished under federal colonialism for sixty-two years. Repeated efforts to join the Union over the decades bore fruit at last in January 1912.[104] As a variation on this lethargic theme, the process of settling its disputed northern line—with almost 170,000 acres at stake—lasted almost another half-century.[105] In 1919 New Mexican authorities sued Colorado to settle the matter once and for all, but the U.S. Supreme Court did not hear arguments until late 1924. The plaintiff state described Darling's line as "inaccurate, defective, and in part a work of pure fiction." Colorado's lawyers retorted that the Darling line remained the legal divide between the two polities after Roosevelt's veto of the Carpenter survey. Justice Edward T. Sanford delivered the court's opinion in January 1925. He admitted that Carpenter's work offered the most accurate demarcation of the thirty-seventh parallel but, hearkening back to the 1849 *Missouri v. Iowa* decision, declared that a boundary already in force took precedence over a more accurate, later survey. The justices ordered a new marking of the Darling line paid for by both states and carried out by a court-appointed commission.[106] Thus the high court affirmed New Mexico's loss of many thousands of acres.

The legal resolution of this boundary dispute met with mixed reception on each side of the line. The Colorado press praised the decision, especially its implications east of the Rockies where the state retained control of valuable oil fields.[107] Newspapers in Santa Fe and Albuquerque, by contrast, printed only brief wire reports of the decision without editorial comment, perhaps having grown accustomed to such disappointments.[108] Carrying out the Supreme Court's directive proved unexpectedly time-consuming, however. A survey team led by Arthur Kidder and funded by both states started work in 1927 to re-mark Darling's 1868 line. Appropriations from each legislature trickled in at a snail's pace, forcing the survey to proceed piecemeal for years. Kidder traveled from his home in Indiana to direct the project only when funds permitted.[109] His team completed the work in 1950, twenty-three years after their infrequent task commenced, but federal officials did not receive Kidder's maps and notes until after his death in 1958.[110] In October 1960, the U.S. Supreme Court accepted the survey results some thirty-five years after requesting them.[111] Just a few months shy of a century had passed since Congress had decreed the New Mexico-Colorado line. The Darling boundary, warts and all, remains the political divide between the two states.

The invisible line that tourist trains chug across with ease today has represented a much more significant obstacle to diverse groups living near it. The thirty-seventh parallel sliced through a cultural pocket of New Mexico and attached it to Colorado, a territory whose predominant Anglo society shared few traits with Hispanos. Good-faith measures to incorporate southern Coloradans into local affairs never quite compensated for tearing them away from New Mexico in the first place. An influx of Anglos to the region introduced challenges of a political rather than ethnic nature and inspired a secessionist movement to redress the newcomers' grievances. Ultimately, however, the ethnic tension inspired by the New Mexico–Colorado boundary proved its most enduring consequence, even as Anglos supplanted Hispanos politically and economically in the region. As one scholar of the San Luis Valley wrote in the 1920s: "The Americans came not to a barren, virgin land, but to find a kindly disposed race, which, far from resenting the invasion by an alien people, received the strangers with an open-handed generosity which has been but illy repaid."[112] Fifty years later another historian observed: "The political separation of most of the Spanish-speaking people in the valley produced a dichotomy of interest which has never been resolved satisfactorily."[113] Although migration and natural population trends made both New Mexico and Colorado more diverse after the mid-nineteenth century, to many people on both sides of the boundary it remains an incomplete and inconvenient barrier between "two distinct civilizations."

Quartzite column along the North Dakota–South Dakota boundary west of the Missouri River

CHAPTER 8

Let Us Divide
The Dakota Boundaries

Roughly a century of state boundary-making drew to a close on the northern Great Plains in the late nineteenth century, a time of great transition for the West in general. The massacre at Wounded Knee Creek in 1890 marked the end of Dakota Sioux authority in the region and represented a late example of the centuries-long weakening of American Indian hegemony in the present-day United States. It also coincided with the federal government's declaration of the frontier's closing, as thousands of settlers established homes across the continent, a process felt especially strongly on the prairies of Dakota Territory. From farmers along the Missouri River to townsfolk on transcontinental railroad lines to argonauts in the Black Hills, the diverse newcomers who made up the territory's American population held disparate visions of the area's future, leading some to demand, "Let us divide."[1] From the 1860s through the 1880s a wide array of boundary proposals emerged, both from locals and federal officials in an attempt to create effective regional government. Their enthusiasm to reshape Dakota Territory veiled a sobering realization—that the opportunity to mold vast tracts of land into new states was ending. For American settlers and politicians, Dakota offered one last boundary-making playground, of which they took full advantage.

The upper Missouri River region experienced more proposals, most of them nonstarters, for the drawing of new political boundaries than almost any other part of the United States. It took forty years for the area known as Dakota Territory to achieve its current organization. The most intense discussions stretched from the territory's organization in 1861 through the end of the 1880s, by which time almost all other western territories had defined their limits. Both on the plains and in the nation's capital, suggestions for redrawing the territorial boundaries cropped up incessantly, with proposals using both geographic and geometric limits. Unlike with most territorial lines drawn earlier through comparatively unknown and unsettled places, the lateness of the Dakota debate allowed both partisan politics and land ownership to play unparalleled roles in the process. The focus upon Dakota intensified the debates over its boundaries beyond those of other states. All parties expressed a sense of desperate dedication as they fought for their preferred division, recognizing that the debate was likely to be the last major one of its kind in the United States. Each faction proved determined to claim the final victory for themselves.

Dakota came into the United States through two acquisitions, the Louisiana Purchase of 1803, which brought in the Missouri River valley, and the British cession of the Red River drainage south of the forty-ninth parallel in 1818. But this naturally divided region possessed an effective authority long before the arrival of Americans. Bands of Native residents united by a common root tongue, Siouan, had dominated the area since the eighteenth century. A year after the Louisiana Purchase, the Lewis and Clark expedition set off up the Missouri River to explore the northern reaches of this new acquisition. In late August 1804 it reached the neighborhood of a large and powerful group of American Indians that the explorers identified in their ever-changing spelling as "Dar ca ter" or "Dar co tar," also called "Scioux" or "Souex" by French traders who had contacted this society decades earlier. Captain William Clark identified twenty different bands "possessing Separate interests" within this Dakota confederacy, a diverse community of some two to three thousand people making up only one segment of the vast regional Sioux culture.[2]

Almost thirty years later artist George Catlin ascended the Missouri River to study and paint its native residents. He spoke highly of the "Dahcotas" as physical specimens, praised their temperate and generous nature, and described them as living and hunting in "a boundless country of green fields." In 1832 Catlin characterized the Dakotas, by which he meant the Sioux generally, as a nomadic group of forty-two bands spanning territory from the Mississippi River to the Rocky Mountains, with an estimated population of forty to fifty thousand.[3] American understanding of the size and complexity of Dakota culture increased exponentially in the following decades. According to Catlin and the Corps of Discovery, therefore, a splintered but effective authority existed in the Missouri River region long before Americans imposed their institutions upon it.

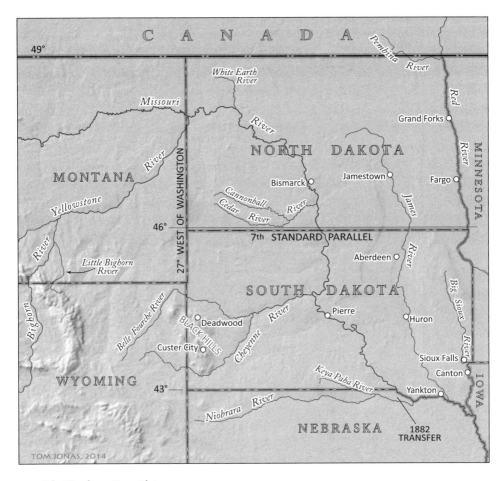

The Northern Great Plains

The first internal American boundaries drawn through the region appeared in the mid-1830s, shortly after Catlin's trek. An amendment to the organic act of Michigan Territory in 1834 extended its jurisdiction westward to the Missouri and White Earth Rivers. Erroneously, the federal government believed that the latter, a tributary of the Missouri, intersected the international line with British Canada.[4] The lack of American settlement on the northern plains at the time prevented this inaccuracy from causing much concern, though. It remained a popularly accepted, if far-flung, boundary for several territories throughout the decade. In 1836 the organic act of Wisconsin Territory adopted a similar western line, which served subsequently as the terminus for the new Iowa Territory two years later.⁵ But with so much focus on Texas and other contentious parts of the trans-Mississippi West at the time, the northern plains attracted comparatively little attention.

In the 1840s American settlement pushed incrementally up the Missouri. As early as 1841 westerners started referring to the area as a nascent "Dacotah Territory," although two more decades passed before such a polity appeared.⁶ With Iowa's promotion to statehood in 1846, the area east of the Missouri again found itself without political organization until the establishment of Minnesota Territory in 1849.⁷ The land west of the river remained unorganized until 1854, when the Kansas-Nebraska Act created the latter territory for the rest of the northern plains.⁸ By the mid-1850s Americans had thus carved the land of the Dakotas and other regional Native cultures into two territories divided by their best-known geographic feature, the Missouri—to the river's east, Minnesota, and to its west, Nebraska.

For a place so connected to the Missouri River, the political fate of the northern plains depended initially upon events involving St. Paul (the capital of the Minnesota Territory since 1849) more than downriver capitals like Omaha and the feuding seats of Kansas Territory. A partisan debate over the final shape of Minnesota swirled before its statehood in May 1858, with Democrats preferring its current vertical alignment. Republicans instead lobbied for a horizontal state between the Mississippi and Missouri Rivers, limited in the north at the forty-sixth parallel, near the present North Dakota–South Dakota line. With Democrats in charge of the White House and Congress, their orientation won the day. In response, land speculators and Republican politicians encouraged settlement west of Minnesota's boundary to make a new territory dominated by their partisans. They hoped to create a new territory, and soon a new state, from the unorganized parcel between the Missouri River and the Minnesota boundary. The Dakota Land Company organized in St. Paul and facilitated settlement in present-day southeastern South Dakota. Within months a meeting for a provisional Dakota Territory convened at the brand-new town of Sioux Falls. As with many local efforts to create a new territory in the trans-Mississippi West, federal leaders expressed little interest at first, especially with their attention diverted to the sectional crisis.⁹

The first serious attempt to organize the northern plains developed in 1860. The House Committee on the Territories proposed creating two horizontal polities from northern Nebraska Territory and the tract left over after Minnesota's statehood. It proposed a line from Big Stone Lake on Minnesota's western boundary running due west to the Missouri River, then north up the river to the forty-sixth parallel, and thence west to the Continental Divide. This would split two new territories—Chippewa to the north and Dakota to the south. A population of some five thousand settlers along the Pembina River, just south of the British border near the Red River, would anchor Chippewa, with about the same number on the Missouri near the southeastern corner of Dakota. Such elongated territories reflected the east–west migration routes across the country. The committee also hoped that they would facilitate railroad construction by sending routes through as few jurisdictions as possible. Several representatives contested the population figures of Chippewa, however, claiming that it lacked sufficient numbers to warrant a territorial government. Eventually the House of Representatives tabled the measure.[10]

As the secession of seven southern states in late 1860 and early 1861 created an unprecedented crisis, the federal government moved to resolve problems in the distant West. Congress created three new territories for mining and agricultural communities, Nevada and Colorado for the former and Dakota for the latter. Limited by the Rocky Mountain crest to the west, the international border to the north, and the states of Minnesota and Iowa to the east, only the southern boundary of the first official Dakota Territory demanded much definition. It followed the Missouri River upstream from its confluence with the Big Sioux River, then continued up the Niobrara and Keya Paha Rivers to the forty-third parallel, and from there due west to the Continental Divide. Within weeks Congress approved and President James Buchanan signed the measure.[11] All of these actions transpired without consulting the Dakotas or the other Sioux factions that still controlled most of the region.

Throughout the 1860s the federal government adjusted Dakota Territory's boundaries to suit the needs of neighboring communities. Idaho Territory's organic act in 1863 pushed Dakota's western limit to the twenty-seventh meridian west of Washington, the present western line for the Dakotas and Nebraska.[12] A year later the creation of Montana Territory and the reorganization of Idaho left a large parcel of land, most of present-day Wyoming, without government. For the time being, Congress attached this tract awkwardly to Dakota.[13] Dakota's leaders, operating from Yankton on the Missouri River, exercised practically no control over this western appendage. Their already overstretched territorial judicial branch, for example, ignored pleas from residents there to hold even rudimentary court sessions.[14] Construction of the Union Pacific Railroad through the southern reaches of this distant segment Dakota Territory provided even more complications for the solons of Yankton. To rid themselves of this burden, the legislature of Dakota Territory memorialized Congress in 1867 to break off that land on the other side of "a broad extent of wild Indian country" from the agricultural settlements on the Missouri.[15] The establishment of Wyoming Territory followed a year later and left Dakota a more manageable, nearly square jurisdiction.[16]

The mid-century American population in Dakota Territory increased rapidly, thanks in large part to homesteading east of the Missouri. Since the late 1850s, American Indians in the region had surrendered millions of square miles of land in the southeastern portion of the territory, present-day eastern South Dakota.[17] By 1870, the federal government had secured title to just over 20 percent of the land encompassed by Dakota's boundaries. Turning the northern plains into townships and sections took more time and effort, though—less than 6 percent of Dakota had been surveyed by that time.[18] But after the Civil War, homesteading expanded dramatically across the northern plains. When the 1870s began, Dakota's American population totaled about fourteen thousand, but it grew to fifty thousand by the middle of the decade—and kept growing.[19] Meanwhile, the roughly twenty-five thousand Native inhabitants found themselves constrained to the southwestern portion of the territory, including the Black Hills.[20] The Fort Laramie Treaty in 1868 defined a Sioux reservation bounded to the south by Nebraska, in the east by the Missouri River, to the north by the forty-sixth parallel, and in the west by the 104th Greenwich meridian.[21] The latter did not correspond to the eastern lines of Wyoming and Montana territories, which rested upon the twenty-seventh meridian west of Washington a few miles to the west of the Greenwich line, leaving a slice of Dakota between Sioux lands and its neighboring polities. Meanwhile, the parallel marking the reservation's northern limit emerged as one of two potential divisions for Dakota.

Competition between Dakota's two American population centers—along the Red River to the north and the Missouri River to the south—intensified with each passing year, and a new transportation route made things even worse. Construction on the Northern Pacific Railroad began in 1870, with crews building west from Minnesota across the northern part of Dakota. Fearing competition from a growing population to their north, southern territorial politicians sought to protect their interests at Yankton, over two hundred miles from the railroad. In early 1871 the territorial legislature sent the first of many memorials asking Congress to divide the territory into northern and southern parts along the forty-sixth parallel, the western half of which served as the northern Sioux reservation boundary already. This proposed division would split Dakota evenly between northern and southern halves. The memorial noted the "broad extent of unoccupied and wild country" separating the northern and southern American settlements, and said that the railroad would connect the north to St. Paul and beyond rather than downriver to Yankton.[22] Congress took no action on the prairie-splitting proposal.

As many Dakotas and other Sioux moved to a reservation west of the Missouri River and American settlers took over the eastern part of the territory in the 1870s, the rectilinear land system spread across the prairie. Surveyors in Minnesota had established the seventh standard parallel, a horizontal line about five miles south of the forty-sixth parallel, to plot land there several years earlier. Standard parallels served as benchmarks for defining and distributing land across the country on the model of the 1785 Land Ordinance. In the early 1870s federal agents carried the seventh standard parallel westward into Dakota Territory. The slight difference between the latitudinal and survey lines complicated division proposals in the years to come. In late 1872, for example, Dakota officials asked for a territorial split using the survey parallel "because the line herein described is already surveyed, marked and established for a considerable distance from the eastern boundary, and will be established hereafter in the prosecution of the public surveys without additional expense for surveying it as a boundary."[23] The seventh standard parallel could thus divide two regions frugally and without affecting the existing property rights of American settlers on the high plains.

The Black Hills in southwestern Dakota provided additional tension to the north-south divide emerging within the territory in the 1870s. As early as 1867 the territorial legislature pleaded with Congress to treat with the Natives for the Black Hills, supposedly rich in mineral and timber resources.[24] Rumors about the region's potential led to an expedition in the summer of 1874 under Lieutenant Colonel George Custer. Custer's command confirmed the presence of gold deposits, among other valuable resources. Later that year Dakota Territory legislators again petitioned Congress to negotiate with the Sioux and open the area to American settlement.²⁵ In the meantime, fortune-seekers violated the treaty rights of the Sioux by sneaking into the Black Hills to build fortified camps and pan for gold. Some ten thousand interlopers had poured into the area by the spring of 1876. The Sioux struck back, but after several months of fierce resistance to both reservation life and the trespassers—culminating in the destruction of Custer's command on the Little Bighorn River that summer—they were forced to accept a new treaty relinquishing their claims to the Black Hills. The reservation boundaries shrank eastward but also expanded slightly across the forty-sixth parallel. The northern reservation limit switched from geometry to geography in 1877, resting instead upon the Cedar and Cannonball Rivers.²⁶ Abandoning the forty-sixth parallel as a reservation boundary coincided with increased debate about its or the seventh standard parallel's suitability as a north-south divide for Dakota.

With native competition pushed out of the way, it did not take long for Americans in the Black Hills to lobby for their own territory. These would-be solons proposed a wide variety of potential boundaries. Less than a month after the Little Bighorn battle in late June 1876, a newspaper at the illegally established town of Deadwood proposed a new polity with geometric boundaries that would include present-day western South Dakota, northeastern Wyoming, and southeastern Montana. Another suggestion called for limits along the Missouri and Bighorn Rivers to the east and west and lines of latitude to the north and south.²⁷ Still others simply proposed attaching the Black Hills to Wyoming while accusing Dakota's government of dismissing the needs of the distant and booming population.²⁸ Many proposals called patriotically for the name "Lincoln Territory," but they met widespread scorn elsewhere in Dakota.²⁹ Yankton's *Daily Press and Dakotaian* objected to "the 'carving-out' being made without much regard for the symmetry of the surrounding portions of the territories," not to mention losing such a valuable region as the Black Hills.³⁰

The not yet surveyed boundary between Wyoming and Dakota Territories caused further problems. Several Wyoming newspapers suggested that flourishing towns like Deadwood and Custer City, not to mention the Black Hills gold itself, stood in their polity.³¹ The uncertain jurisdiction affected the establishment of law and order in the mining camps. In one famous episode of the period, Jack McCall, accused of murdering James Butler "Wild Bill" Hickok in August 1876, called unsuccessfully for a dismissal of his charges since his trial was taking place in a Dakota court, arguing that Deadwood actually stood in Wyoming.³² To settle the matter, a survey team marked the contested line in the summer of 1877 and located almost all of the Black Hills in Dakota. Their work reinforced the jurisdiction of Yankton over the turbulent region, and affirmed the conviction of McCall—an academic point since he had already been hanged in the territorial capital that spring.[33]

With their position in Dakota Territory established, supporters of Lincoln Territory resumed their efforts to break away in the fall of 1877. A Deadwood newspaper claimed that fully three-quarters of the American population in the Black Hills wanted their own territory, and Congress took note.[34] That October, Senator Alvin Saunders of Nebraska introduced a measure to create Lincoln from parts of Dakota, Wyoming, and Montana Territories, but it stalled in committee.[35] The next year Saunders introduced another unsuccessful bill to split Dakota, this time vertically along the 100th meridian to make Lincoln out of the western half.[36] The Lincoln movement percolated for a few more years, but as transportation and communication links between the Black Hills and Yankton improved in the late 1870s, it "finally died of inanition," in the words of historian Howard Roberts Lamar.[37]

Dakota Territory in the 1870s thus contained three distinct American population centers, all straining in different directions. Indeed, the debates over Lincoln Territory did not distract Dakotans from pushing for a north-south divide as well. Legislators in Yankton supported a failed effort to make Pembina Territory out of their northern reaches in 1874.[38] A similar effort several years later, under the name of Huron Territory, did not specify what line to use for a boundary, but insisted only that the break occur.[39] The desire to split grew so intense that Dakota's largest settlements, far removed from the proposed dividing lines, expressed little concern for the effects that break would have on those living near the potential boundaries. As the *Dakota Herald* grumbled: "The advocacy of a division on the 7th standard parallel simply because if made on the 46th parallel it will intersect some farmer's potato patch or throw his hog-yard into two different territories, seems to us to be puerile. In so grave a matter as the dismemberment of a great Territory like this the trivial question of the personal accommodation of a few individuals or even communities, should not be allowed to enter. The 'greatest good to the greatest number' should be the incentive."[40] The *Herald* failed to explain how the forty-sixth parallel represented a superior choice to the already established land survey line, however. Its dismissive attitude boded poorly for those living in the marginal area. Over the years, Dakota's boundary debate stirred many passions, not all of which drew upon logic or fairness.

The division debate grew more complicated with the addition of statehood proposals in the late 1870s and the escalating role of political partisanship. In 1877 the two leading newspapers of Yankton, the Republican *Daily Press and Dakotaian* and the Democratic weekly *Dakota Herald*, took opposing sides on the matter. The Republican sheet declared: "To be a state [south of forty-six degrees] is to be Dakota indeed and with all the advantages increased. To be a territory or two divided territories is to remain helpless and impotent."[41] Meanwhile, the Democratic paper remained "rather indifferent" to the debate. The *Herald* viewed southern Dakota's statehood as a scheme by Republican politicians, "and all of the star-spangled banner, E Pluribus Unum clap-trap, and unfettered freedom and great commanding influence bombast . . . is solely in the interest of those persons who expect to fill the offices—for the tax-eaters and against the tax-payers."[42] Holding true to its stance, the *Herald* boasted two years later that "the Territory of Dakota is a big thing and its citizens are all proud of it."[43]

The fate of Dakota changed when a newly appointed territorial governor, Nehemiah G. Ordway, arrived in Yankton in the summer of 1880. The *Herald*, of the opposite political persuasion to the new executive, remarked that "he has it in his power to become the worthy, impartial and popular governor, or the political, selfish and designing officer. It is to be hoped for the good of the people and Territory we will find in our new governor the former officer, when he will find an appreciative constituency."[44] But when lobbying President Rutherford Hayes for the post the previous year, Ordway had foreshadowed the *Herald*'s coming disappointment. He explained his suitability by noting that no other native of New Hampshire had ever been appointed a territorial governor, and therefore the Granite State—through himself—merited some largesse.[45] Having won over the president with his infallible logic, over the course of four years Ordway proved a divisive, objectionable politician whose name remains infamous on the northern plains to this day.

More than a year into his tenure, Ordway waded into the territory's long-simmering political debate. In September 1881 he called a meeting of territorial officials in Fargo to consider dividing Dakota.[46] A statehood bill had appeared in Congress the previous winter for everything south of the forty-sixth parallel, but it had failed after objections from Red River valley residents.[47] Ordway believed the idea a worthwhile one, though, and used his authority to explore it. The Fargo meeting attracted little popular attention, and the *Herald*, which opposed division and supported statehood for all of Dakota Territory, believed that "Ordway's attempt to stick his facial projection into concerns that did not belong to him proved a disastrous failure."[48] But Republicans like Dakota's territorial delegate Richard F. Pettigrew and Minnesota's Senator William Windom agreed with the governor. Both men introduced bills in Congress in late 1881 to break Dakota along the forty-sixth parallel and put the southern half on the path to statehood.[49] They received support at a meeting of some four hundred division advocates in Sioux Falls in early 1882. Resolutions from that meeting called for a split along the forty-sixth parallel, retaining the name "Dakota" for themselves, and bestowing "North Dakota" on their new neighbor, in the vein of Virginia and the broken-away West Virginia.[50] In response, a Grand Forks newspaper saw the "shadowy hand of Gov. Ordway" trying to create a position for himself as the appointed executive of the new northern territory.[51]

In the early months of 1882, both houses of Congress considered the bills to divide Dakota Territory along the forty-sixth parallel and admit the southern half to the Union. The partisanship already apparent in Dakota intensified in the chambers of the U.S. Capitol. Republicans, then the majority party in the House of Representatives, supported the plan enthusiastically in the Committee on the Territories. But their Democratic colleagues did not think southern Dakota possessed enough people to justify a seat in the House. The Democratic majority in the Senate agreed, and the plan fizzled.[52] From 1882 on, the intense national party competition of the Gilded Age shaped what many Dakotans considered a local matter.

Although the plan to split Dakota Territory failed in 1882, Congress revised the polity's external boundaries that year. In February, the legislative branch debated a transfer of land from Dakota to Nebraska. It affected a skinny parcel between the Niobrara and Keya Paha Rivers to the south and the forty-third parallel to the north, which Nebraskans wished to annex. The federal government had declared most of the affected land as part of a reservation for the Poncas years earlier, although by 1882 the Poncas had relocated to Indian Territory "at the point of a bayonet," as Senator George Edmunds of Vermont noted bluntly. Senator Henry Dawes of Massachusetts, whose later legislation dramatically affected Indians in the West, responded: "The Ponca tribe have been driven off; no one inhabits the reservation now, and their dwellings, those that have not been carried over this river by enterprising settlers in Nebraska, have been swept down the river" by seasonal floods. Brushing the objections of Edmunds aside, Congress pushed the legislation through in early 1882.[53] With President Chester Arthur's signature and the assent of Nebraska's legislature, the Cornhusker State achieved its present boundaries and limited Dakota to the land that would be split between two states by the decade's end.[54]

Congressional refusal to divide Dakota in 1882 did not slow the movement within the territory. Mass meetings emerged for both supporters and opponents in the 1880s. A significant one took place in Canton, just across the Big Sioux River from Iowa, in the summer of 1882. About two hundred people gathered there to advocate the forty-sixth parallel as the proper northern limit for their state. It also declared that "we strongly deprecate the addition of any part of our territory to the State of Nebraska."[55] Northern Dakotans dismissed the division talk generally. The *Bismarck Herald* compared the idea with the fate of Virginia and West Virginia, claiming: "Neither of the Virginia states amount to much since the grand old State was divided."[56] Although the towns along the Northern Pacific Railroad shared little with those on the Missouri River or in the Black Hills, the residents there seemed satisfied with the status quo.

Congress returned to the division question in February 1883. The House of Representatives, with its Republican majority, considered a bill to separate Dakota using the forty-sixth parallel. William Grout, a Republican from Vermont, described the northern and southern population centers as possessing vastly different interests. Commerce for the former traveled through Minnesota to the Great Lakes, while the latter dealt with St. Louis and Chicago. The fastest route between northern and southern Dakota involved a train trip through St. Paul, hundreds of miles out of the way. By contrast, Samuel Randall, a Pennsylvania Democrat, suggested that if Dakota needed a new boundary at all it should fall on the Missouri River. A fellow Democrat, James Blount of Georgia, observed the lack of a geographic barrier along the suggested geometric boundary: "There are no mountain ranges between these proposed divisions of the State. There are no natural barriers of any kind." Republican Julius Burrows of Michigan pointed to "a barren space of lands extending for a hundred miles which will probably never be brought under cultivation" between the proposed halves of Dakota.[57] But the Republicans failed to muster enough support to end the debate, and once again a division bill stalled in Congress.

A more immediate issue distracted Dakotans from division in early 1883—that of moving the territorial capital. The lack of a clear alternative had left the seat of government tethered at Yankton since 1861, but tensions within the immense polity brought the capital's location increasingly into question. Suggestions for moving it first appeared in 1878 with several proposals to send the seat of government up the Missouri River to Bismarck.[58] That town served as the western terminus of the Northern Pacific Railroad from 1873 to 1882, as the line struggled to recover after the Panic of 1873 and waited for federal troops to subdue American Indians in the railroad's path. Following the completion of the transcontinental line in 1883, the prominence of Bismarck seemed assured. In addition, other towns along the Northern Pacific dreamed of claiming the capital, including Jamestown, the largest settlement between Bismarck and the Red River.[59] In response, southern Dakotans proposed relocating the capital to a spot near the center of their proposed state, not far from the present capital city of Pierre.[60]

The territorial legislature passed a capital removal bill by a narrow margin in March 1883. It allowed Governor Ordway to appoint a nine-member commission to select a new center of authority for Dakota. Although the territorial council confirmed the commission's membership, the bill gave Ordway an astonishingly free hand to shape the territory's political and economic future. The *Dakota Herald*, already opposed to Ordway on partisan grounds, reported that citizens of Yankton now referred to the governor and his commission "as the vilest rogues and thieves." Nonetheless, the newspaper found some solace "that by the act for which they are condemned they rid Yankton of a very undesirable class of people, and this is a compensation which should be taken into account."[61] Newspapers from all over the territory lampooned the capital removal bill. For once even the Democratic *Herald* and the Republican *Daily Press and Dakotaian* found themselves on the same side of an argument, fearing for the future of their shared town of Yankton. Another territorial press compared Ordway to the era's most famous corrupt politician, New York City's William M. "Boss" Tweed, while one even cast Ordway in the same light as the Germanic tribes that sacked the Roman Empire.[62]

In reaction to the capital removal bill, demands for Ordway's replacement headed to Washington, D.C., as fast as the telegraph lines could carry them. President Arthur took no immediate action, however, and Ordway remained in power. Within months the commission selected a hilltop site in Bismarck, where many, including Ordway, had already speculated in local real estate. Alexander McKenzie, a Bismarck businessman and the Republican kingpin of northern Dakota, exerted economic and political pressure in favor of the deal. The commission approved designs quickly for an impressive capitol building, an immense structure complete with a tower and dome that promised to place Dakota's edifice in the same league with other Gilded Age statehouses.[63] Former President Ulysses S. Grant—no stranger to accusations of political corruption—traveled to Bismarck to help lay the building's cornerstone on September 5, 1883.[64]

Losing the capital to northern Dakota fueled the fire of "divisionists" in the south. Several hundred concerned residents met at Huron, up the James River from Yankton, in late June 1883. They resolved in favor of breaking on the forty-sixth parallel, with statehood for the south and territorial status for the north.[65] Three months later an even more zealous group assembled in Sioux Falls to compose a state constitution in the hopes of speeding along their progress toward membership in the Union. The document also called for a northern boundary of the state of Dakota along the forty-sixth parallel.[66] That November, voters south of the line approved the proposed constitution, with only 25 percent of the electorate casting ballots.[67] Noting the poor turnout, Congress debated but did not pass an enabling act for the region south of that latitude in early 1884, with the Republican Senate supporting and the Democratic House opposing the measure. Only the long-awaited replacement of Governor Ordway alleviated the frustration of many Dakotans in the summer of 1884.[68]

In December 1884, Congress considered the political fate of Dakota Territory more extensively than ever before. A bill by the Senate's majority Republicans split the region along the forty-sixth parallel, naming the southern half "Dakota" and the northern half "Lincoln," with an enabling act for statehood and a new territorial organic act, respectively.[69] In opposition to the measure, Senator George Vest, a Democrat from Missouri, saw no need to break "a territory across [in] which there is no natural dividing line, no range of impassable mountains, no deep or navigable river, a country which God has cemented together, and which is as much homogenous as He has made any portion of the American continent."[70] In response Senator Benjamin Harrison, a Republican from Indiana, the most vocal supporter of the bill and chairman of the Senate Committee on the Territories, praised the nation's many "certain and unfluctuating" geometric political barriers.[71] He also observed that Dakota's legislature had, over the years, established two normal schools, two agricultural colleges, and other such paired institutions with an eye toward an eventual division. The Senate approved the bill and passed it along to the Democratic-controlled House Committee on the Territories, where many such measures had gone to die. This one was no exception.[72]

Once again the burden of division shifted westward. The territorial legislature in Bismarck approved by a wide margin a bill for a southern constitutional convention, essentially agreeing to its own dismemberment, in March 1885. It also sent a memorial to Congress for a division of the territory along either the forty-sixth or seventh standard parallels, at the federal government's discretion. Ninety-five delegates assembled in Sioux Falls that September to draft a set of laws for their half of Dakota. The convention selected the forty-sixth parallel as the northern limit of their new state.[73] In early November voters south of that line approved the proposed constitution almost five to one, but as in 1883 the low turnout—some thirty thousand out of about four hundred thousand eligible voters—cast a pall. Observers from as far away as New York City described the resistance to division in northern Dakota and among Democrats in the territory, some of whom boycotted the election in protest.[74] Regardless of the election's shortcomings, leaders of Dakota organized with the expectation that the territory would be welcomed into the Union by Congress. Prominent division advocate Arthur Mellette won election as governor, and he often traveled to Washington, D.C., to lobby for his "state's" interests.[75]

In the spring of 1886, members of Congress introduced three different bills to reorganize Dakota. Two Senate measures, both from the Republican majority, divided the territory in half, one along the forty-sixth parallel and the other on the seventh standard parallel. Both named the southern half "Dakota" and the northern half "North Dakota." In the House of Representatives, William Springer of Illinois offered a bill on behalf of the Democratic Party that controlled the chamber. Springer insisted that an increasing population living near the proposed divisions, as well as the promise of north–south transportation routes, negated the need for a split. His bill authorized statehood for Dakota as a whole, a notion sure to meet the same fate in the Republican Senate that Republican division bills had met in the Democratic House.[76] "If Dakota should be admitted into the Union as one state," Springer argued, "it will be but a few years until State pride, local interests, and its diversified resources will be such as to demonstrate beyond all peradventure the wisdom of single Statehood."[77]

When the Dakota territorial legislature convened in early 1887, advocates for a state south of forty-six degrees attempted to get support from the north. They expressed willingness to concede a boundary on the seventh standard parallel to inconvenience as few property owners as possible, and to share the "Dakota" name.[78] In addition, they called for constitutional conventions for both halves to exert twice the pressure for statehood on the federal government.[79] Although the solons of Bismarck did not acquiesce, they did permit a referendum on whether "to divide Dakota on the seventh standard parallel, an established line running alongside the imaginary line of the forty-sixth parallel of latitude" to appear on the ballot the following November.[80]

Other legislation reflected the tension that would ultimately rend Dakota in two. One law in particular demonstrated the unique problems facing the territory. The *Daily Press and Dakotaian* of Yankton, a staunch supporter of dividing Dakota, described a measure banning the shooting of prairie chickens before the first of September each year. The newspaper argued that by then "the bird is over-ripe in the sunny southern portion of the territory" and too adept at hiding from hunters, whereas in the colder northern reaches the birds had not grown as skittish. The *Press and Dakotaian* concluded: "With our wide area we cannot have a homogenous prairie chicken law. What is one man's full fledged meat is a mere spring chicken to another man with so vast a range of latitude between the two. There has been an earnest endeavor upon our part to reconcile the antagonistic phases of the situation, but every little while something comes to convince us that Dakota should divide up its effects and dissolve partnership. The prairie chicken law should convince the most skeptical. United it can never be made satisfactory to both sections. Let us divide."[81] This opinion hints at the myriad factors involved in splitting Dakota. Indeed, who could have anticipated the boundary-making consequences of such *fowl* legislation?

A convention of some four hundred delegates assembled in Huron in July 1887 to plan for the upcoming boundary ballot measure.[82] The delegates rejected the admission of the whole territory as a state and for called a split along the seventh standard parallel. Northern Dakotans, not invited to the Huron meeting, looked on with scant interest. The *Bismarck Daily Tribune*, a supporter of statehood but not division, worried about the effect of splitting the territory: "Dakota has the greatest of everything. We could still shout if divided, but only half as loud. We would still be great but not greatest."[83] With the matter officially in the hands of the federal government anyway, the Bismarck newspaper knew that the stalemate would hold until one party took control of both houses of Congress and the presidency.[84] The expectation of further transportation and communication links between the two halves of the territory further made the matter moot. The *Tribune* concluded: "Five years from now there will be no divisionists in Dakota."[85] Its prophecy proved accurate, but not for the reasons the *Tribune* hoped, as the territory had split by then and the argument ended.

As the 1887 vote on separation drew nearer, the measure's supporters grew increasingly confident. A meeting in Fargo in October united divisionists from both parts of the territory for the first time.[86] The assembled crowd agreed to a memorial that Dakota "declares for division on the seventh standard parallel—which obviates the danger of dividing counties, townships and farms, with the border lines."[87] But when the polls closed on November 8, once again the voter turnout stood far lower than anticipated, with just over seventy thousand ballots cast out of a territorial population of more than six hundred thousand. Ballots in some counties, like Bismarck's Burleigh, did not even include the question, and in others only the choice to approve division appeared on the ballot, forcing opponents to write in their opinion about how to do it.[88] As a whole the territory approved division by some forty-five hundred votes, but when broken down by section the north rejected it by almost ten thousand while the south approved it by over fourteen thousand.[89] With opinion so split, competing conventions were held, one to rally for a one-state solution, the other for a two-state solution. The first assembled in mid-December 1887 at Aberdeen, not far from the forty-sixth parallel, and agreed on a memorial to Congress asking permission to hold a constitutional convention for the whole territory.[90] Divisionists met in Huron in early January 1888 to insist on two states broken by the seventh standard parallel.[91]

When Congress assembled in the winter of 1887–88 it returned to the matter of Dakota. A bill drafted by the Republican majority in the Senate's Committee on the Territories used both the seventh standard and forty-sixth parallels to split the territory, the former east of the Missouri River and the latter through the sparsely populated area west of the river. But once again the proposal died in the Democratic-controlled House Committee on the Territories.[92] Republican territorial delegate Oscar Gifford introduced legislation in the House of Representatives to divide his constituency along the seventh standard parallel and provide enabling acts to both.[93] Representative Springer, still the chair of the Territorial Committee, countered with an "omnibus bill" that proposed to enable Dakota, Washington, Montana, and New Mexico Territories all to apply for statehood. He hoped that the Democratic Party could compete more successfully in the other three territories, having long before acknowledged their minority on the northern plains. Springer's idea was met with scorn—the *Bismarck Daily Tribune*, for example, referred to it as "That Bologna Bill" and objected to the attachment of any other polity's fate to their own.[94] Once again all of these measures died in opposing chambers, leading the *Tribune* to gasp: "Poor Dakota! How she has been buffetted about by the high-fed, high-stepping horde of politicians who strut about the national capital in the mask of statesmen."[95]

Debate on dividing Dakota all but disappeared from the national and local scene in the months leading up to the 1888 elections, as if everyone recognized that its outcome would settle the matter. The Republicans backed Senator Harrison for president and adopted a plank admitting two Dakotas into the Union.[96] Democrats renominated President Grover Cleveland, who had kept his distance from the hot-button Dakota issue over the previous four years. A few small gatherings of division advocates met in Dakota during the intervening months, but generally the region adopted a wait-and-see attitude.[97] Following Harrison's victory in November, as well the election of a comfortable Republican majority in both houses of Congress, divisionists in Dakota rejoiced. As the *Daily Press and Dakotaian* gloated: "With a [R]epublican presidential and congressional triumph comes certainty of statehood for two Dakotas," both of which it expected to support that party in future elections.[98]

With the matter of division practically assured after the 1888 election, attention turned instead to the process of statehood. Southern Dakotans expected to gain admission under the 1885 Sioux Falls constitution, and "Governor" Mellette met with President-elect Harrison several times before his inauguration in March 1889.[99] Meanwhile, northern Dakotans assembled at Jamestown in December to plan for their impending promotion.[100] Back in Washington, D.C., the despairing Democrats, led by Representative Springer, looked to the potential admission of Republican-dominated states with dread. They considered pushing through a bill to admit southern Dakota but leave the northern half as a territory, or to approve Springer's omnibus bill with Indian Territory added, before the end of President Cleveland's term.[101] The lame duck House of Representatives and Senate hammered away on the matter of western statehood in early 1889. They eventually reached a compromise on the omnibus bill to provide enabling acts for Washington, Montana, and two Dakotas, split along the seventh standard parallel.[102] Thus Congress protected the interests of property owners rather than dividing houses, farms, and land by using a more traditional latitudinal line.

The omnibus bill reached President Cleveland's desk on February 22, 1889, only ten days before his term expired. Upon receiving word that he had signed the bill, newspapers on the northern plains changed their mastheads—somewhat prematurely—to "North Dakota" and "South Dakota."[103] This pride and perhaps relief at the beginning of the end of this interminable debate, felt on both sides of the seventh standard parallel, resonated in diverse ways. That summer the *Washington Post* noted that travelers from the northern plains would sign hotel registers with the name of their "state," unlike most visitors, such was their pride in the polity from whence they came. The *Post* compared the actions of such guests with those of an eager boy wearing "his first pair of pants or first morocco-topped, copper-toed boots."[104] Such benign press paternalism failed to mask a sympathetic welcoming of more states to the Union, the ultimate object of a century's political work in the West.

On Independence Day 1889, amid much fanfare—including a visit by Sitting Bull to the statehouse-in-progress in Bismarck—twin constitutional conventions organized at Bismarck and Sioux Falls. In a telling contrast, South Dakotans placed their polity's name and boundary along the seventh standard parallel prominently in the first article of their constitution, while North Dakotans mentioned both in a seventeenth article, blandly titled "Miscellaneous," near the end of their constitution.[105] Division had always been more popular in the south, and the constitutions reflected that attitude. In addition, the conventions recognized the unique challenges inherent in making two states from one territory. A joint commission of fourteen members assembled in Bismarck to divide the territorial records and the debt inherited from that temporary government. Any documents pertaining specifically to one half were parceled out appropriately, while civil servants copied any that dealt with matters spanning the common boundary.[106] With their long-anticipated split nearly accomplished, a remarkable collegiality replaced animosity across the seventh standard parallel.

Without revealing which bill he signed first, Harrison penned the statehood declarations for both North and South Dakota on November 2, 1889, bringing to a close the last great western boundary battle.[107] Its controversies had drawn attention not only from the northern plains and marbled halls of Washington, D.C., but from Americans at large. Two articles from that month's Portland *Morning Oregonian* reflected that curiosity. On November 15 the newspaper responded to "several inquiries" it had received from readers regarding the difference between a standard parallel and a line of latitude, with a concise description of the former as a uniquely American phenomenon.[108] Eight days later the *Morning Oregonian* reprinted a story that had appeared initially in the *Chicago Herald*. It noted that a clerical error in North Dakota's statehood bill omitted the word "standard" from its southern boundary, and that "a single word probably never made a bigger difference." The article continued: "By describing its southern boundary as the seventh parallel, North Dakota was represented as spreading its wheat fields, blizzards and prohibition principles down through the country into upper South America. It is hardly necessary to say that the mistake was quickly noticed and corrected, and the ambitious young state put back where she belongs."[109] The appearance of such stories, whether mundane or amusing, in newspapers nationwide illustrated the widespread interest Americans held in both the divide between the two new states and boundary-making generally in the nineteenth century.

Following the admission of North and South Dakota to the Union, the need to mark the long-debated division took center stage. South Dakota's new members of Congress introduced legislation to establish the boundary in the waning days of 1889. Some of their colleagues objected to spending any more time or money on the division of Dakota, but the bill cleared the chambers and was signed by President Harrison by the following fall. A team led by Charles H. Bates of Yankton marked the line with over seven hundred ruddy quartzite columns—"the most superb scratching posts for cattle that were ever seen," in Bates's words—between September 1891 and August 1892.[110] Wagon teams hauling two pillars each carried these loads more than sixty miles south from the Northern Pacific tracks over the course of the year.[111] These monuments provide an appropriate tribute to the passionate, almost interminable, debate about how to divide Dakota. Setting the boundary along the seventh standard parallel struck many at the time as just and appropriate for the two "great commonwealths of farmers." As one 1889 account claimed, "every farmer, tax-collector, surveyor, and county official within fifty miles of the dividing line will have a hundred reasons for thanking Congress for making this division as it is, and in the end they, and not the geographers, are the ones that should be consulted in the matter."[112] The divide resonates into the twenty-first century in myriad ways. Since 2004, for example, the winning team of the annual North Dakota State University–South Dakota State University football game has received the Dakota Trophy, a miniature replica of one of the quartzite boundary markers. It seems appropriate that the two land grant universities should meet on the gridiron to capture a symbol of their common line.[113]

The stresses of population and geography pulled Dakota in different directions. The repeated proposals for dividing the large territory testified to its complicated political, social, and economic realities. Yet historians have not always recognized that tension. Many South Dakotans perceive an east river–west river divide in attitudes of politics, economics, and society, a decades-old tension within the state. This perceived split became more apparent after the earlier political break with North Dakota, which left the southern state divided almost perfectly along the Missouri River's course.[114] Modern attitudes are often based on inaccurate views of early Dakota tensions. In historian Mark Wahlgren Summers's work on Gilded Age politics, he suggests that "from settlement patterns" a division along the Missouri River would have made as much sense as the geometric split achieved in 1889.[115] But the facts do not bear this out. From the earliest days of American settlement on the northern plains, a vast expanse existed between northern and southern population centers in Dakota. Aside from the aberration of the Black Hills gold rush in the 1870s, which inspired several abortive plans to split the territory vertically, in the long run it made more sense to accommodate the stable settlements along the Northern Pacific Railroad and the Missouri River near Nebraska and Iowa with their own political communities.

Summers makes a broader yet similarly problematic claim: "The fluidity of territorial lines was a reality of the Gilded Age."[116] In fact, compared with the dynamic redrawing of the trans-Mississippi West from the 1840s through the late 1860s, boundaries in the Gilded Age proved remarkably stable. Only four significant changes to the U.S. continental map occurred between 1868 and 1907, dates representing the creation of Wyoming Territory and Oklahoma's statehood respectively. Aside from splitting Dakota, the others consisted of altering Nebraska's northern line in 1882, creating Oklahoma Territory out of western Indian Territory in 1890, and a dispute between Texas and Oklahoma over their Red River boundary, which was settled by the U.S. Supreme Court in 1896.[117] Drawing the North Dakota–South Dakota boundary commanded the most local and national interest of the four. Both Republicans and Democrats saw consequences on the northern plains as they wrestled for control of Congress and the presidency. The competitive politics of the era prevented fluidity in boundary matters—the stakes for both parties stood far too high.

In addition, the popular criticism of Republicans for allowing partisanship to determine the outcome of the Dakota matter seems disingenuous. Considering the close elections of the Gilded Age, both parties hoped to pad their slim margins in Congress and the Electoral College. Republican control of Congress and the White House from 1889 to 1891 gave them a chance that the Democrats would have used to their own ends had the 1888 vote turned out differently. If the Democrats had retained the presidency and House of Representatives and had taken the Senate, almost assuredly they would have admitted Dakota as a single state for partisan reasons of their own. The Democratic Party could not forever resist the pressure for statehood and would have preferred one Republican-dominated state to two. In the presidential election of 1892—the first after the statehood of the Dakotas, Washington, Montana, Idaho, and Wyoming—Republican Benjamin Harrison won four of the six new states, Populist James Weaver won one, Democrat Grover Cleveland won none, and the three candidates split North Dakota's electoral votes.[118] If Republicans admitted these states to ensure their hold on the presidency, Cleveland's victory in 1892 demonstrated that elections depended on more than a few western polities.

The division of Dakota Territory held significance both for the northern plains and the nation. It provided separate polities for two fast-growing population centers that had struggled for decades to cooperate. It also permitted them to share their admired "Dakota" moniker. In addition, the new line accommodated American settlers—and their potato patches, hog yards, and prairie chickens—by using the land survey boundary rather than the forty-sixth parallel. While the latter would have corresponded to the other latitudes used for most east-west boundaries in the trans-Mississippi West, the former recognized the reality of boundary-making in settled regions. Indeed, the Dakota debate offers an example of the changing character of the West, and the United States in general, at the end of the nineteenth century. Dakota provided one last opportunity to pare down immense tracts into new states, an activity that engrossed several generations of Americans. The chance to create just one more boundary proved irresistible at the end of the nineteenth century, as the long game of drawing lines in the West concluded. In addition, the split illustrated competitive party politics in the Gilded Age at both the local and national levels. In the end, the debates over dividing Dakota represented more than a logical response to the tensions of a vast region and its separate American populations. They also demonstrated the changing importance of the West to the nation at large, as the region emerged from a colonial tutelage and its polities gained equality with the older, more established states.

A man and his ass on the New Mexico–Arizona boundary, mid-twentieth-century postcard

Conclusion
A Broader View of Borderlands

While contemplating the 1839 pseudo-war between the state of Missouri and Iowa Territory, a New York newspaper observed: "There is no country in the world like ours for boundary questions."[1] The preceding chapters demonstrate that notion repeatedly across time, place, and circumstance. Throughout the nineteenth century, Americans carved up the trans-Mississippi West with lines that shaped states modeled upon the older members of the Union. These political divides restructured enormous tracts of land that had expanded the United States through purchase, negotiation, and conquest. Invisible lines slicing through prairies, forests, mountains, and deserts created a special framework that the newcomers used to claim the region as their own. The process established new states that, after having endured territorial adolescence, would exist (in theory) on an equal basis with the older ones. It involved the active participation of American settlers living in the regions divided, politicians in the national capital, and interested parties across the country. Unique factors contributed to the evolution and ultimate shape of these lines, yet their existence fulfilled an essential goal for nineteenth century Americans. State boundaries remade a diverse, complicated landscape into a distinctly American West.

The final boundary-making in the late nineteenth century western United States is not the end of the story, however. Consequences of these invisible lines resonated long after the region's map took on its familiar shape. These boundaries complicated the always contentious issue of water rights, for example, as residents on either side of an invisible line sought to protect their access to that precious resource in a plethora of courtroom battles. But the modern significance of state lines does not rest solely with the rivers that cross them and sometimes form them. As a reporter quipped in reference to a dispute between Colorado and Utah in 1962: "It's never too late in Western U.S. history for a state boundary squabble."[2] The last major boundary change in the continental United States involved the deletion of one rather than the addition of one, when Congress erased the invisible line between Oklahoma and Indian Territories to admit the state of Oklahoma in 1907.[3] After that, the federal government and its constituents made only a few slight alterations to accommodate local circumstances. Following a century of line-drawing, since the early twentieth century most Americans have accepted the existence of state boundaries with scarcely a thought. Yet the trans-Mississippi region's forty-eight state boundaries represent realities far more significant than their accompanying highway signage might suggest. Two illustrative phenomena bear this out—the use of invisible divides to defend against perceived threats, and the lengths to which those who live near such lines will go to rectify their sense of marginalization. Both echo nineteenth-century experiences and demonstrate the ongoing importance of state boundaries.

The desire to use boundaries to protect a state from outsiders, even fellow Americans, resonated long after Missourians and Iowans shouldered arms. During the Great Depression, two western states used their boundaries to defend local industries and jobs from foreign competition, whether defined as outside the state or outside the country. In this they drew inspiration from Governor David Sholtz of Florida, who ordered highway patrol officers to guard each major road leading into the state in late 1935 and bar any transient who might place further strain on Florida's glutted labor pool.[4] Following Florida's lead, in February 1936 the Los Angeles, California, police chief acted on his own to defend the Golden State from a similar onslaught. For three weeks officers under his control stationed themselves along roads leading from Oregon, Nevada, and Arizona until pressure both within and outside California forced the chief to recall his officers.[5] In April 1936, Governor Edwin C. Johnson of Colorado attempted a similar blockade, this time using state troops. The Colorado National Guard patrolled the boundary between Colorado and New Mexico for two weeks in an ethnocentric and ultimately failed effort to prevent the entry of Latino migrant laborers.[6] In all three cases, states viewed their boundaries as a way to defend against the economic crisis of the 1930s. By guarding them like an international border, politicians and officials sought to protect their own workers. Such efforts demonstrated the protective potential ascribed to these invisible lines by viewing boundaries as barriers.

Three quarters of a century later, during the "Great Recession"—the bleakest economy in the United States since the Great Depression—the use of state lines to defend against outside labor resurfaced. This time, perceived federal failures to enforce immigration rules inspired Republican-dominated state governments to take up the matter within their own boundaries. Arizona led the way in 2010, with the Support Our Law Enforcement and Safe Neighborhoods Act, which Governor Jan Brewer described as a response to "a crisis we did not create and the federal government has refused to fix." The act authorized state and local law enforcement to investigate suspected cases of illegal immigration and turn over violators to federal agents for deportation. Brewer declared that along with boosting defenses on the state's southern boundary shared with Mexico, "today . . . Arizona strengthens its security *within* our borders."[7] Arizona's stand inspired legislatures in Alabama, Georgia, Indiana, South Carolina, and Utah to approve similar and in some cases even stricter laws the following year. Eight other states considered but did not adopt measures in the same vein in 2011.[8] Although none of these proposals went as far as establishing a border patrol, the insulating potential of invisible state barriers remained great.

Like those attempted during the Great Depression, the twenty-first-century state defense measures faced challenges almost immediately. A lawsuit challenging Arizona's measure appeared before the U.S. Supreme Court in 2012. The state found sympathetic ears among some justices; as conservative Antonin Scalia asked: "If, in fact, somebody who does not belong in this country is in Arizona, Arizona has no power? What does sovereignty mean if it does not include the ability to defend your borders?"[9] Republican presidential candidate Mitt Romney echoed this sentiment: "I believe that each state has the duty—and the right—to secure our borders and preserve the rule of law, particularly when the federal government has failed to meet its responsibilities."[10] Ultimately a majority of justices ruled that most of the law infringed upon federal powers, although they supported the right of state and local law enforcement to investigate the immigration status of individuals they might detain or arrest.[11] The debate over which invisible barriers can serve which purposes, one of the myriad issues of American federalism over the centuries, simmers on across the United States.

Additionally, some Americans who live near state boundaries feel neglected as a result of their location. As in southern Colorado in 1877, some of these marginal denizens propose secession to make their own polities. Perhaps the most famous such a move took place along the Oregon-California boundary, where in late November 1941 a half-dozen counties declared the state of Jefferson. Its supporters described themselves in "patriotic rebellion" against California and Oregon, and they insisted that "Jeffersonians intend to secede each Thursday until further notice." After only a couple weeks, though, the attack on Pearl Harbor and U.S. entry into World War II deprived Jefferson of its media attention. Yet the idea lingers, expressed most recently by frustrated citizens in Siskiyou County, California, in the summer of 2013.[12] Aggrieved residents of southwestern Kansas proposed a similar revolution in 1992. Angered by that state's efforts to raise taxes but reduce their area's share of revenue, seven counties placed the question of political divorce before voters that spring. The movement drew interest from other Kansans as well as dissatisfied Oklahomans and Coloradans nearby. Eighty-four percent of the local electorate supported secession at the ballot box that April, and a constitutional convention for West Kansas assembled in September 1992. Ridicule from the press helped undermine the proposal, however, and secessionism receded into the prairie grasses from whence it came.[13] As recently as 2013, commissioners in Colorado counties along the northern and eastern state boundaries discussed forming their own state, called either North Colorado or New Colorado. These predominantly Republican regions felt slighted by a Democrat-dominated state government, especially after passage of locally unpopular laws regarding energy development and gun control. Proponents, frustrated with decades-old tensions between rural and urban communities, pointed an accusatory finger toward political lines in the West. One secession advocate stated: "We are at a point where we need a national discussion about representation and state boundaries. Some of the reasons why borders were drawn when states were created no longer exist. This is a healthy and necessary discussion."[14] The struggles of secessionists illustrate the difficulty of breaking from long-established polities and also the lengths to which frustrated Americans will go to rectify their sense of weakened sovereignty.

Controversies like ones inspired by state boundaries and described throughout the preceding pages evoke a thriving and compelling field within academic history: borderlands studies. Nearly a century old now, borderlands scholarship emerged not long after the state-centered study of American history had all but vanished. Herbert Eugene Bolton provided the field its name as well as its seminal work in 1921's *The Spanish*

Borderlands.[15] Unlike Frederick Jackson Turner's emphasis on the frontier process of Anglo-American westward expansion, Bolton focused upon a contested place—the northern fringe of Spain's American empire. Percolating in academia ever since, work on borderlands history in North America has flourished since the 1990s. Borderlands studies have expanded to incorporate other imperial edges in North America as well as the evolving borders between the United States and its neighbors. The field's blossoming historiography took place without a significant intellectual structure to define it, however. One might characterize borderlands as a field saturated with scholarship yet lacking in theory.

Since the late twentieth century, a few scholars have attempted to define the skeleton of this burgeoning body of work. None commands more attention than a 1999 article in the *American Historical Review* written by Jeremy Adelman and Stephen Aron. They defined North American borderlands as "the contested boundaries between colonial domains," suggesting that the give-and-take nature of such amorphous places existed until borders between Euro-American claims were formalized via treaties. Therefore, borderlands could only exist in regions split eventually by official borders and did not exist after the completion of those lines. To Adelman and Aron, borderlands in North America disappeared after the early nineteenth century, coincidentally the same time the United States started drawing state lines outside its original limits. In perhaps their most oft-quoted line, Adelman and Aron declared: "This shift from inter-imperial struggle to international coexistence turned borderlands into *bordered* lands."[16] In the following *American Historical Review*, several scholars critiqued their attempt to create borderlands theory. Perhaps the most thorough deconstruction came from John R. Wunder and Pekka Hämäläinen, who expressed their frustration at a lack of American Indian agency in this "game of imperial chess."[17] Yet the sweeping claims of Adelman and Aron all but ensured their appearance in any historiographic discussion of borderlands, including this one.

A dozen years later, an article coauthored by one of Adelman and Aron's most strident critics reassessed borderlands scholarship. Hämäläinen and Samuel Truett offered their own take on theory in a special borderlands issue of the *Journal of American History* in 2011. They emphasized the existence of borderlands between Native cultures and Euro-American empires, not to mention between Native groups themselves, in addition to the contested zones between imperial claims long accepted in the field. Yet the authors struggle, as do so many others, to discern what to do with this broadened field. "We now find borderlands everywhere," according to Hämäläinen and Truett, "but our ability to interweave their stories—and use them to contest older narratives and transcend older boundaries—is as limited as ever." Like Adelman and Aron, Hämäläinen and Truett suggest a similar time frame for the field, with the heyday of borderlands winding down in the early nineteenth century.[18]

The lack of a clear definition or direction in borderlands studies as explored by Hämäläinen and Truett concerns many historians who contribute to it. Their article quotes Alan Taylor, who notes the "problem that scholars find borderlands everywhere—leaving us with no space, at least in North America, that is not a borderland." Hämäläinen and Truett describe the phenomenon thus: "The breadth of the field is its virtue, and also its chief vulnerability."[19] In this vein, a 2011 volume of the *Western Historical Quarterly* included an article by Kelly Lytle Hernández on the status of borderlands research. She points out diverse disciplines that illustrate the consequences of boundaries beyond the traditional purview of borderlands historians. This broader view incorporates any number of imposed divisions. Hernández observes both the challenges and benefits of a broader view of borderlands, one that "may bleed the concept into irrelevance but, along the way . . . also highlight that colonial disputes and national borders are not the only way to meaningfully bisect and order communities."[20] Taking advantage of such an incomplete wall surrounding borderlands studies, I intend to lead internal boundaries—state lines in particular—through the breach. By doing so, I hope to rectify more than a century of historiographic dismissal and demonstrate the benefits of exploring many kinds of borderlands, with as little irrelevant bleeding as possible.

"Too often," Adelman and Aron write, "students of borderlands neglect the power politics of territorial hegemony."[21] Yet too many scholars in the field also fail to recognize that this process did not end with the drawing of international borders, as if a diplomat's signature on a treaty flipped some magic switch that caused all competition to cease. As historical geographer D. W. Meinig notes, "a boundary line creates a border zone that becomes an important fact of life, day by day, for all who live within it."[22] Meinig's observation proves just as accurate for state boundaries as international borders, as the preceding pages illustrate over and over. Yet many scholars would accept instead Glen M. Leonard's interpretation of state lines as "boundaries once alive and vibrant [that] have become mere administrative niceties."²³ Belittling state lines as little more than crusty relics ignores their lingering effects, especially for those living closest to them. Whether Adelman and Aron's "power politics of territorial hegemony" take place within a country or between it and another matters little. In either case, a borderland results. The process of internal organization in the American West inspired a plethora of contested zones, battled over throughout the nineteenth century and with consequences that resonate into the twenty-first. These frontiers within the United States I term *intranational borderlands*.

An intranational borderland originates with any division within a country, around which emerges a region whose inhabitants experience both the obstacles and opportunities inherent in any marginal place where two or more authorities meet and often overlap. These borderlands can develop before, during, or after the formalization of that division. For those who live within them, intranational borderlands offer unique complications. At times their contested nature provides benefits to residents, while at other times they find themselves frustrated by their position on the edge of not one but two or more political communities. As historian Richard White suggests, "The boundaries of the American West are a series of doors pretending to be walls."²⁴ Although White was referring to external limits of the West, internal divisions within the West also represent both portals and barricades for the region's inhabitants and have done so for well over two centuries.

I proposed the concept of intranational borderlands in a paper delivered at the Western History Association meeting in Scottsdale, Arizona, in October 2005. In the session, my paper instigated many thoughtful comments and suggestions, in particular from the late David Weber. He encouraged me to submit a proposal for a "Comparative Perspectives on North American Borderlands" conference he was co-organizing at the Filson Institute in Louisville, Kentucky. A year later I traveled to the falls of the Ohio River for an intimate and thought-provoking meeting. Yet the news that borderlands theorist Stephen Aron would chair my session inspired a stomach full of butterflies. Responding thoughtfully to my concept, Aron pondered if the dilution of borderlands studies by incorporating state lines would lead to a debate about the term in general, turning it into a "b-word" like the 1990s battle over "frontier" as the "f-word."²⁵ And where would it end? Counties? Cities? School districts? As Aron asked the audience, where do we draw the line? (I believe he intended the pun.)

Weber addressed Aron's concerns in his conference-ending remarks. In them he called for a broad, inclusive field to investigate the myriad implications of boundaries of all kinds. After all, Weber asked, what sense does it make to draw a border around borderlands history and define what is in the field and what is out?

Excluding new interpretations of borderlands dismisses the vital interactions along any number of imposed divisions and denigrates many promising paths of inquiry. Quixotic efforts to encapsulate a field that thus far has eluded all attempts at definition reflect the contentious nature of borderlands themselves. Moreover, the need to defend borderlands scholarship against perceived dilution strikes me as unwarranted if not laughable. One envisions a handful of academic ideologues staffing the toll booth on the sagebrush flats in the film *Blazing Saddles*, hoping that no one realizes that they can—and should—walk around it.[26] Academia offers several examples of this potential for scholarly evolution, including the Association for Borderlands Studies, a professional organization established in 1976 focusing on the U.S.-Mexico border region. It now boasts a membership from across the globe and facilitates conversations on "a variety of borders, borderlands, and border regions worldwide" through the *Journal of Borderlands Studies*.[27] The primary scholarly journal of American West history, the *Western Historical Quarterly*, also recognizes the malleability of the field. Each edition includes a "Recent Articles" roundup on scholarship of interest to readers. Through 2002 the *Quarterly* hearkened back to the ghost of Bolton through its listings under "Spanish Borderlands." It acknowledged an increasing diversity of research, however, by switching the category to "International Borderlands" in the spring 2003 volume.[28] This simple action suggests that some historians consider borderlands studies a flexible field, able to accommodate new points of view.

To demonstrate the existence of nineteenth-century intranational borderlands, I return to chapters 3–8. Each of the state boundaries outlined in these chapters experienced borderlands conditions. The western limit of Arkansas evolved through a great amount of give-and-take involving two groups of migrants—American settlers and American Indians. The Choctaws and Cherokees of the Deep South, longtime neighbors of Euro-Americans, recognized the value their competitors placed upon boundaries and used that insight to their advantage in the trans-Mississippi West. Throughout the 1820s, Americans and Native residents of the borderlands alike found their livelihoods at times protected and at other times threatened by the negotiations over it. The same conditions that protected Tom Graves from prosecution also forced American settlers from their farms in Lovely's Purchase. Similarly, a deal struck between the federal government and the Osages instigated the contentious borderland between Missouri and Iowa. Its erroneous survey led nearly to war in the 1830s as both polities sought to defend their claim to the disputed region. For some residents within it, however, the lack of clarity provided an opportunity. The "Hairy Nation" played officials off one another to avoid paying taxes while the line separating the two polities remained uncertain. After both states resolved the question of jurisdiction, this crafty community dispersed in search of greener pastures.

Native cultures in the Pacific Northwest also inspired borderlands conditions, even if they did not play a significant role in the creation of Oregon Country boundaries. When American officials chose to ignore the Oregon-Washington line during a conflict with Native groups in the 1850s, they inspired political and military feuding that spanned the continent. In this case, volunteer troops and territorial officials created their own borderland. A similar situation emerged between California and Utah in 1860, as violence between American settlers and the Paiutes proved impossible for the responsible authority to defuse and illegal for the other to affect. The federal government contributed to this borderland repeatedly by declaring an overlapping jurisdiction between California and the new territory of Nevada. But American settlers in the Honey Lake valley provided the most dramatic challenge. Like the "Hairy Nation," Isaac Roop and his neighbors used vague divides for their own purposes. Beyond Utah's legal control and too isolated for California to exercise authority, they sought several times to create their own territory in between. Roop remains ensconced in this borderland a century and a half after his death. Even though his home stood in California, modern Nevadans honor Roop as their first—albeit extralegal—executive, with a portrait in the Carson City capitol. His body rests on a California mountainside overlooking the Honey Lake valley, under a large gravestone declaring him "governor of the provisional territory of Nevada."

Ethnic divisions dominated the borderland between New Mexico and Colorado in the 1860s and 1870s. Hispano settlers of northern New Mexico learned the power of American boundary-making the hard way when Colorado's southern line cleaved some three thousand of them away from their cultural and political ties. Even good-faith efforts on the part of Colorado's Anglos to accommodate their Hispano neighbors rankled, for without the line such accommodation would not have been necessary. The experience of these Hispanos brings new meaning to a popular saying in the Southwest: "We didn't cross the border. The border crossed us." Americans coming into the region experienced their own borderland, as they plotted secession from both polities to make their own in between. While the San Juan movement failed, like other western secession efforts to date, it illustrates the feeling of neglect experienced by many marginalized groups. Politics more than ethnicity affected the final specific borderland in this book, the one located between two divergent population centers in Dakota Territory. This contested zone represents perhaps the best-defined, not to mention narrowest, of them all—the five-mile-wide strip between the forty-sixth parallel and the seventh standard parallel. While politicians on the northern plains and in Washington, D.C., argued over which line should split the territory, the farmers within it awaited word of their fate. By selecting the seventh standard parallel, the only land survey line in the country used also as a state boundary, their agrarian property remained whole and packaged neatly into either North or South Dakota.

In contrast to Adelman and Aron's claim of defined borders signifying the end of borderlands, international divides remain contested regions long after the ink dries on their treaties—witness recent ongoing arguments over policing illegal immigration and narcotics-related bloodshed along the U.S.-Mexico border. Similarly, the consequences of their intranational counterparts linger following the formal declaration and survey of those lines. The examples of boundary defense during the Great Depression and simmering secessionist efforts illustrate this ongoing reality. For good measure, I offer three additional instances of intranational borderlands today. The residents of Texarkana experience divergent economic opportunities depending on whether they live on the Texas or Arkansas side of State Line Road. This street follows not only a state boundary, but also what served from 1819 to 1845 as part of the southwestern border of United States. To level the playing field in this bifurcated town, in 1977 the state government of Arkansas ceased collecting income tax from their half, balancing Texas's lack of an income tax. While the situation evened out somewhat after that decision, the Texas side of the city has long been the faster-growing one. As one journalist commented in 2007, "to the casual observer [State Line Road] is merely another road separating two states. But for those in the neighborhood, the road represents something of a great divide."[29] No

A BROADER VIEW OF BORDERLANDS 225

A man and his ass in the "border town" of Texarkana, mid-twentieth-century postcard

international border slices through Texarkana these days, but its residents know firsthand that their state boundary can still mean complications.

On the other side of the Lone Star State, another intranational borderland lurks. According to the Compromise of 1850, the longitudinal limit of the Texas panhandle should be the 103rd Greenwich meridian, but a piecemeal survey nine years later located the boundary several miles too far west. Increasing American settlement on the staked plains exposed the error, which Congress accepted as the legal limit between Texas and New Mexico Territory in 1891, twenty-one years before the latter joined the Union at long last.[30] The inconsistency has rankled many New Mexicans ever since, leading to several memorials from the state legislature to redress the situation. In 1953 officials adopted a sarcastic declaration claiming large chunks of Texas for New Mexico, which they would surrender for a quarter of all mineral rights in their neighboring state.[31] Less tongue-in-cheek measures passed the New Mexico legislature in 1987 and 1991. The state's attorney general promised to raise the issue with Texas and the federal government, but to no avail.[32] Some residents in the contested zone liked the notion of joining New Mexico, with its lower tax rates and vehicle fees and "prettier" license plates. Others, including a newspaper columnist in Amarillo, threatened to organize a militia to defend against any encroachment.[33] The postmaster of Texline, perched on the inaccurate boundary, complained: "It would change the ZIP code. It would change the time zone. It would change the state. I don't like it. Besides, I like being a Texan."[34] The debate resumed in the new millennium as the New Mexico legislature tried three more times to reclaim the half million acres lost long ago, as well as the "subsurface mineral rights, oil and gas royalties and income, property taxes and grazing privileges" thus denied to them. One politician suggested New Mexico's legislature should meet their counterparts to the east in the disputed area and settle the matter with an all-out melee.[35] Such a clash has yet to transpire, perhaps a good thing for New Mexico as the Texas legislature outnumbers its neighbor's by almost seventy.

Like Texarkana, another internal set of "border towns" also represent intranational borderlands, much to their residents' chagrin. The municipalities of Wendover, Utah, and West Wendover, Nevada, suffer from what amounts to a single community paying for the infrastructure and management of two, as the majority of the former's residents work in the latter. West Wendover enjoys a greater revenue stream, the result of legalized gambling, while neighboring Wendover struggles to make ends meet. Blending the two communities into one by altering the state boundary between them seemed the most sensible option, not to mention a popular one, as a 2001 poll indicated near-unanimous support for the idea on the Utah side of the line.[36] The next year members of Congress from both states introduced a bill in the House of Representatives to pave the way for a slight shift of the line, placing both communities in Nevada. Representative James Hansen of Utah observed, "[K]nowing where the boundaries lie, it seems that some of the practical challenges faced by every small town [are] amplified by this particular area because of the unique mix of circumstances."[37] The Republican-controlled House passed the bill but the Democratic leadership in the Senate chose not to act on the measure. In any case, voters in both towns approved a merger referendum in November 2002, and over the next several years Nevada legislators held public meetings to discuss the next step.[38] The effort petered out in 2006 for several reasons—Wendover's county objected to the loss of its airport, Nevada did not relish accepting responsibility for Wendover's spiraling debt, and its strongest advocates no longer held their local political offices.[39] Yet the borderland frustrations that inspired the talk of boundary-shifting linger there in the heart of the Great Basin.

Lest I develop a reputation as an academic iconoclast, I do not suggest that intranational borderlands are more important than those defined by

Adelman and Aron or Hämäläinen and Truett. Nor do I suggest that they are less important. The experiences of one provide insight into the other. International borders and intranational boundaries are not mutually exclusive but instead offer complementary ways to conceive the challenges faced by marginalized people. Hämäläinen and Truett worry about the ability of borderlands scholars to incorporate their work into broader historical narratives. Inclusion of diverse kinds of borderlands offers a way to combat this problem. Alan Taylor suggests that seeing borderlands everywhere is a problem, but I disagree. It is an opportunity. Further studies might well demonstrate borderlands conditions along all manner of imposed divisions throughout North America and around the world.[40] Ignoring them neglects many people whose day-to-day realities center upon their own vitally important borderlands, as D. W. Meinig argues. Some historians might dismiss this interpretation as chaotic or parochial, but if borderlands were really as rare as they claim, we would not keep finding new ones.

Instead of treating borderlands and frontiers as either Boltonian places or Turnerian processes, we should portray the contested nature of marginal places as a concept, a distinct interpretive framework. In so doing, historians can develop a compelling aspect to their profession. As Weber suggested, there cannot be a border around borderlands history, or the field endangers its future and makes irrelevant its contributions by refusing to acknowledge new possibilities, especially as exceptions to accepted scholarship increase exponentially. By engaging with diverse kinds of boundary-inspired competition, historians can transform borderlands into a theoretical structure commanding just as much attention as the "holy trinity" of class, gender, and ethnicity. Environmental history offers a useful model—like borderlands, it boasts strong ties to American West studies and has evolved in thought-provoking ways to affect interpretations of communities around the world. Bolton presaged just such an expansion of the field in 1933: "Borderland zones are vital not only in the determination of international relations, but also in the development of culture. . . . By borderland areas not solely geographical regions are meant; borderline studies of many kinds are similarly fruitful."[41] Perhaps one day the *Western Historical Quarterly* can complete its evolution from "Spanish Borderlands" to "International Borderlands" to simply "Borderlands."

Bolton laid the foundation for a dynamic field rife with possibilities with his 1921 book *The Spanish Borderlands*. His work appeared a generation after the successful efforts of Woodrow Wilson and Frederick Jackson

Turner to disparage state-centered narratives of American history. The same year that Bolton's book hit the shelves, President Warren G. Harding—Wilson's less academic successor in the White House—drove a symbolic nail into the coffin of state boundaries. Speaking to a gathering of newspaper reporters in New York City, Harding declared: "Our Constitution was adopted in order to form a more perfect union, and as the national life has developed under it that union has been so perfected that State lines have well-nigh ceased to have more than geographical and political significance."[42] Yet the lingering effects of state boundaries as exemplified by intranational borderlands belie Harding's claim. In contrast to their common interpretation as stale anachronisms, the lines that both divide and unite the country from within remain essential, consequential elements of American life.

Not only is it never too late for a state boundary feud, it's never too late for historians to acknowledge the essential role they played in creating the American West. They transformed places identified by Europeans as *La Louisiane*, the Oregon Country, and *El Norte*, and which American Indian cultures called simply "home," into functional members of the Union. Each state line boasts a unique creation story, yet when viewed together over the course of the nineteenth century the boundary-making process illustrates the Americanization of a vast, complicated region. In addition, it's never too late for borderlands scholars to embrace a more open approach to the field. Examples throughout these pages demonstrate the existence of intranational borderlands along state boundaries throughout the West. By treating borderlands as a general interpretive structure, the concept can shed light on the consequences of marginalization among communities large and small, all worthy of having their story told. Far from bleeding borderlands into irrelevancy, such an approach would make its scholarship even more useful. After more than a century of neglect, state boundaries deserve greater attention, both for their own merits and defects and as ways to conceive of a broader view of borderlands. It's never too late.

Appendix
"The Honey War"

On October 26, 1839, the *Missouri Whig and General Advertiser* published a satirical poem in honor of the boundary conflict between Missouri and Iowa Territory. Several sources attribute it to John I. Campbell of Palmyra, Missouri, but the poem as printed in the newspaper does not contain a byline.[1] Intended to be sung to the tune of "Yankee Doodle," the complete poem reads:

The Honey War

Ye freemen of the happy land,
Which flows with milk and honey,
Arouse to arms, your poneys [sic] mount,
Regard not *blood* or *money*.
Old Governor Lucas, tiger like,
Is prowling round our borders,
But Governor Boggs is wide awake,
Just listen to his orders.*

Three *bee trees* stand about the line
Between our state and Lucas,
Be ready all the trees to fall,
And bring things to a focus.
We'll show old Lucas how to brag,
And claim our precious *honey*,

---

* We understand the Governor has ordered Gen. [Daniel] Willock [of the state militia and sheriff of Marion County, Missouri] to be in readiness with several thousand men.

He also claims, I understand,
Of us three bits in money.†

The dog which *barks* will seldom bite,
Then let him rave and splutter;
How impudent must be the wight
Who can such vain words utter.
But he will learn before he's done,
Missouri is not Michigan.
Our *bee trees* stand on our own land,
Our *honey* then we'll bring in.

Conventions, boys, now let us hold,
Our *honey* trade demands it,
Likewise the *three bits* all in gold,
We all must understand it.
Now in conventions let us meet,
In *peace* this thing to *settle*,
Let not the tiger's war-like words
Now raise too high our metal [*sic*].

Now if the Governors want to fight,
Just let them meet in person,
For Governor Boggs can Lucas flog,
And teach the *brag* a lesson.
And let the victor cut the trees,
And have three bits in *money*,
And wear a crown from town to town,
Anointed with pure honey.

And then no widows will be made,
No orphans unprotected,
Old Lucas will be nicely flogg'd,
And from our line ejected.
Our *honey trade* will then be placed
Upon a solid basis,
And Governor Boggs, where'er he goes,
Will meet with smiling faces.

† Taxes.

Notes

INTRODUCTION

1. "Western and Northern Boundary," *Missouri Argus*, March 4, 1836, 3.
2. See Hermann von Holst, *The Constitutional and Political History of the United States*, 7 volumes (Chicago: Callaghan and Company, 1877); George Bancroft, *History of the Formation of the Constitution of the United States of America*, 2 volumes (New York: D. Appleton and Company, 1882); Hubert Howe Bancroft, *The Works of Hubert Howe Bancroft*, 39 volumes (San Francisco: A. L. Bancroft, 1882–90).
3. Susan Schulten, *Mapping the Nation: History and Cartography in Nineteenth-Century America* (Chicago: University of Chicago Press, 2012), 185.
4. Woodrow Wilson, "The Making of the Nation," *Atlantic Monthly* 80, no. 477 (July 1897), 2.
5. Frederick Jackson Turner, "The Middle West," in *The Frontier in American History* (1920; repr., New York: Dover Publications, 1996), 127.
6. For monographs on specific boundaries or boundary-related legislation, see Jay Monaghan, *Civil War on the Western Border, 1854–1865* (Lincoln: University of Nebraska Press, 1955); Mark J. Stegmaier, *Texas, New Mexico, and the Compromise of 1850: Boundary Dispute and Sectional Crisis* (Kent, OH: Kent State University Press, 1996); Nicole Etcheson, *Bleeding Kansas: Contested Liberty in the Civil War Era* (Lawrence: University of Kansas Press, 2004); Michael F. Holt, *The Fate of Their Country: Politicians, Slavery Extension, and the Coming of the Civil War* (New York: Hill and Wang, 2004); Jeremy Neely, *The Border between Them: Violence and Reconciliation on the Kansas-Missouri Line* (Columbia: University of Missouri Press, 2007); and John R. Wunder and Joann M. Ross, eds., *The Nebraska-Kansas Act of 1854* (Lincoln: University of Nebraska Press, 2008). For encyclopedic works, see Henry Gannett, *Boundaries of the United States and of the Several States and Territories with an Outline of the History of All Important Changes of Territory* (Washington, D.C.: Government Printing Office, 1904); Franklin K. van Zandt, *Boundaries of the United States and the Several States* (Washington, D.C.: Government Printing Office, 1966); Jim Feldman, *The Shape of the Nation: Why the States Are Shaped Like That* (Madison, WI: Feldman, 1999); Andro Linklater, *Measuring America: How the United States Was Shaped by the Greatest Land Sale in History* (New York: Walker, 2002); Gary Alden Smith, *State and National Boundaries of the United States* (Jefferson, NC: McFarland, 2004); Andro Linklater, *The Fabric of America: How Our Borders and Boundaries Shaped the Country and Forged Our National Identity* (New York: Walker, 2007); Tegan Hansen and Jerry Hansen, *State Boundaries of America: How, Why and When American State Lines were Formed* (Westminster, MD: Heritage Books, 2007); Mark Stein, *How the States Got Their Shapes* (New York: Smithsonian Books/Collins, 2008); Kathy Guyton, *U.S. State Names: The Stories of How Our States Were Named* (Nederland, CO: Mountain Storm Press, 2009); Bill Hubbard, Jr., *American Boundaries: The Nation, the States, the Rectangular Survey* (Chicago: University of Chicago Press, 2009); Mark Stein, *How the States Got Their Shapes Too: The People Behind the Borderlines* (New York: Smithsonian

Books/Collins, 2011). Entertaining works on alternative state boundaries include G. Etzel Pearcy, *A 38 State USA* (Redondo Beach, CA: Plycon Press, 1973) and Michael J. Trinklein, *Lost States: True Stories of Texlahoma, Transylvania, and Other States that Never Made It* (Philadelphia: Quirk Books, 2010).

    7. *Congressional Globe*, 28th Cong., 2nd sess. (February 11, 1845), 274.

    8. Schulten, *Mapping the Nation*, 7, 12–14, 25, 57, 75, 184.

    9. Albert D. Richardson, *Our New States and Territories, Being Notes of a Recent Tour of Observation through Colorado, Utah, Idaho, Nevada, Oregon, Montana, Washington Territory and California* (New York: Beadle, 1866), 14.

    10. *Boundary Line between Colorado and Utah*, 45th Cong., 2nd sess., April 30, 1878, H. Rep. 708, serial 1825.

    11. *Congressional Record*, 47th Cong., 2nd sess. (February 5, 1883), 2109.

    12. Quotation in "Troubles in Colorado," *Daily Press and Dakotaian*, June 15, 1877, 2.

    13. *Resolution of the Legislature of Texas, in Favor of the Passage of an Act, Extending the Jurisdiction of that State over the Sabine Pass, the Sabine Lake, and the Sabine River*, 30th Cong., 1st sess., April 17, 1848, S. Misc. Doc. 123, serial 511; *Resolution of the Legislature of Louisiana, in Favor of the Extension of the Jurisdiction of that State to the Western Bank of the Sabine*, 30th Cong., 1st sess., April 28, 1848, S. Misc. Doc. 135, serial 511.

    14. Matthew A. Byron, "Crime and Punishment: The Impotency of Dueling Laws in the United States" (PhD diss., University of Arkansas, 2008), 79–104.

    15. Montana State Constitution, 1972; Charles S. Johnson, "Saying 'Sovereignty in Jeopardy,' Stevensville Woman Wants Borders Back in Montana Constitution," *Missoulian*, February 10, 2010, http://missoulian.com/ (accessed July 24, 2012); Charles S. Johnson, "3 Ballot Measures Qualify for November," *Missoulian*, July 20, 2010, http://missoulian.com/ (accessed July 24, 2012).

    16. Jeremy Adelman and Stephen Aron, "From Borderlands to Borders: Empires, Nation-States, and the Peoples in Between in North American History," *American Historical Review* 104, no. 3 (June 1999), 840 (italics original).

    17. Carl Abbott, *Political Terrain: Washington, D.C. from Tidewater Town to Global Metropolis* (Chapel Hill: University of North Carolina Press, 1999), 174.

    18. Kristen Leigh Painter, "Ceremony Marks Completion of History Colorado Center Construction," *Denver Post*, October 8, 2011, www.denverpost.com/ (accessed October 11, 2011).

    19. William D. Pattison, *Beginnings of the American Rectangular Land Survey System, 1784–1800* (1957; repr., Chicago: Department of Geography, University of Chicago, 1964), 1. For a more recent comprehensive interpretation see Hubbard, *American Boundaries*.

    20. Andro Linklater, *Measuring America: How the United States Was Shaped by the Greatest Land Sale in History* (New York: Walker, 2002), 6.

    21. William E. Riebsame, ed., *Atlas of the New West: Portrait of a Changing Region* (New York: W. W. Norton, 1997), 53.

    22. "California Legislature," *Daily Alta California*, March 19, 1862, 1.

    23. Session Laws of Nevada, 5th Legislature, March 2, 1871, 187–88.

    24. Quotation in Herman J. Deutsch, "The Evolution of Territorial and State Boundaries in the Inland Empire of the Pacific Northwest," *Pacific Northwest Quarterly* 51, no. 3 (July 1960), 118.

    25. Malcolm G. Comeaux, "Attempts to Establish and Draw a Western Boundary," *Annals of the Association of American Geographers* 72, no. 2 (June 1982), 254.

    26. D. W. Meinig, *The Shaping of America: A Geographical Perspective on 500 Years of History*, vol. 3, *Transcontinental America, 1850–1915* (New Haven: Yale University Press, 1998), 34–35.

27. Derek R. Everett, *The Colorado State Capitol: History, Politics, Preservation* (Boulder: University Press of Colorado, 2005), 1.

28. *Resolution of the General Assembly of Missouri, in Regard to Ceding a Portion of Nemaha County, in the State of Nebraska, to the State of Missouri*, 44th Cong., 1st sess., December 14, 1875, H. Misc. Doc. 15, serial 1698.

29. *In the Senate of the United States*, 51st Cong., 1st sess., December 4, 1889, S. Misc. Doc. 10, serial 2697. See also *Letter from the Secretary of the Interior, Transmitting in Response to Senate Resolution of December 5, 1889, Report on the Boundary Line between the States of Iowa and Nebraska*, 51st Cong., 1st sess., January 6, 1890, S. Ex. Doc. 20, serial 2682; Daniel Henry Ehrlich, "Problems Arising from Shifts of the Missouri River on the Eastern Border of Nebraska," *Nebraska History* 57, no. 2 (Fall 1973): 341–63.

30. *Report of the Secretary of the Interior*, 51st Cong., 2nd sess., 1890, H. Ex. Doc. 1, pt. 5, serial 2844, 250–57.

31. "The Eastern Line of the State—An Hour among the Snowy Mountains," *Daily Evening Bulletin*, April 1, 1863, 3.

32. Robert B. Houston, "On the Face of the Earth: Marking Colorado's Boundaries, 1868–1925," *Colorado History* 10 (2004): 101.

33. Quotation in Thomas L. Karnes, *William Gilpin: Western Nationalist* (Austin: University of Texas Press, 1970), 264.

34. R. B. Townshend, *A Tenderfoot in Colorado* (Boulder: University Press of Colorado, 2008), 40–41.

35. Samuel Bowles, *The Parks and Mountains of Colorado: A Summer Vacation to the Switzerland of America, 1868*, ed. James H. Pickering (1869; repr., Norman: University of Oklahoma Press, 1991), 64.

36. *Congressional Record*, 48th Cong., 2nd sess. (December 11, 1884), 184.

37. See *American Prime Meridian*, 31st Cong., 1st sess., May 2, 1850, H. Rep. 286, serial 584; *Congressional Globe*, 31st Cong., 1st sess. (May 2, 1850), 891–92; Joseph Hyde Pratt, "American Prime Meridians," *Geographical Review* 32, no. 2 (April 1942), 233–44. Matthew Fontaine Maury, a cartographer world-renowned for his oceanic charts, served as the observatory's superintendent at the time of the meridian's creation. As befit his position and his observatory's new status, Maury produced an unprecedented, intricate national map replete with statistics illustrating the supposed supremacy of the United States in 1860, even as sectional tension threatened to dissolve the country. See Schulten, *Mapping the Nation*, 108–10.

38. "The Proposed Territory of Dakota," *Chicago Daily Tribune*, March 1, 1878, 7.

39. Quotation in "The Saunders Bill," *Daily Press and Dakotaian*, March 15, 1878, 2.

40. *Boundaries of the State of Oregon*, 44th Cong., 1st sess., July 18, 1876, H. Rep. 764, serial 1712, 3.

41. *How the West Was Won* (DVD), directed by John Ford, Henry Hathaway, and George Marshall, 1962 (Burbank, CA: Warner Home Video, 2008).

42. Versions of these chapters appeared respectively in the following journals: Derek R. Everett, "On the Extreme Frontier: Crafting the Western Arkansas Boundary," *Arkansas Historical Quarterly* 67, no. 1 (Spring 2008), 1–26; "To Shed Our Blood for Our Beloved Territory: The Iowa-Missouri Borderland," *Annals of Iowa* 67, no. 4 (Fall 2008), 269–97.

1. A LITTLE PATCH OF GROUND

1. William Shakespeare, *The Tragedy of Hamlet, Prince of Denmark*, act IV, scene 4, lines 9–28, in *William Shakespeare: Four Tragedies*, ed. T. J. B. Spencer (London: Penguin Books, 1994), 228–29.

2. See Carlton J. H. Hayes and James H. Hanscom, *Ancient Civilizations: Prehistory to the Fall of Rome* (New York: Macmillan, 1983), 165; Norman Bancroft Hunt, *Historical Atlas of Ancient Mesopotamia* (New York: Checkmark Books, 2004), 98–99; Eric H. Cline and Mark W. Graham, *Ancient Empires: From Mesopotamia to the Rise of Islam* (New York: Cambridge University Press, 2011), 50.

3. Alan Taylor, *American Colonies: The Settling of North America* (New York: Penguin Books, 2001), 47.

4. D. W. Meinig, *The Shaping of America: A Geographical Perspective on 500 Years of History*, vol. 1, *Atlantic America, 1492–1800* (New Haven: Yale University Press, 1986), 80–83; William Cronon, *Changes in the Land: Indians, Colonists, and the Ecology of New England*, rev. ed. (New York: Hill and Wang, 2003), 54–81; James D. Drake, *The Nation's Nature: How Continental Presumptions Gave Rise to the United States of America* (Charlottesville: University of Virginia Press, 2011), 242–47.

5. Francis Newton Thorpe, *The Federal and State Constitutions, Colonial Charters, and Other Organic Laws of the States, Territories, and Colonies Now or Heretofore Forming the United States of America*, vol. 1 (Washington, DC: Government Printing Office, 1909), 53.

6. See David Stick, *Roanoke Island: The Beginnings of English America* (Chapel Hill: University of North Carolina Press, 1983); Lee Miller, *Roanoke: Solving the Mystery of the Lost Colony* (New York: Penguin Books, 2000); Giles Milton, *Big Chief Elizabeth: How England's Adventurers Gambled and Won the New World* (London: Hodder and Stoughton, 2000).

7. Francis Newton Thorpe, *The Federal and State Constitutions, Colonial Charters, and Other Organic Laws of the States, Territories, and Colonies Now or Heretofore Forming the United States of America*, vol. 7 (Washington, DC: Government Printing Office, 1909), 3783–85.

8. Ibid., 3795. Plymouth lost its "Sea to Sea" charter in 1630 when a new document limited its claims to the area around the town of Plymouth and Cape Cod, accommodating Massachusetts Bay's "Sea to Sea" charter of 1629.

9. *St. Louis Enquirer*, September 22, 1819, 3.

10. "Virginia Claims Pilgrims and Even Reaches Out for Wyoming," *Wyoming State Tribune*, April 21, 1936, 12; "'Pilgrim Dads' Landed in Virginia, He Finds," *Miles City Daily Star*, April 22, 1936, 1.

11. Francis Newton Thorpe, *The Federal and State Constitutions, Colonial Charters, and Other Organic Laws of the States, Territories, and Colonies Now or Heretofore Forming the United States of America*, vol. 3 (Washington, DC: Government Printing Office, 1909), 1847.

12. Oliver Perry Chitwood, *A History of Colonial America* (New York: Harper & Brothers Publishers, 1931), 98.

13. Francis Newton Thorpe, *The Federal and State Constitutions, Colonial Charters, and Other Organic Laws of the States, Territories, and Colonies Now or Heretofore Forming the United States of America*, vol. 5 (Washington, DC: Government Printing Office, 1909), 2534.

14. Ibid., 2762.

15. Thomas Jefferson, *Notes on the State of Virginia*, ed. David Waldstreicher (1787; repr., Boston: Bedford/St. Martin's, 2002), 81.

16. The best work on the Mason-Dixon line, from which this narrative is derived, is Edwin Danson's *Drawing the Line: How Mason and Dixon Surveyed the Most Famous Border in America* (New York: John Wiley & Sons, 2001).

17. The "English" empire transformed into the "British" empire in 1707 after the Act of Union blended England, Scotland, and Wales into the country of Great Britain. The addition of Ireland (reduced to Northern Ireland in 1922) via the 1801 Act of Union created the United Kingdom.

18. See Fred Anderson, *Crucible of War: The Seven Years' War and the Fate of Empire in British North America, 1754–1766* (New York: Vintage Books, 2000), 565–80, 634–37; Colin G. Calloway, *The Scratch of a Pen: 1763 and the Transformation of North America* (Oxford, UK: Oxford University Press, 2006), 92–100; R. Douglas Hurt, *The Ohio Frontier: Crucible of the Old Northwest, 1720–1830* (1996; repr., Bloomington: Indiana University Press, 1998), 55–56; Drake, *The Nation's Nature*, 115–18. In the afterword to the 2003 edition of *Changes in the Land*, William Cronon describes his futile attempts to track down examples of colonial outrage at the Proclamation Line, suggesting that the barrier caused little stir in the 1760s. See Cronon, *Changes in the Land*, 177–78. Although the Quebec Act of 1774 inspired anger in North America by extending the ethnically French colony's jurisdiction southward to the Ohio River, placing another objectionable barrier against trans-Appalachian growth, the American Revolution prevented the act's territorial clauses from coming into effect. See Meinig, *The Shaping of America*, vol. 1, 279–80, 288; Drake, *The Nation's Nature*, 119, 147.

19. Drake, *The Nation's Nature*, 70, 80, 112–13, 153–54, 158–59, 162–67, 194–96, 204.

20. Articles of Confederation, 1777, Article IX.

21. Quotation in Merrill Jensen, *The Articles of Confederation: An Interpretation of the Social-Constitutional History of the American Revolution, 1774–1781* (Madison: University of Wisconsin Press, 1963), 200.

22. *Journals of the Continental Congress*, vol. 11, June 25, 1778, 649, 650.

23. Gary Alden Smith, *State and National Boundaries of the United States* (Jefferson, NC: McFarland, 2004), 96–97; William A. Degregorio, *The Complete Book of U.S. Presidents* (1984; repr., New York: Wings Books, 1993), 106–107; Rachel M. Kochmann, *Presidents: Birthplaces, Homes, and Burial Sites: A Pictorial Guide* (1976; repr., Detroit Lakes, MN: Midwest Printing, 1993), 24.

24. "Jackson Issue Due a Kickin' Around," *News and Courier* (Charleston, South Carolina), August 16, 1979, 12-B.

25. "Old Hickory's Birthplace Settled in Football Fray," *Free Lance-Star* (Fredericksburg, Virginia), August 18, 1979, 15.

26. Jensen, *The Articles of Confederation*, 196–210, 238.

27. Drake, *The Nation's Nature*, 218–29.

28. Henry Gannett, *Boundaries of the United States and of the Several States and Territories, with an Outline of the History of All Important Changes of Territory* (Washington, DC: Government Printing Office, 1904), 9–10.

29. *Journals of the Continental Congress*, vol. 25, September 13, 1783, 558, 559–60.

30. Julian P. Boyd (ed.), *The Papers of Thomas Jefferson*, vol. 6, *21 May 1781–1 March 1784* (Princeton: Princeton University Press, 1952), 577–80; *Journals of the Continental Congress*, vol. 26, March 1, 1784, 112–17.

31. *Journals of the Continental Congress*, vol. 26, March 1, 1784, 118.

32. Julian P. Boyd (ed.), *The Papers of Thomas Jefferson*, vol. 1, *1760–1776* (Princeton: Princeton University Press, 1950), 363.

33. Boyd, *The Papers of Thomas Jefferson*, vol. 6, 590, 601–602.

34. Ibid., 581.

35. *Journals of the Continental Congress*, vol. 26, March 1, 1784, 118.

36. Boyd, *The Papers of Thomas Jefferson*, vol. 6, 603–605.

37. *Journals of the Continental Congress*, vol. 26, March 1, 1784, 118–20.

38. Ibid., April 23, 1784, 274–75.

39. Boyd, *The Papers of Thomas Jefferson*, vol. 6, 604, 613–16; *Journals of the Continental Congress*, vol. 26, April 23, 1784, 275–79. See also Robert F. Berkhofer, Jr., "Jefferson, the

Ordinance of 1784, and the Origins of the American Territorial System," *William and Mary Quarterly* 29, no. 2 (April 1972), 231–62; Richard P. McCormick, "The 'Ordinance' of 1784?," *William and Mary Quarterly* 50, no. 1 (January 1993), 112–22.

40. *Journals of the Continental Congress*, vol. 26, May 28, 1784, 446; Peter S. Onuf, *Statehood and Union: A History of the Northwest Ordinance* (Bloomington: Indiana University Press, 1987), 22–24; James C. Scott, *Seeing Like a State: How Certain Schemes to Improve the Human Condition Have Failed* (New Haven: Yale University Press, 1998), 49–51.

41. For more information on the early decades of the rectilinear land distribution system, see Malcolm J. Rohrbough, *The Land Office Business: The Settlement and Administration of American Public Lands, 1789–1837* (Belmont, CA: Wadsworth, 1990).

42. D. J. Waldie, *Holy Land: A Suburban Memoir* (1996; repr., New York: W. W. Norton, 2005), 3.

43. As with New York and Virginia, cession and completion of the transfer took some time in each case. Massachusetts gave up its interest to parts of present-day Illinois, Michigan, and Wisconsin in 1784, a deal completed in 1785. In 1786, Connecticut surrendered its strip of territory except present-day northeastern Ohio, which it held until 1800. North Carolina gave up its trans-Appalachian claim in 1790, and that area joined the Union as Tennessee in 1796. Georgia did not relinquish its holdings over most of present-day Alabama and Mississippi until 1802. See Hurt, *The Ohio Frontier*, 164–66; Dave Foster, *Tennessee: Territory to Statehood* (2000; repr., Johnson City, TN: Overmountain Press, 2002), 32–33; Buddy Sullivan, *Georgia: A State History* (Charleston, SC: Arcadia Publishing, 2003), 46–47.

44. Stanislaus Murray Hamilton, *The Writings of James Monroe, Including a Collection of His Public and Private Papers and Correspondence Now for the First Time Printed* (1898; repr., New York: AMS Press, 1969), 117–18.

45. Ibid., 126–27.

46. Ibid., 140–41.

47. The entire Northwest Ordinance of 1787 appears in Onuf, *Statehood and Union*, 60–64. For scholarly interpretations of the Northwest Ordinance, see Onuf, *Statehood and Union*, 59; Merrill Jensen, *The New Nation: A History of the United States during the Confederation, 1781–1789* (New York: Vintage Books, 1950), 354; Meinig, *The Shaping of America*, vol. 1, 391–92.

48. *Annals of the Congress of the United States*, 9th Cong., 1st sess. (January 29, 1806), 78; ibid., January 30, 1806, 79. See also Malcolm J. Rohrbough, *Trans-Appalachian Frontier: People, Societies, and Institutions, 1775–1850* (Bloomington: Indiana University Press, 2008), 121–25.

49. *The Debates and Proceedings in the Congress of the United States*, 16th Cong., 1st sess. (January 20, 1820), 213. See also ibid., January 17, 1820, 140; January 25, 1820, 239.

50. Ibid., January 19, 1820, 163.

51. Ibid., January 28, 1820, 310.

52. Ibid., January 25, 1820, 248.

53. Session Laws of Dakota Territory, 12th Legislative Session, 1877, iii.

54. Catherine Drinker Bowen, *Miracle at Philadelphia: The Story of the Constitutional Convention, May to September 1787* (New York: Little, Brown, 1966), 174.

55. James Madison, *Journal of the Federal Convention*, vol. 1, ed. E. H. Scott (Chicago: Albert, Scott, 1893), 63.

56. Drake, *The Nation's Nature*, 297–98.

57. Madison, *Journal of the Federal Convention*, vol. 1, 148, 169, 263.

58. Ibid., 195.

59. Ibid., 298.

60. Drake, *The Nation's Nature*, 304.

61. Madison, *Journal of the Federal Convention*, vol. 1, 327, 346.

62. For more on Morris's objections, see James Madison, *Journal of the Federal Convention*, vol. 2, ed. E. H. Scott (Chicago: Albert, Scott, 1893), 460, 631.

63. United States Constitution, 1787, Article IV, section 3.

64. The Constitution does not describe the legality of adding territory to the United States by negotiation or force either, a problematic issue beginning with the Louisiana Purchase and extending for over a century as the country grew larger. For a thorough investigation of this problem, see Gary Lawson and Guy Seidman, *The Constitution of Empire: Territorial Expansion and American Legal History* (New Haven: Yale University Press, 2004).

65. Earl S. Pomeroy, *The Territories and the United States, 1861–1890: Studies in Colonial Administration* (1947; repr., Seattle: University of Washington Press, 1969), 5.

66. Drake, *The Nation's Nature*, 275; Andrew R. L. Cayton, *The Frontier Republic: Ideology and Politics in the Ohio Country, 1780–1825* (Kent, OH: Kent State University Press, 1986), 34–35.

67. *The Debates and Proceedings in the Congress of the United States*, 1st Cong., 2nd sess. (April 2, 1790), 2269.

68. Ibid., May 26, 1790, 2286.

69. William Maclay, *The Journal of William Maclay, United States Senator from Pennsylvania, 1789–1791* (1890; repr., New York: Frederick Ungar, 1965), 378. Neither the proposal nor the discussion around it appears in the skimpy records of debate in the Senate that day. See *The Debates and Proceedings in the Congress of the United States*, 1st Cong., 3rd sess. (February 11, 1791), 1799.

70. Lowell H. Harrison, *Kentucky's Road to Statehood* (Lexington: University Press of Kentucky, 1992), 12.

71. Ibid., 7–9; Foster, *Tennessee*, 3.

72. Drake, *The Nation's Nature*, 257.

73. Forrest McDonald, *States' Rights and the Union: Imperium in Imperio, 1776–1876* (Lawrence: University Press of Kansas, 2000), 9–11.

74. Hurt, *The Ohio Frontier*, 249–54; Cayton, *The Frontier Republic*, 52, 68–80.

2. THE FAIREST PORTION OF OUR UNION

1. *Arkansas Gazette*, October 22, 1822, 3.

2. Peter J. Kastor, *The Nation's Crucible: The Louisiana Purchase and the Creation of America* (New Haven: Yale University Press, 2004), 15.

3. *Annals of the Congress of the United States*, 7th Cong., 2nd sess. (October 21, 1803), 1005–1006.

4. *Message of President Jefferson*, 8th Cong., 1st sess., October 17, 1803, American State Papers 01, Foreign Relations, vol. 1, 61.

5. *Annals of the Congress of the United States*, 8th Cong., 1st sess. (February 24, 1804), 1058.

6. Ibid., March 26, 1804, 1293–98; S. Charles Bolton, "Jeffersonian Indian Removal and the Emergence of Arkansas Territory," in *A Whole Country in Commotion: The Louisiana Purchase and the American Southwest*, ed. Patrick G. Williams, S. Charles Bolton, and Jeannie M. Whayne (Fayetteville: University of Arkansas Press, 2005), 80–81.

7. *Remonstrance of the People of Louisiana against the Political System Adopted by Congress for Them*, 8th Cong., 2nd sess., December 31, 1804, American State Papers 037, Miscellaneous, vol. 1, 396–99.

8. *Revision of the Political System Adopted for Louisiana*, 8th Cong., 2nd sess., January 25, 1805, American State Papers 037, Miscellaneous, vol. 1, 417–18.

9. *Remonstrance of the People of Louisiana*, December 31, 1804, 399.

10. Ibid., 400. This argument appeared in later congressional debates over the perils of subdividing the West. In 1856, for example, one wag suggested that the United States could through territorial policy create its own version of "the rotton-borough system of England, which they have found necessary to do away with." See *Congressional Globe*, 34th Cong., 1st sess. (June 23, 1856), 1447.

11. *Appendix to the Annals of the Congress of the United States*, 8th Cong., 1st sess., 1610.

12. Gerald T. Hanson and Carl H. Moneyhon, *Historical Atlas of Arkansas* (Norman: University of Oklahoma Press, 1989), 27.

13. Stephen Aron, *How the West Was Lost: The Transformation of Kentucky from Daniel Boone to Henry Clay* (Baltimore: Johns Hopkins University Press, 1996), 64–70; Dave Foster, *Tennessee: Territory to Statehood*, 2nd ed. (Johnson City, TN: Overmountain Press, 2002), 30; Malcolm J. Rohrbough, *Trans-Appalachian Frontier: People, Societies, and Institutions, 1775–1850* (Bloomington: Indiana University Press, 2008), 29–30. See also Stanislaus Murray Hamilton, ed., *The Writings of James Monroe, Including a Collection of His Public and Private Papers and Correspondence Now for the First Time Printed*, vol. 1 (1898; repr., New York: AMS Press, 1969), 89.

14. *Description of Louisiana*, 8th Cong., 1st sess., November 14, 1803, American State Papers 037, Miscellaneous, vol. 1, 344.

15. James D. Drake, *The Nation's Nature: How Continental Presumptions Gave Rise to the United States of America* (Charlottesville: University of Virginia Press, 2011), 233–34.

16. *Exploration of Louisiana*, 8th Cong., 1st sess., March 8, 1804, American State Papers 037, Miscellaneous, vol. 1, 390.

17. In "State Boundaries," *Arkansas Gazette*, January 22, 1820, 3.

18. *Description of Louisiana*, 8th Cong., 1st sess., November 14, 1803, American State Papers 037, Miscellaneous, vol. 1, 345.

19. In "State Boundaries," *Arkansas Gazette*, January 22, 1820, 3.

20. *St. Louis Enquirer*, December 1, 1819, 2. Taking into account the altered boundaries of the Louisiana Purchase following the 1819 treaty with Spain, the remaining land eventually became all or part of thirteen states, half the *Enquirer*'s proposed number.

21. *St. Louis Enquirer*, September 25, 1819, 3.

22. *Annals of the Congress of the United States*, 16th Cong., 1st sess. (December 29, 1819), 43.

23. *Revision of the Act "For the Government of the Territory of Missouri,"* 12th Cong., 2nd sess., January 29, 1813, American State Papers 038, Miscellaneous, vol. 2, 202.

24. *St. Louis Enquirer*, September 25, 1819, 3.

25. *St. Louis Enquirer*, September 4, 1819, 3. See also Walter B. Stevens, *Centennial History of Missouri (The Center State): One Hundred Years in the Union, 1820–1921* (Chicago: S. J. Clarke, 1921), 34.

26. *Annexation of a Part of the Territory of Nebraska to Kansas*, 35th Cong., 2nd sess., February 9, 1859, H. Misc. Doc. 50, serial 1016.

27. *Annexation of a Portion of the Territory of Arizona to the Territory of Utah*, 38th Cong., 2nd sess., February 24, 1865, H. Misc. Doc. 53, serial 1232. See also Malcolm L. Comeaux, "Attempts to Establish and Change a Western Boundary," *Annals of the Association of American Geographers* 72, no. 2 (June 1982), 254–71.

28. "States West of the Mississippi," in *Arkansas Gazette*, February 12, 1820, 3.

29. *Annals of the Congress of the United States*, 16th Cong., 1st sess. (January 14, 1820), 118.

30. *St. Louis Enquirer*, December 1, 1819, 2.

31. *Annals of the Congress of the United States*, 16th Cong., 1st sess. (December 29, 1819), 43.

32. See Hanson and Moneyhon, *Historical Atlas of Arkansas*, 27; Stevens, *Centennial History of Missouri*, 71; Milton D. Rafferty, *Historical Atlas of Missouri* (Norman: University of

Oklahoma Press, 1982), 2. A letter in the *National Intelligencer* (Washington, D.C.) described the boundaries of Arkansas in late 1824 and expressed the territory's remorse at not holding the Bootheel Region. The *Intelligencer*'s article was reprinted in the *Arkansas Gazette*, January 4, 1825, 3.

33. *Memorial of the General Assembly of Missouri*, February 28, 1831, 1.

34. *Documents Relating to the Extension of the Northern Boundary Line of the State of Missouri*, 24th Cong., 1st sess., February 26, 1836, S. Doc. 206, serial 281, 4.

35. Stephen Aron, *American Confluence: The Missouri Frontier from Borderland to Border State* (Bloomington: Indiana University Press, 2006), 229–33.

36. *Limits of Missouri*, 22nd Cong., 1st sess., July 14, 1832, H. Rep. 512, serial 228.

37. *In the Senate of the United States*, 24th Cong., 1st sess., March 16, 1836, S. Doc. 251, serial 281, 4.

38. *Western Boundary of Missouri*, 24th Cong., 2nd sess., January 17, 1837, H. Ex. Doc. 83, serial 303. For other political debates about the "Platte Purchase," see *In the Senate of the United States*, 23rd Cong., 1st sess., April 8, 1834, S. Doc. 263, serial 240; *Boundary Line between Land of the United States and Missouri*, 23rd Cong., 2nd sess., January 28, 1835, H. Doc. 107, serial 273; *Amendments—Constitution—Missouri*, 23rd Cong., 2nd sess., February 21, 1835, H. Doc. 162, serial 274; *Missouri—Extension of Western Boundary*, 24th Cong., 1st sess., February 26, 1836, H. Rep. 379, serial 294; Session Laws of the State of Missouri, 1837, Ninth General Assembly, 28–29. For a recent scholarly treatment of the boundary shift, see Christopher M. Paine, "The Platte Earth Controversy: What Didn't Happen in 1836," *Missouri Historical Review* 91, no. 1 (October 1996), 1–23.

39. See H. Jason Combs, "The Platte Purchase and Native American Removal," *Plains Anthropologist* 47, no. 182 (August 2002), 265–74.

40. *Missouri—Extension of Western Boundary*, 24th Cong., 1st sess., February 26, 1836, H. Rep. 379, serial 294, 5.

41. *In the Senate of the United States*, 36th Cong., 1st sess., January 26, 1860, S. Misc. Doc. 9, serial 1038.

42. *Memorial of the Legislature of Missouri, relative to the Extension of the Southwestern Boundary of the State, and the Extinguishments of the Indian Title to Certain Lands*, 31st Cong., 1st sess., February 25, 1850, H. Misc. Doc. 34, serial 581.

43. *Congressional Globe*, 28th Cong., 2nd sess. (February 10, 1845), 269. See also ibid., February 11, 1845, 273–74.

44. D. W. Meinig, *The Shaping of America: A Geographical Perspective on 500 Years of History*, vol. 2, *Continental America, 1800–1867* (New Haven: Yale University Press, 1993), 437.

45. *Congressional Globe*, 29th Cong., 1st sess. (June 8, 1846), 938.

46. *Appendix to the Congressional Globe*, 29th Cong., 1st sess. (June 8, 1846), 668–69. See also *Congressional Globe*, 29th Cong., 1st sess. (June 8, 1846), 938–41.

47. "The Admission of Iowa," *Burlington Hawk-Eye*, March 6, 1845, 2. See also "Who Is the Wolf?," *Burlington Hawk-Eye*, March 13, 1845, 2; "To the People of Iowa Territory," *Burlington Hawk-Eye*, June 12, 1845, 1.

48. See Franklin K. van Zandt, *Boundaries of the United States and the Several States*, Geological Survey Bulletin 1212 (Washington, DC: Government Printing Office, 1966), 212–13.

49. *Western Boundary of Iowa*, 34th Cong., 1st sess., July 28, 1856, H. Misc. Doc. 139, serial 867. See also *Iowa Northern Boundary*, 32nd Cong., 1st sess., February 12, 1852, H. Ex. Doc. 662, serial 694, 2; *Annals of Iowa* 4, no. 8 (January 1901), 609.

50. *Western and Northern Boundary of Iowa*, 34th Cong., 1st sess., August 14, 1856, H. Rep. 347, serial 870. See also *Resolution of the General Assembly of the State of Iowa, in Favor of the*

Enlargement of the Boundaries of that State, 34th Cong., 3rd sess., January 28, 1857, S. Misc. Doc. 14, serial 890.

51. *St. Louis Enquirer*, December 4, 1819, 2.

52. *Annals of the Congress of the United States*, 16th Cong., 1st sess. (January 13, 1820), 86; ibid., January 14, 1820, 110, 118.

53. Ibid., January 25, 1820, 239.

54. Ibid., January 14, 1820, 107.

55. Ibid., January 18, 1820, 158.

56. Ibid., February 1, 1820, 337.

57. Ibid., January 21, 1820, 232–33.

58. Ibid., February 3, 1820, 363.

59. See ibid., February 15, 1820, 417.

60. Ibid., February 16, 1820, 424. See also ibid., February 17, 1820, 426–28; ibid., March 2, 1820, 469.

61. Ibid., March 2, 1820, 1580.

62. Ibid., January 17, 1820, 135.

63. Ibid., January 19, 1820, 184.

64. Ibid., March 2, 1820, 1578.

65. Ibid., February 17, 1820, 426–28; ibid., March 2, 1820, 1587.

66. *Message from the President of the United States, Notifying the House of Representatives of His Approval of the Bill "to Establish the Territorial Government of Oregon,"* 30th Cong., 2nd sess., December 6, 1848, H. Ex. Doc. 3, serial 538.

67. See *Resolutions of the Legislature of Rhode Island, Requesting Their Representatives to Vote against the Repeal of the Missouri Compromise*, 33rd Cong., 1st sess., January 30, 1854, H. Misc. Doc. 10, serial 741; *Resolutions of the Legislature of New York of the Subject of the Proposed Organization of the Territory of Nebraska*, 33rd Cong., 1st sess., February 17, 1854, H. Misc. Doc., 16, serial 741; *Resolutions of the Legislature of Massachusetts, Remonstrating against the Passage of the Nebraska Bill*, 33rd Cong., 1st sess., February 24, 1854, H. Misc. Doc. 20, serial 741; *Resolutions of the General Assembly of Connecticut, Remonstrating against the Passage of the Bill Organizing the Territories of Kansas and Nebraska, and against the Repeal of the Missouri Compromise*, 33rd Cong., 1st sess., May 22, 1854, H. Misc. Doc, 77, serial 741; *Resolutions of the Legislature of Connecticut, in Favor of Restoring the Missouri Compromise to Full Force and Effect; in Favor of an Amendment to the Fugitive Slave Law, to Secure to Every Person Claimed under Said Law the Right of Trial by Jury; and Censuring the Hon. Isaac Toucey for His Vote on the Final Passage of the Bill for the Organization of the Territories of Kansas and Nebraska*, 33rd Cong., 1st sess., July 6, 1854, S. Misc. Doc. 70, serial 705; *Resolutions of the Legislature of Rhode Island, on the Subject of the Nebraska Bill*, 33rd Cong., 1st sess., July 15, 1854, H. Misc. Doc. 94, serial 741; *Resolutions of the General Assembly of the State of Connecticut, Protesting against the Repeal of the Missouri Compromise, and Urging the Repeal of the Fugitive-Slave Law of 1850*, 33rd Cong., 1st sess., July 25, 1854, H. Misc. Doc. 96, serial 741; *Resolutions of the Legislature of Michigan Respecting Slavery*, 33rd Cong., 2nd sess., February 5, 1855, S. Misc. Doc. 11, serial 772; *Resolutions of the Legislature of Massachusetts, concerning the Repeal of the Missouri Compromise*, 34th Cong., 1st sess., March 7, 1856, H. Misc. Doc. 41, serial 866; *Resolutions of the Legislature of Ohio, in Favor of the Prohibition of Slavery in the Territories of the United States, and the Immediate Admission of Kansas into the Union as a State*, 34th Cong., 1st sess., May 5, 1856, S. Misc. Doc. 49, serial 835; *Resolutions of the Legislature of New Hampshire, Relative to Slavery in the Territories, and the Repeal of the Missouri Compromise*, 34th Cong., 1st sess., May 25, 1856, H. Misc. Doc. 115, serial 867; *Resolutions of the Legislature of the State of Connecticut, Instructing the Senators and Requesting the Representatives of that State in Congress to Vote against the Admission of Another Slaveholding State into the Union*, 35th Cong., 1st sess., March 8, 1858, S.

Misc. Doc. 188, serial 936; *Resolutions of the Legislature of the State of Massachusetts, in Relation to the Decision of the Supreme Court in the Case of Scott versus Sanford*, 35th Cong., 1st sess., April 21, 1858, serial 963.

68. *Resolutions of the Legislature of Maine, against the Passage of the Nebraska Bill*, 33rd Cong., 1st sess., March 6, 1854, H. Misc. Doc. 23, serial 741; *Resolutions of the Legislature of the State of Maine, against Slavery in the United States, in Favor of the Abolition of Slavery in the District of Columbia, and the Repeal of the Fugitive Slave Law*, 34th Cong., 1st sess., January 3, 1856, S. Misc. Doc. 11, serial 835; *Resolutions of the Legislature of Maine, Relative to Slavery*, 34th Cong., 1st sess., June 20, 1856, H. Misc. Doc. 121, serial 867; *Resolutions of the Legislature of the State of Maine, Relative to the Decision of the Supreme Court of the United States in the Case of Dred Scott*, 35th Cong., 1st sess., January 20, 1858, H. Misc. Doc. 31, serial 961; *Resolutions of the Legislature of the State of Maine, in Relation to Kansas and Slavery*, 35th Cong., 1st sess., March 25, 1858, S. Misc. Doc. 206, serial 936.

69. *Resolutions of the Legislature of Missouri Relative to the Subject of Slavery in the Organization of New Territories or States out of Territory Now Belonging to, or Hereafter to Be Acquired by, the United States*, 30th Cong., 1st sess., December 21, 1847, H. Misc. Doc. 2, serial 523.

70. *Resolutions of the Legislature of Virginia, in Relation to Slavery*, 30th Cong., 2nd sess., February 5, 1849, S. Misc. Doc. 48, serial 533.

71. *Resolutions of the Legislature of Arkansas, Approving the Kansas and Nebraska Act, and the Repeal of the Missouri Compromise*, 33rd Cong., 2nd sess., February 9, 1855, H. Misc. Doc. 26, serial 807; *Resolutions of the Legislature of the State of Texas, Relative to the Acts of Congress Admitting California into the Union, Fixing the Boundary of Texas, Establishing Territorial Governments in Utah and New Mexico, the Fugitive-Slave Law, and the Kansas-Nebraska Act*, 34th Cong., 1st sess., February 6, 1856, S. Misc. Doc. 15, serial 835; *Joint Resolutions of the Legislature of the State of Tennessee, Instructing Her Senators and Requesting Her Representatives to Vote for the Admission of Kansas under the Lecompton Constitution, and in Reference to the Hon. John Bell*, 35th Cong., 1st sess., March 15, 1858, H. Misc. Doc. 81, serial 963.

72. *Resolutions of the General Assembly of Missouri, on the Subject of Slavery in the Territories, District of Columbia, and States*, 31st Cong., 1st sess., December 31, 1849, H. Misc. Doc. 5, serial 581.

73. *In the Senate of the United States*, 36th Cong., 2nd sess., December 28, 1860, S. Misc. Doc. 7, serial 1089. For another peace proposal involving the Missouri Compromise line, see "Domestic Intelligence," *Harper's Weekly*, March 9, 1861, 151.

74. For a comprehensive interpretation of the creation and consequences of the Missouri Compromise, see Robert Pierce Forbes, *The Missouri Compromise and its Aftermath: Slavery and the Meaning of America* (Chapel Hill: University of North Carolina Press, 2007).

75. Ibid., 226.

76. Kastor, *The Nation's Crucible*, 228.

77. Woodrow Wilson, "The Making of the Nation," *Atlantic Monthly* 80, no. 477 (July 1897), 9.

78. James N. Woolworth, *Nebraska in 1857* (New York: A. S. Barnes, 1857), 14.

3. ON THE EXTREME FRONTIER

1. *Resolutions of a Number of Citizens of Washington County, Arkansas, in Relation to the Defence of the Western Frontier*, 25th Cong. 2nd sess., April 5, 1838, S. Doc. 358, serial 317, 1.

2. See Henry Thompson Malone, *Cherokees of the Old South: A People in Transition* (Athens: University of Georgia Press, 1956); David H. Corkran, *The Cherokee Frontier: Conflict and Survival, 1740–1762* (Norman: University of Oklahoma Press, 1962); Grace Steele Woodward,

The Cherokees (Norman: University of Oklahoma Press, 1963); Arthur H. DeRosier, *The Removal of the Choctaw Indians* (Knoxville: University of Tennessee Press, 1970); Jesse O. McKee and John A. Schlenker, *The Choctaws: Cultural Evolution of a Native American Tribe* (Jackson: University Press of Mississippi, 1980); Carolyn Reeves, *The Choctaw before Removal* (Jackson: University Press of Mississippi, 1985).

3. *Annals of the Congress of the United States*, 8th Cong., 1st sess., 1299–1300; S. Charles Bolton, "Jeffersonian Indian Removal and the Emergence of Arkansas Territory," in *A Whole Country in Commotion: The Louisiana Purchase and the American Southwest*, ed. Patrick G. Williams, S. Charles Bolton, and Jeannie M. Whayne (Fayetteville: University of Arkansas Press, 2005), 77–90.

4. Charles J. Kappler, ed., *Indian Affairs: Laws and Treaties* (Washington, DC: Government Printing Office, 1904), 96.

5. See Robert A. Myers, "Cherokee Pioneers in Arkansas: The St. Francis Years, 1785–1813," *Arkansas Historical Quarterly* 56, no. 2 (Summer 1997): 127–57.

6. Gerald T. Hanson and Carl H. Moneyhon, *Historical Atlas of Arkansas* (Norman: University of Oklahoma Press, 1989), 19.

7. "Organic Law," *Arkansas Gazette*, November 27, 1819, 2.

8. *Arkansas Gazette*, 11 March 1820, 4.

9. "Indian Hostilities," *Arkansas Gazette*, December 30, 1820, 3; *Arkansas Gazette*, January 13, 1821, 4. See also Robert W. Frazer, *Forts of the West: Military Forts and Presidios and Posts Commonly Called Forts West of the Mississippi River to 1898* (1965; repr., Norman: University of Oklahoma Press, 1972, 16–17.

10. Ina Gabler, "Lovely's Purchase and Lovely County," *Arkansas Historical Quarterly* 19, no. 1 (Spring 1960): 31–33; Brad Agnew, "The Cherokee Struggle for Lovely's Purchase," *American Indian Quarterly* 2, no. 4 (Winter 1975), 347.

11. See Edwin C. Bearss and Arrell M. Gibson, *Fort Smith: Little Gibraltar on the Arkansas*, 2nd ed., (Norman: University of Oklahoma Press, 2002), 20–65; Kathleen DuVal, *The Native Ground: Indians and Colonists in the Heart of the Continent* (Philadelphia: University of Pennsylvania Press, 2006), 208–26.

12. See S. Charles Bolton, *Territorial Ambition: Land and Society in Arkansas, 1800–1840* (Fayetteville: University of Arkansas Press, 1993), 64–66.

13. Quotation in *Arkansas Gazette*, October 21, 1820, 3.

14. *Arkansas Gazette*, October 7, 1820, 3. See also ibid., December 2, 1820, 3.

15. "Choctaw Treaty," *Arkansas Gazette*, December 16, 1820, 2.

16. "Choctaw Treaty," *Arkansas Gazette*, December 9, 1820, 3.

17. "Choctaw Treaty," *Arkansas Gazette*, December 16, 1820, 3.

18. *Arkansas Gazette*, January 6, 1821, 3.

19. Ibid., January 13, 1821, 4.

20. Ibid., January 20, 1821, 3.

21. Quotation in "Arkansas Territory," *Arkansas Gazette*, February 3, 1821, 2.

22. *Arkansas Gazette*, January 13, 1821, 4.

23. Ibid., January 27, 1821, 2, 4. No record of this resolution appears in the session laws of Louisiana, although the *Gazette* reported that it passed with an amendment on January 10, 1821, indicating that it might have passed the state house of representatives but not the state senate. The Louisiana legislature did resolve to have Louisiana's geometric boundaries between the Sabine and Mississippi Rivers surveyed and marked in early 1820. See Session Laws of Louisiana, 4th General Assembly, 2nd sess., March 16, 1820, 126.

24. "Choctaw Treaty," *Arkansas Gazette*, January 6, 1821, 3.

25. "The Choctaw Treaty Ratified," *Arkansas Gazette*, February 24, 1821, 3.

26. "Choctaw Treaty," *Arkansas Gazette*, April 7, 1821, 3.

27. *Arkansas Gazette*, March 10, 1821, 3.

28. Ibid., March 17, 1821, 3.

29. Ibid., June 2, 1821, 3. The *Gazette* nonetheless reprinted an article from the *St. Louis Enquirer* describing the glories of Mexican Texas on June 23, 1821 (3), but briefly noted on June 25, 1822, that those who had moved south of the border were "heartily sick of Texas" (3). See also *Arkansas Gazette*, October 15, 1822, 3.

30. *Execution of the Treaty with the Choctaws, of the 18th October, 1820*, 17th Cong., 2nd sess., February 6, 1823, American State Papers 8, Indian Affairs, vol. 2, 394.

31. See *Arkansas Gazette*, May 12, 1821, 3; ibid., May 19, 1821, 3; "Arkansas," *St. Louis Enquirer*, June 23, 1821, 2; *Arkansas Gazette*, June 30, 1821, 3.

32. "The Choctaw Treaty—Again," *Arkansas Gazette*, September 1, 1821, 3.

33. *Arkansas Gazette*, May 28, 1822, 3.

34. W. David Baird, "Arkansas's Choctaw Boundary: A Study of Justice Delayed," *Arkansas Historical Quarterly* 28, no. 3 (Autumn 1969): 206–207.

35. *Arkansas Gazette*, April 14, 1821, 1. See also "Movements of the Cherokees," *Arkansas Gazette*, April 8, 1823, 3.

36. *Arkansas Gazette*, June 2, 1821, 3. In the late spring of 1821, Congress considered a bill to provide public land elsewhere in Arkansas to settlers displaced by the Cherokee grant in 1817, but the idea stalled with the end of the legislative session that summer. See *Arkansas Gazette*, June 9, 1821, 3.

37. "Lovely's Purchase," *Arkansas Gazette*, July 30, 1822, 3.

38. Gabler, "Lovely's Purchase and Lovely County," 33.

39. *Arkansas Gazette*, November 12, 1822, 4.

40. In the summer of 1823, Arkansas Territory experienced a brief panic when rumors of the loss of Lovely's Purchase to the Cherokees surfaced. Indeed, the federal government revised the western boundary of the Cherokee reserve, but it affected only a small portion of the Purchase in present-day western Arkansas. See "Unwelcome News," *Arkansas Gazette*, June 17, 1823, 3; *Arkansas Gazette*, July 1, 1823, 3; ibid., July 12, 1823, 4; ibid., November 4, 1823, 3.

41. "Trial for Murder," *Arkansas Gazette*, April 29, 1823, 3. See also Joseph Patrick Key, "Indians and Ecological Conflict in Territorial Arkansas," Arkansas Historical Quarterly 59, no. 2 (Summer 2000), 143–44; Stanley Hoig, *The Cherokees and Their Chiefs: In the Wake of Empire* (Fayetteville: University of Arkansas Press, 1998), 135–36.

42. *Arkansas Gazette*, May 13, 1823, 4. See also ibid., May 6, 1823, 3.

43. *Arkansas Gazette*, May 20, 1823, 3; ibid., May 27, 1823, 3. See also ibid., January 13, 1824, 3.

44. *Message from the President of the United States, Transmitting Information in Relation to the Western Boundary Line of the Territory of Arkansas*, 18th Cong., 1st sess., February 23, 1824, H. Doc. 84, serial 97.

45. *In the Senate of the United States*, 18th Cong., 1st sess., March 23, 1824, S. Doc. 58, serial 91, 1–2. Delegate Conway expressed a similar opinion in a letter to the *Arkansas Gazette*, August 17, 1824, 3.

46. *Annals of the Congress of the United States*, 18th Cong., 1st sess., 3241–42.

47. "From Fort Smith," *Arkansas Gazette*, April 13, 1824, 3; Frazier, *Forts of the West*, 18, 120.

48. *Extinguishment of Title to Quapaw Indians to Land in Arkansas*, 18th Cong., 1st sess., December 8, 1823, American State Papers 31, Public Lands, vol. 4, 1; "Treaty with the Quapaws," *Arkansas Gazette*, July 13, 1824, 3; "The Quapaw Treaty," *Arkansas Gazette*, November

23, 1824, 3; *Treaty with the Quapaw Tribe*, 18th Cong., 2nd sess., January 19, 1825, American State Papers 8, Indian Affairs, vol. 2, 529–31; "Ratification of the Quapaw and Choctaw Treaties," *Arkansas Gazette*, March 29, 1825, 3; *Arkansas Gazette*, April 5, 1825, 1.

49. *Lovely's Purchase—Arkansas*, 20th Cong., 1st sess., April 30, 1828, H. Doc. 268, serial 174, 10. See also "The Cherokee Boundary," *Arkansas Gazette*, January 19, 1825, 3; *Arkansas Gazette*, February 22, 1825, 3.

50. *Treaty with the Choctaws*, 18th Cong., 2nd sess., January 27, 1825, American State Papers 8, Indian Affairs, vol. 2, 549.

51. Ibid., 550–54.

52. Ibid., 555–57.

53. Ibid., 557–58; "New Choctaw Treaty," *Arkansas Gazette*, March 3, 1825, 3.

54. *Arkansas Gazette*, April 5, 1825, 1.

55. "Ratification of the Quapaw and Choctaw Treaties," *Arkansas Gazette*, March 29, 1825, 3. The Choctaws specifically insisted on the permanence of the new boundary during negotiations; see *Encroachments on Choctaw Lands*, 19th Cong., 2nd sess., January 8, 1827, H. Doc. 39, serial 150, 6.

56. *Arkansas Gazette*, April 19, 1825, 3. By early 1826 the *Gazette* reported that Conway had secured permission to let the American settlers remain in their homes on the Choctaw reserve for another year. See "Important to Persons Residing on the Choctaw Lands," *Arkansas Gazette*, February 7, 1826, 3.

57. Clarence Carter, ed., *The Territorial Papers of the United States*, vol. 20, *The Territorial Papers of Arkansas, 1825–1829* (Washington, DC: Government Printing Office, 1954), 164, 328.

58. *Arkansas Gazette*, December 20, 1825, 3. At the same time, the federal government negotiated a treaty with the Osages to remove them from the last portions of land they claimed in Missouri and Arkansas to a reserve beyond the western political borders of both polities. See *Treaties with the Osage, Kanzas, and Shawanee Tribes*, 19th Cong., 1st sess., December 14, 1825, American State Papers 8, Indian Affairs, vol. 2, 588–89, 591–92.

59. *Arkansas Gazette*, May 10, 1825, 3; "Western Boundary of Arkansas," *Arkansas Gazette*, May 24, 1825, 3.

60. "Lovely's Purchase," *Arkansas Gazette*, July 4, 1826, 3; *Arkansas Gazette*, November 28, 1826, 3.

61. Ina Gabler, "Lovely's Purchase and Lovely County," *Arkansas Historical* Quarterly 19, no. 1 (Spring 1960), 34. See also *Lovely's Purchase—Arkansas*, April 30, 1828, 21.

62. *Arkansas Gazette*, April 10, 1827, 3.

63. *Encroachments on Choctaw Lands*, January 8, 1827, 7–8. See also *Settlers on the Choctaw Lands in Arkansas*, 19th Cong., 2nd sess., February 27, 1827, American State Papers 31, Public Lands, vol. 4, 958–60.

64. "Intruders on Choctaw Lands," *Arkansas Gazette*, June 26, 1827, 2.

65. *Gales & Seaton's Register of Debates in Congress*, 19th Cong., 2nd sess. (January 23, 1827), 73.

66. *Arkansas Gazette*, June 12, 1827, 3.

67. Agnew, "The Cherokee Struggle for Lovely's Purchase," 354.

68. *Lovely's Purchase—Arkansas*, April 30, 1828, 31.

69. See Myers, "Cherokee Pioneers in Arkansas," 150–53, 157; Key, "Indians and Ecological Conflict," 143–44; Hoig, *The Cherokees and Their Chiefs*, 106; DuVal, *The Native Ground*, 217–18.

70. Gabler, "Lovely's Purchase and Lovely County," 34–35, 36–37; Agnew, "The Cherokee Struggle for Lovely's Purchase," 353–54.

71. "Very Late from Washington," *Arkansas Gazette*, June 11, 1828, 3; *Arkansas Gazette*, June 18, 1828, 3.

72. "Cherokee Treaty," *Arkansas Gazette*, June 25, 1828, 1.

73. Quotation in Gabler, "Lovely's Purchase and Lovely County," 38.

74. "Cherokee Treaty," *Arkansas Gazette*, June 25, 1828, 1.

75. *Arkansas Gazette*, March 11, 1829, 3.

76. Matthew A. Byron, "Crime and Punishment: The Impotency of Dueling Laws in the United States" (PhD diss., University of Arkansas, 2008), 98–100.

77. See *Arkansas Gazette*, September 9, 1828, 3; September 16, 1828, 3; October 14, 1828, 3; November 25, 1828, 3; December 2, 1828, 3; January 13, 1829, 3.

78. Ibid., July 2, 1828, 3. See also Bolton, *Territorial Ambition*, 66.

79. "Settlers' Grants," *Arkansas Gazette*, July 23, 1828, 3.

80. *Arkansas Gazette*, October 7, 1828, 3; "Proclamation," *Arkansas Gazette*, November 25, 1828, 4.

81. "Cherokee Treaty," *Arkansas Gazette*, June 25, 1828, 3; *Arkansas Gazette*, July 16, 1828, 3; October 28, 1828, 3.

82. "Cherokee Treaty," *Arkansas Gazette*, June 25, 1828, 1. See also *Arkansas Gazette*, December 23, 1828, 3.

83. "Cherokee Improvements," *Arkansas Gazette*, March 13, 1829, 3.

84. "Donations to Settlers," *Arkansas Gazette*, September 30, 1828, 3.

85. *Arkansas Gazette*, January 13, 1829, 3; "Highly Important," *Arkansas Gazette*, January 27, 1829, 3.

86. "Miller Claims—All Aback Again," *Arkansas Gazette*, February 25, 1829, 3; *Arkansas Gazette*, March 25, 1829, 2.

87. "Donation Claims—Again," *Arkansas Gazette*, April 29, 1829, 3.

88. See *Application of Arkansas for a Donation of Land to Settlers Who Were Removed by the Treaty with the Choctaw Indians*, 24th Cong., 1st sess., December 16, 1835, American State Papers 35, Public Lands, vol. 8, 245.

89. *Appendix to Gales & Seaton's Register of Debates in Congress*, 21st Cong., 1st sess., xxxii–xxxiii.

90. See *Memorial of the Inhabitants of Chester County, Pennsylvania, Praying that the Act Passed at the Last Session, for the Removal of the Indians beyond the Mississippi, May Be Repealed, and that No Treaty Made under that Law Be Confirmed*, 21st Cong., 2nd sess., December 31, 1830, S. Doc. 16, serial 203; *Memorial of Inhabitants of Woodbridge, New Jersey, for Protection to the Indians and Repeal of the Act of Last Session on the Subject*, 21st Cong., 2nd sess., January 14, 1831, S. Doc. 31, serial 203; *Memorial of Inhabitants of Andover, Essex Co., Massachusetts, that the Indians Be Protected in Their Rights, &c.*, 21st Cong., 2nd sess., January 21, 1831, S. Doc. 34, serial 203; *Massachusetts*, 21st Cong., 2nd sess., February 14, 1831, H. Doc. 106, serial 209-1; *Memorial from Sundry Citizens of Pennsylvania, Praying that the Cherokee Indians May Be protected in Their Rights, &c.*, 22nd Cong., 1st sess., February 14, 1832, S. Doc. 61, serial 213. Not all petitions expressed sympathy for the Native people; see *Maine*, 21st Cong., 2nd sess., February 9, 1831, H. Doc. 89, serial 208; *Inhabitants of Surry, State of Maine*, 21st Cong., 2nd sess., February 21, 1831, H. Doc. 138, serial 209-1.

91. See *Application of Arkansas for a Donation of Land to Actual Settlers for the Protection of the Frontiers*, 22nd Cong., 1st sess., December 19, 1831, American State Papers 33, Public Lands, vol. 6, 318; *Application of Arkansas for a Donation of Land to Actual Settlers on the Western Frontier of that State*, 23rd Cong., 1st sess., December 11, 1833, American State Papers 33, Public Lands, vol. 6, 623.

92. William E. Unrau, *The Rise and Fall of Indian Country, 1825–1855* (Lawrence: University Press of Kansas, 2007), 2–3.

93. *Application of Arkansas for a Removal of the Troops from Fort Gibson to Fort Smith*, 23rd Cong., 1st sess., January 13, 1834, American State Papers 20, Military Affairs, vol. 5, 242–43. See also *Defence of Frontiers of Arkansas*, 23rd Cong., 1st sess., February 10, 1834, H. Rep. 255, serial 261, 1; *On the Expediency of Removing the Troops from Fort Gibson to the Western Boundary of Arkansas*, 24th Cong., 1st sess., March 21, 1836, American State Papers 21, Military Affairs, vol. 6, 181–84.

94. *Memorial, Choctaw Indians—Boundary of Arkansas*, 24th Cong., 1st sess., April 18, 1836, H. Doc. 220, serial 291.

95. Grant Foreman, ed., *A Traveler in Indian Territory: The Journal of Ethan Allen Hitchcock, Late Major-General in the United States Army* (Cedar Rapids, IA: Torch Press, 1930), 23. Hitchcock traveled from Fort Smith through Indian Territory in the late weeks of 1841.

96. *Resolutions of a Number of Citizens of Washington County, Arkansas*, April 5, 1838, 1. Settlers on the western edges of other states felt similarly—see, for example, "The Protection of the Frontier" in *The Far West* (Liberty, Missouri), August 11, 1836, 3. See also Richard L. Trotter, "For the Defense of the Western Border: Arkansas Volunteers on the Indian Frontier, 1846–1847," *Arkansas Historical Quarterly* 60, no. 4 (Winter 2001), 394–410.

97. *Report of the Secretary of War, with Plans for the Defence and Protection of the Western Frontiers of the United States, and Statements of the Number of Indians and Warriors on Those Frontiers*, 25th Cong., 2nd sess., January 3, 1838, American State Papers 22, Military Affairs, vol. 7, 778, 779.

98. *Report of the Secretary of War*, January 3, 1838, 782. The residents of Washington County, Arkansas, expressed their strong support for the Cross plan in their request for frontier defenses; see *Resolutions of a Number of Citizens of Washington County, Arkansas*, April 5, 1838, 2.

99. *Report of the Secretary of War*, January 3, 1838, 784–86.

100. John W. Morris and Edwin C. McReynolds, *Historical Atlas of Oklahoma* (Norman: University of Oklahoma Press, 1965), 19; *Boundary of Choctaw and Chickasaw Country*, 40th Cong., 2nd sess., April 21, 1868, H. Ex. Doc. 359, serial 1343. See also Hanson and Moneyhon, *Historical Atlas of Arkansas*, 28.

101. *Memorial of the Choctaw Nation of Indians*, 41st Cong., 2nd sess., March 18, 1870, S. Misc. Doc. 90, serial 1408, 12.

102. Morris and McReynolds, *Historical Atlas of Oklahoma*, 19. See also *In the Senate of the United States*, 45th Cong., 3rd sess., February 4, 1879, S. Rep. 714, serial 1838; Baird, "Arkansas's Choctaw Boundary," 212–22.

103. *Cherokee Advocate*, January 20, 1892, 2.

104. United States Statutes at Large, Public Law 67, 58th Cong., 3rd sess. (February 10, 1905), 714–15; Morris and McReynolds, *Historical Atlas of Oklahoma*, 55.

105. Session Laws of Arkansas, 35th General Assembly, 1st sess., January 25, 1905, 839–40; ibid., January 30, 1905, 837–38.

4. BLOOD WILL BE SHED

1. *Boundary—Missouri and Iowa*, 26th Cong., 1st sess., December 31, 1839, H. Ex. Doc. 36, serial 364, 3.

2. See Elliott West, "Lewis and Clark: Kidnappers," in *A Whole Country in Commotion: The Louisiana Purchase and the American Southwest*, ed. Patrick G. Williams, S. Charles Bolton, and Jeannie M. Whayne (Fayetteville: University of Arkansas Press, 2005), 3–20.

3. Zebulon M. Pike, *An Account of Expeditions to the Sources of the Mississippi, and through the Western Parts of Louisiana, to the Sources of the Arkansaw, Kans, La Platte, and Pierre Jaun, Rivers; Performed by Order of the Government of the United States during the Years 1805, 1806, and 1807* (Philadelphia: C. & A. Conrad, 1810), 4.

4. *The History of Van Buren County, Iowa, Containing a History of the County, its Cities, Towns, &c.* (Chicago: Western Historical Company, 1878), 428–29.

5. Pike, *An Account of the Expeditions to the Sources of the Mississippi*, 4.

6. Stephen Aron, *American Confluence: The Missouri Frontier from Borderland to Border State* (Bloomington: Indiana University Press, 2006), 115.

7. *The Osages*, 11th Cong., 2nd sess., March 14, 1810, American State Papers 07, Indian Affairs, vol. 1, 766.

8. *The Osages*, 11th Cong., 2nd sess., January 16, 1810, American State Papers 07, Indian Affairs, vol. 1, 763; Aron, *American Confluence*, 139–48; Robert W. Frazer, *Forts of the West: Military Forts and Presidios and Posts Commonly Called Forts West of the Mississippi River to 1898* (1965; repr., Norman: University of Oklahoma Press, 1972), 75–76.

9. Landon Y. Jones Jr., "The Council that Changed the West: William Clark at Portages des Sioux," *Gateway Heritage* 24, nos. 2 & 3 (Fall 2003 & Winter 2004): 88–95; Aron, *American Confluence*, 156–58.

10. *Boundary between Missouri and Iowa*, 25th Cong., 3rd sess., January 30, 1839, H. Doc. 128, serial 347, 3. See also Frank E. Landers, "The Southern Boundary of Iowa," *Annals of Iowa* 1, no. 8 (January 1895): 642–43.

11. See Milton D. Rafferty, *Historical Atlas of Missouri* (Norman: University of Oklahoma Press, 1982), 2.

12. "Constitution of the State of Missouri," *Arkansas Gazette*, 23 September 1820, 1.

13. *Treaties with the Ioway, Sac, and Fox Tribes*, 18th Cong., 2nd sess., December 15, 1824, American State Papers 08, Indian Affairs, vol. 2, 525. See also *Proposition to Extinguish Indian Title to Lands in Missouri*, 18th Cong., 1st sess., May 14, 1824, American State Papers 08, Indian Affairs, vol. 2.

14. Thomas M. Spencer, "'Demand Nothing but What Is Strictly Right and Submit to Nothing that Is Wrong': Governor Lilburn Boggs, Governor Robert Lucas, and the Honey War of 1839," *Missouri Historical Review* 103, no. 1 (October 2008): 36.

15. *Memorial of the Legislature of Missouri, Praying That Improvements May Be Made in the Navigation of the Mississippi River; That an Alteration Be Made in the Northern Boundary Line of That State; and That Certain Indian lands Be Purchased by the United States*, 20th Cong., 2nd sess., S. Doc. 88, serial 182.

16. *Memorial of the General Assembly of Missouri, That the N. and N.W. Boundary May Be Enlarged, and a Mounted Force Granted for the Protection of the Frontier of the State, and Its Trade with Mexico and the Indians*, 21st Cong., 2nd sess., February 28, 1831, S. Doc. 71, serial 204, 4.

17. See *In the Senate of the United States*, 23rd Cong., 1st sess., April 8, 1834, S. Doc. 263, serial 240, 2; *Northern Boundary of Missouri*, 25th Cong., 2nd sess., April 6, 1838, H. Rep. 768, serial 335. Technically not part of an Indian reserve, the "half-breed" lands were opened to settlement in the late 1830s by both Wisconsin and Iowa Territories. See Session Laws of Iowa, 1st Territorial Assembly, January 24, 1839, 224–25; ibid., January 25, 1839, 225; "Trial Relative to the Half Breed Tract," *Burlington Hawk-Eye*, April 16, 1846, 1.

18. "Northern Boundary Line of Missouri," *Missouri Argus*, February 17, 1837, 1.

19. Session Laws of Missouri, 9th General Assembly, 1st session, December 21, 1836, 26–28.

20. *Missouri Argus*, August 23, 1837, 1.

21. William Salter, "Iowa in Unorganized Territory of the United States, August 10, 1821–June 28, 1834," *Annals of Iowa* 6, no. 3 (July 1903): 195.

22. Ibid., 196. For a concise narrative of the Black Hawk War, see James E. Davis, *Frontier Illinois* (Bloomington: Indiana University Press, 1998), 193–98.

23. Frazer, *Forts of the West*, 48–49. See also *Missouri Republican*, August 31, 1839, 2.

24. *Amendments—Constitution—Missouri*, 23rd Cong., 2nd sess., February 21, 1835, H. Doc. 162, serial 274.

25. John Plumbe, Jr., *Sketches of Iowa and Wisconsin, Embodying the Experience of a Residence of Three Years in Those Territories* (1839; repr., Iowa City: Athens Press, 1948), 2.

26. A. R. Fulton, "Van Buren County," *Annals of Iowa* 3 (April 1884): 43; Salter, "Iowa in Unorganized Territory," 204–205; Landers, "The Southern Boundary of Iowa," 644.

27. *Proceedings of the Legislature of Wisconsin Territory, in Relation to the Boundary between That Territory and the State of Missouri*, 25th Cong., 2nd session, January 2, 1838, S. Doc. 63, serial 314.

28. *Memorial of the General Assembly of Missouri*, February 28, 1831, 3.

29. Session Laws of Missouri, December 21, 1836, 27.

30. *Iowa Territory*, 25th Cong., 2nd session, February 6, 1838, H. Rep. 585, serial 334. No mention of the dispute over the southern boundary of the new Iowa Territory appeared in its organic act. See Session Laws of Iowa, 1st Territorial Assembly, June 12, 1838, 31–40.

31. *Boundary between Missouri and Iowa*, 25th Cong., 3rd sess., January 30, 1839, H. Doc. 128, serial 347, 11–12; *Missouri Courier*, December 1, 1838, 2.

32. *Boundary between Missouri and Iowa*, January 30, 1839, 5. Before investigating the competing boundary claims of Missouri and Iowa Territory, Lea surveyed land in present-day Iowa and Minnesota and served as Tennessee's chief engineer. The town of Albert Lea, Minnesota—located within the area he scouted in the early 1830s—bears his name.

33. Ibid., 7–10, 16–24. Missouri's legislature approved a law in 1839 that reinforced the state's claim to the 1837 line as their official northern border; see Session Laws of Missouri, 10th General Assembly, 1st session, February 11, 1839, 14. See also "Boundary Question," *Territorial Gazette and Burlington Advertiser*, March 2, 1839, 3; "Missouri and Iowa," *Missouri Republican*, March 14, 1839, 2.

34. Plumbe, *Sketches of Iowa and Wisconsin*, 101, italics original.

35. "Iowa Lands," *Missouri Argus*, August 13, 1839, 2.

36. *Missouri Republican*, October 28, 1839, 2.

37. Plumbe, *Sketches of Iowa and Wisconsin*, 3.

38. *The History of Van Buren County, Iowa*, 361–62, 467–509.

39. Session Laws of Iowa, 2nd Territorial Assembly, December 31, 1839, 150–51.

40. Fulton, "Van Buren County," 43.

41. "The Hairy Nation," *Western Gazette*, May 13, 1854, 2.

42. Alfred Hebard, "The Border War between Iowa and Missouri, on the Boundary Question," *Annals of Iowa* 1, no. 8 (January 1895): 652.

43. *Missouri Republican*, July 30, 1839, 2.

44. Quotation in *Missouri Republican*, May 22, 1839, 2.

45. Quotation in *Missouri Whig and General Advertiser*, August 10, 1839, 4.

46. For a history of Governor Boggs's extermination campaign, see Stephen C. LeSueur, *The 1838 Mormon War in Missouri* (Columbia: University of Missouri Press, 1987).

47. For a narrative of the Ohio-Michigan boundary conflict, see Willard V. Way, *The Facts and Historical Events of the Toledo War of 1835* (Toledo, OH: Daily Commercial Steam Book and Job Printing House, 1869).

48. Charles S. Larzelere, "Notes and Documents: The Iowa-Missouri Disputed Boundary," *Mississippi Valley Historical Review* 3, no. 1 (June 1916): 80.

49. Quotation in Spencer, "Demand Nothing but What Is Strictly Right," 26.

50. *Message from the President of the United States, in Relation to the Disputed Boundary Line between the State of Missouri and Territory of Iowa*, 26th Cong., 1st sess., December 24, 1839, S. Doc. 4, 3–5; *Missouri Republican*, August 15, 1839, 2.

51. *Missouri Whig and General Advertiser*, September 7, 1839, 3. See also *Missouri Republican*, August 30, 1839, 2.

52. *Missouri Republican*, August 31, 1839, 2.

53. Ibid., October 8, 1839, 2, italics original.

54. Ibid., September 30, 1839, 2.

55. Ibid.

56. Duane Meyer, *The Heritage of Missouri—A History* (St. Louis: State Publishing Co.), 185.

57. "The Honey War," *Missouri Whig and General Advertizer*, October 26, 1839, 3.

58. *Boundary—Missouri and Iowa*, December 31, 1839, 3, 4; Spencer, "Demand Nothing but What Is Strictly Right," 28–30.

59. "Border Difficulties," *Missouri Argus*, November 1, 1839, 3. See also *The History of Van Buren County, Iowa*, 365.

60. *Missouri Republican*, November 7, 1839, 2.

61. "The Border Trouble," *Missouri Argus*, November 23, 1839, 1.

62. "Iowa Difficulties," *Missouri Whig and General Advertiser*, November 9, 1839, 1. See also *Boundary—Missouri and Iowa*, 26th Cong., 1st sess., December 31, 1839, H. Ex. Doc. 36, serial 364, 5–8, 10–12.

63. Spencer, "Demand Nothing but What Is Strictly Right," 30.

64. *Missouri Republican*, November 12, 1839, 2.

65. *Boundary—Missouri and Iowa*, December 31, 1839, 14.

66. "Missouri and Iowa," *Missouri Whig and General Advertiser*, November 30, 1839, 2; "Boundary War," *Missouri Republican*, December 5, 1839, 2.

67. "The Border War," *Missouri Republican*, December 7, 1839, 2; Spencer, "Demand Nothing but What Is Strictly Right," 31.

68. "The Border War," *Missouri Republican*, December 7, 1839, 2. See also "Border Trouble," *Missouri Republican*, December 4, 1839, 2.

69. Boundary—Missouri and Iowa, December 31, 1839, 3, 20.

70. *Ibid.*, 15. See also Meyer, *The Heritage of Missouri*, 183, Carroll J. Kraus, "A Study in Border Controversy: The Iowa-Missouri Boundary Dispute," *Annals of Iowa* 40, no. 2 (Fall 1969): 93.

71. *Boundary—Iowa and Missouri*, 26th Congress, 1st session, February 12, 1840, H. Ex. Doc. 97, serial 365, 3.

72. See Eric McKinley Erickson, "The Honey War," *Palimpsest* 6 (September 1924), 346–48.

73. J. M. D. Burrows, "Rumors of War," *Palimpsest* 24, no. 2 (February 1943), 72.

74. George Earle Shankle, a scholar of American nicknames, found at least two explanations for the "Puke" moniker. One refers to the prevalence of Missourians from Pike County who emigrated to California during the gold rush, and who became known as "Pukes" as a corruption of the county name. Since the nickname was used in the late 1830s, this explanation seems off chronologically. A more likely story comes from northern Illinois in the 1820s, when Missourians flocked to the Galena lead mines in such great numbers "that those already

there declared the State of Missouri had taken a 'puke.'" See George Earle Shankle, *American Nicknames: Their Origin and Significance* (New York: H. W. Wilson, 1937), 355.

75. Spencer, "Demand Nothing but What Is Strictly Right," 32.

76. Quotation in Walter B. Stevens, *Centennial History of Missouri (The Center State): One Hundred Years in the Union, 1820–1921* (Chicago: S. J. Clarke, 1921), 65.

77. "Important and Cheering Intelligence!!," *Missouri Whig and General Advertiser*, December 14, 1839, 2; "Border War with Iowa," *Missouri Republican*, December 17, 1839, 2; "Public Meeting," *Missouri Whig and General Advertiser*, December 21, 1839, 2; "To the Citizens of Marion County," *Missouri Whig and General Advertiser*, December 21, 1839, 2; Spencer, "Demand Nothing but What Is Strictly Right," 33. See also *Memorial of the Legislative Assembly of the Territory of Iowa, Praying the Adjustment of the Boundary Line between That Territory and the State of Missouri*, 26th Cong., 1st sess., January 8, 1840, S. Doc. 53, serial 355.

78. *Missouri Whig and General Advertiser*, December 21, 1839, 3. See also "The Border War Ended," *Daily National Intelligencer*, December 30, 1839, 3.

79. Ibid., 2. See also *Missouri Whig and General Advertiser*, January 18, 1840, 3; "To 'A Citizen of Marion,'" *Missouri Whig and General Advertiser*, January 25, 1840, 1.

80. "Iowa War," *Missouri Whig and General Advertiser*, January 4, 1840.

81. Quotation in Spencer, "Demand Nothing but What Is Strictly Right," 35–36.

82. "Governor's Message," *Hawkeye and Iowa Patriot*, November 12, 1840, 1. Missouri's legislature appropriated fifteen hundred dollars to Sheriff Gregory in early 1841 to compensate for legal expenses incurred as he brought suit against the Van Buren County officials who had arrested him. See Session Laws of Missouri, 11th General Assembly, 1st sess., February 16, 1841, 128; Charles Negus, "The Southern Boundary of Iowa," *Annals of Iowa* 4, no. 4 (October 1866), 749–50.

83. Spencer, "Demand Nothing but What Is Strictly Right," 35.

84. *Missouri Whig and General Advertiser*, December 21, 1839, 2.

85. Ibid., January 4, 1840, 3. See also *Missouri Republican*, February 25, 1840, 2; "Governor's Message," *Missouri Whig and General Advertiser*, November 28, 1840, 1; Spencer, "Demand Nothing but What Is Strictly Right," 36.

86. *Congressional Globe*, 26th Cong., 1st sess. (January 10, 1840), 110–12.

87. *Boundary of Missouri and Iowa*, 26th Cong., 1st sess., February 4, 1840, H. Rep. 2, serial 370.

88. *Missouri Whig and General Advertiser*, February 29, 1840, 3; ibid., March 14, 1840, 3.

89. "Southern Boundary," *Territorial Gazette and Burlington Advertiser*, December 5, 1840, 3.

90. Session Laws of Missouri, 11th General Assembly, 1st sess., February 16, 1841, 106–107.

91. *Iowa Militia*, 26th Cong., 1st sess., March 6, 1840, H. Doc. 123, serial 365.

92. *Iowa Militia Claim*, 26th Cong., 1st session, May 25, 1840, H. Rep. 543, serial 372; "Military," *Territorial Gazette and Burlington Advertiser*, August 8, 1840, 2–3.

93. See "Iowa and Missouri War," *Iowa Territorial Gazette and Advertiser*, March 27, 1841, 3; *Iowa—Expenses Southern Boundary*, 27th Cong., 2nd sess., February 12, 1842, H. Doc. 84, serial 402; *Pay of Iowa Militia*, 27th Cong., 2nd sess., July 4, 1842, H. Rep. 896, serial 410; *Congressional Globe*, 27th Cong., 3rd sess. (February 20, 1843), 311; *Iowa Militia*, 28th Cong., 1st sess., March 28, 1844, H. Rep. 371, serial 446; *Iowa Militia*, 28th Cong., 2nd sess., January 31, 1845, H. Rep. 88, serial 468; *Congressional Globe*, 28th Cong., 2nd sess. (February 10, 1845), 268; *Iowa Militia*, 29th Cong., 1st sess., May 19, 1846, H. Rep. 678, serial 490; *Congressional Globe*, 29th Cong., 1st sess. (June 9, 1846), 949; *Pay for Militia—Called Out by U.S. Marshal*, 33rd Cong., 2nd sess., February 20, 1855, H. Misc. Doc. 31, serial 807.

94. Regardless of the controversies surrounding the tenure of Governor Lucas, the *Territorial Gazette and Burlington Advertiser* printed a flowery farewell to him and thanked him profusely for his services as the first territorial executive. See "Gov. Lucas," *Territorial Gazette and Burlington Advertiser*, April 17, 1841, 2.

95. Session Laws of Missouri, 11th General Assembly, 1st sess., December 19, 1840, 23–24, 127–28. See also *Letter from the Governor of Missouri, Transmitting an Act of the General Assembly of the State of Missouri for Ascertaining and Settling the Northern Boundary Line of that State; Together with a Memorial of the Legislature of Said State on the Same Subject*, 26th Cong., 2nd sess., January 5, 1841, S. Doc. 40, serial 376.

96. *Hawkeye and Iowa Patriot*, March 4, 1841, 2; March 11, 1841, 2, 3. See also "Iowa and Missouri War," *Territorial Gazette and Burlington Advertiser*, March 27, 1841, 3; "The Matter of Our Southern Boundary Line Set Right," *Territorial Gazette and Burlington Advertiser*, May 1, 1841, 2. Missouri's legislature passed a similar bill in early 1843, but again Congress took no action. See *Boundary with Iowa*, 28th Cong., 1st sess., December 22, 1843, H. Doc. 26, serial 441.

97. *Northern Boundary of Missouri*, 27th Cong., 3rd sess., December 31, 1842, H. Doc. 38, serial 420, 13–20.

98. *Northern Boundary of Missouri*, 27th Cong., 2nd sess., January 20, 1842, H. Doc. 48, serial 402; *Northern Boundary of Missouri*, 27th Cong., 3rd sess., 11 February 1843, H. Doc. 138, serial 421.

99. *Boundary between Missouri and Iowa*, 27th Cong., 2nd sess., May 26, 1842, H. Rep. 791, serial 410, 11.

100. *Congressional Globe*, 27th Cong., 2nd sess. (July 20, 1842), 770–71.

101. *Appendix to the Congressional Globe*, 27th Cong., 2nd sess. (July 20, 1842), 248.

102. Ibid., 945.

103. *Burlington Hawk-Eye*, October 17, 1844, 3.

104. *Memorial of the General Assembly of Missouri, Praying that the Southern Boundary Line of the Proposed State of Iowa May Be Made to Conform to the Northern Boundary Line of the State of Missouri*, 28th Cong., 2nd sess., February 19, 1845, S. Doc. 110, serial 456, 1.

105. Floyd Calvin Shoemaker, "Some Historic Lines in Missouri," *Missouri Historical Review* 3, no. 4 (July 1909), 267.

106. "Governor's Message," *Burlington Hawk-Eye*, December 11, 1845, 2.

107. *Boundary Line between Iowa and Missouri*, 29th Cong., 1st sess., February 5, 1846, H. Doc. 104, serial 483, 2.

108. Shoemaker, "Some Historic Lines in Missouri," 267–68; Charles Negus, "Southern Boundary of Iowa," *Annals of Iowa* 5, no. 1 (January 1867), 786–87.

109. *North American and Daily Advertiser* (Philadelphia), June 7, 1845, 1.

110. *Boundary Line between Iowa and Missouri*, 29th Cong., 1st sess., February 17, 1846, H. Doc. 126, serial 483.

111. "The Constitution," *Burlington Hawk-Eye*, June 4, 1846, 1.

112. Landers, "The Southern Boundary of Iowa," 646.

113. United States Reports, January 1849 term, 663–66.

114. Ibid., 666–77.

115. Ibid., 679.

116. "Iowa and Missouri Boundary," *New York Herald*, May 1, 1849, 1.

117. United States Reports, December 1850 term, 1–54; Negus, "Southern Boundary of Iowa," 789–93.

118. "The Hairy Nation," *Western Gazette*, May 13, 1854, 2.

119. "To Settle a Boundary Line," *Daily Inter-Ocean* (Chicago), August 2, 1895, 3; "Missouri-Iowa Boundary Line," *Daily Inter-Ocean* (Chicago), December 27, 1895, 1; United States Reports, December 1896 term, 118–44.

120. David Gebhard and Gerald Mansheim, *Buildings of Iowa* (New York: Oxford University Press, 1993), 107.

121. Troy L. Hayes, "Missouri/Iowa Boundary Line Investigation," *American Surveyor* 3, no. 2 (March/April 2006): 33–37.

122. Quotation in Clarence M. Conkling, "Look at the State They're In!," *Saturday Evening Post*, October 20, 1951, 194.

123. See also Ben Hur Wilson, "The Southern Boundary," *Palimpsest* 20 (1938) 413–24; John Ira Barrett, "The Legal Aspects of the Iowa-Missouri Boundary Dispute, 1839–1851," master's thesis, Drake University, 1959; Craig Hill, "The Honey War," *Pioneer America* 14, no. 2 (July 1982): 81–88.

5. NATURE HAS MARKED OUT THE BOUNDARIES

1. Joel Palmer, *Journal of Travels over the Rocky Mountains, to the Mouth of the Columbia River; Made during the Years 1845 and 1846* (Cincinnati: J. A. and U. P. James, 1847), 116.

2. *Great Britain—Convention of October 20, 1818*, 15th Cong., 2nd sess., December 29, 1818, American State Papers 04, Foreign Relations, vol. 4, 348–407.

3. *Convention with Great Britain for Continuing in Force the Commercial Convention of the Third of July, 1815. —Convention with Great Britain for Continuing in Force the Third Article of the Convention of the 20th of October, 1818, in Relation to the Territories Westward of the Rocky Mountains. —Convention with Great Britain for the Reference to a Friendly Sovereign the Points of Difference Relating to the Northeastern Boundary of the United States*, 20th Cong., 1st session, December 12, 1827, American State Papers 06, Foreign Relations, vol. 6, 639–706; *Conventions with Great Britain —1. Continuing in Force the Commercial Convention of July 8, 1815; 2. Continuing in Force the Convention for the Joint Occupation of the Territory West of the Rocky Mountains of October 20, 1818; and 3. Convention for Carrying into Effect the Fifth Article of the Treaty of Ghent Relative to the Northeastern Boundary of the United States*, 20th Cong., 1st sess., May 19, 1828, American State Papers 06, Foreign Relations, vol. 6, 999–1002.

4. See Robert W. Frazer, *Forts of the West: Military Forts and Presidios and Posts Commonly Called Forts West of the Mississippi River to 1898* (1965; repr., Norman: University of Oklahoma Press, 1972), 176–77.

5. *Memorial of a Number of Citizens of the Oregon Territory, Praying Congress to Take Possession of, and Extend Their Jurisdiction over, the Said Territory*, 25th Cong., 3rd sess., January 28, 1839, S. Doc. 154, serial 340, 2.

6. *Territory of Oregon*, 25th Cong., 3rd sess., February 16, 1839, H. Rep. 101, serial 351, 2.

7. *Petition of a Number of Citizens of the Oregon Territory, Praying the Extension of the Jurisdiction and Laws of the United States over that Territory*, 26th Cong., 1st sess., June 4, 1840, S. Doc. 514, serial 360; *Petition of a Number of Citizens of the Territory of Oregon, Praying the Extension of the Jurisdiction of the United States over that Territory*, 28th Cong., 1st sess., February 7, 1844, S. Doc. 105, serial 433.

8. *Oregon Territory*, 28th Cong., 1st sess., March 12, 1844, H. Rep. 308, serial 445.

9. John D. Unruh, Jr., *The Plains Across: The Overland Emigrants and the Trans-Mississippi West, 1840–1860* (1979; repr., Urbana: University of Illinois Press, 1993), 119.

10. *Oregon*, 29th Cong., 1st sess., December 2, 1845, H. Doc. 3, serial 480, 3. See also *Congressional Globe*, 29th Cong., 1st sess. (December 8, 1845), 24.

11. *Amended Organic Laws of the Territory of Oregon, as Adopted by the People of that Territory on the Last Saturday of July, 1845*, 29th Cong., 1st sess., May 21, 1846, S. Doc. 353, serial 476. Several earlier movements to create local government had transpired in Oregon in the early 1840s; see Gustavus Hines, *Oregon: Its History, Condition and Prospects* (Buffalo, NY: Geo. H. Derby, 1851), 417–37.

12. See Frederick Merk, *The Oregon Question: Essays in Anglo-American Diplomacy and Politics* (Cambridge, MA: Harvard University Press, 1967), 49–50, 195, 205, 234–54.

13. *The Times*, January 3, 1846, 4. See also ibid., March 16, 1846, 4; ibid., June 30, 1846, 4. For a thorough study of the British press reaction to the Oregon issue, see Thomas C. McClintock, "British Newspapers and the Oregon Treaty of 1846," *Oregon Historical Quarterly* 104, no. 1 (Spring 2003), 96–109.

14. *Message from the President of the United States, Communicating a Copy of the Convention for the Settlement of the Oregon Question, Concluded the 15th of June, 1846, between the United States and Great Britain, and Recommending to Congress the Adoption of Such Measures for Facilitating the Occupation and Settlement of that Territory*, 29th Cong., 1st sess., August 6, 1846, S. Doc. 476, serial 478.

15. Hall J. Kelley, *A Geographical Sketch of that Part of North America, called Oregon* (Boston: J. Howe, 1830), 18–25.

16. *Territory of Oregon*, February 16, 1839, 7–8.

17. Lansford W. Hastings, *The Emigrant's Guide to Oregon and California, Containing Scenes and Incidents of a Party of Oregon Emigrants; a Description of Oregon; Scenes and Incidents of a Party of California Emigrants; and a Description of California; with a Description of the Different Routes to Those Countries; and All Necessary Information Relative to the Equipment, Supplies, and the Method of Traveling* (1845; repr., Bedford, MA: Applewood Books, 1994), 24–26. In 1850 a correspondent for a Washington, D.C., newspaper provided an extensive description of the Oregon Territory, with geographic divisions practically copied from Hastings. See "Oregon," *Daily National Intelligencer*, February 1, 1850, 2. Residents of The Dalles, Oregon Territory, petitioned Congress unsuccessfully in 1855 and 1857 to reorganize the Northwest by limiting both Oregon and Washington Territories west of the Cascade Mountains and creating the rest of the region on both sides of the Columbia River as a new territory. See D. W. Meinig, *The Great Columbia Plain: A Historical Geography, 1805–1910* (1968; repr., Seattle: University of Washington Press, 1995), 170.

18. Alexander Ross, *Adventures of the First Settlers on the Oregon or Columbia River* (London: Smith, Elder, 1849), 100–101.

19. Palmer, *Journal of Travels over the Rocky Mountains*, 88, 113, 116.

20. *Oregon*, 30th Cong., 1st sess., August 10, 1848, H. Misc. Doc. 98, serial 523, 2. See also *Petition of Citizens of Oregon, Praying that the Laws of the United States May Be Extended over that Territory, &c.*, 30th Cong., 1st sess., May 8, 1848, S. Misc. Doc. 136, serial 511, italics original.

21. David Lavender, "Fort Vancouver and the Pacific Northwest," in *Fort Vancouver*, National Park Handbook 113 (Washington, DC: Government Printing Office, 2001), 106.

22. *Congressional Globe*, 30th Cong., 1st sess. (August 12, 1848), 1078–80. See also *Approval of the Oregon Bill*, 30th Cong., 2nd sess., December 6, 1848, H. Ex. Doc. 3, serial 538.

23. Unruh, *The Plains Across*, 119–20.

24. James W. Scott and Roland L. DeLorme, *Historical Atlas of Washington* (Norman: University of Oklahoma Press, 1988), 21. According to the son of the first territorial governor, the population north of the Columbia River had grown to just under four thousand by 1853, when Washington Territory was organized. See Hazard Stevens, *The Life of Isaac Ingalls Stevens*, vol. 1 (Boston: Houghton, Mifflin, 1900), 411.

25. Robert E. Ficken, *Washington Territory* (Pullman: Washington State University Press, 2002), 16–17; Thomas Wickham Prosch, *McCarver and Tacoma* (Seattle: Lowman & Hanford, 1906), 151–52.

26. *Oregon Territory*, 32nd Cong., 1st sess., February 19, 1852, H. Misc. Doc. 14, serial 652, 1.

27. "Introductory," *Columbian*, September 11, 1852, 2; "The New Territory—Northern Oregon," *Columbian*, October 9, 1852, 2; "What Northern Oregon Wants," *Columbian*, October 16, 1852, 2; "Matters and Things in General," *Columbian*, October 30, 1852, 2.

28. "Northe[r]n Oregon—Abuses, &c.," *Columbian*, October 23, 1852, 2.

29. "Public Meeting," *Columbian*, November 13, 1852, 1. See also "Prepare! Prepare!," *Columbian*, November 6, 1852, 2; "Turn Out! Turn Out!," *Columbian*, November 13, 1852, 2. Monticello, no longer extant, stood where the city of Longview, Washington, is today. A historical marker stands in Longview to commemorate the "Monticello Convention."

30. *Congressional Globe*, 32nd Cong., 2nd sess. (February 8, 1853), 541. D. W. Meinig criticizes the Monticello convention's choice of the upper Columbia River as a boundary, considering its service as a unifying factor in the basin of western Washington, in contrast to the river's more imposing character closer to the Pacific Ocean. See Meinig, *The Great Columbia Plain*, 168–70.

31. "Rally! Rally!," *Columbian*, November 20, 1852, 2. See also *Columbian*, February 12, 1853, 2.

32. "The New Territory Convention," *Columbian*, December 4, 1852, 2. See also "Meeting of Congress—Northern Oregon—Its Wants, Ggrievances [sic], Interests, &c., &c.," *Columbian*, December 11, 1852, 2. While Congress debated the fate of Columbia Territory, Olympians prepared for a second convention to take place in their city in May 1853, a meeting that never transpired since Congress established Washington Territory in the meantime. See "Division of the Territory—The New Territory Convention to Meet at Olympia, on the 11th of May, Next," *Columbian*, March 26, 1853, 2.

33. *Congressional Globe*, 32nd Cong., 2nd sess. (February 8, 1853), 539–42; ibid., February 10, 1853, 555; ibid., March 2, 1853, 1020; *Appendix to the Congressional Globe*, 32nd Cong., 2nd sess. (March 2, 1853), 338–40. For more information on the debate over Washington Territory's name, see Robert E. Ficken, "Columbia, Washington or Tacoma: The Naming and Attempted Renaming of Washington Territory," *Columbia Magazine* 17, no. 1 (Spring 2003), 25–30.

34. "Washington Territory," *Daily National Intelligencer*, March 26, 1853, 3. The *Intelligencer* recognized the problem of naming the territory after the city within the District of Columbia, suggesting that Congress named it "with singular inappropriateness, and as contributing fresh confusion to our already confused nomenclature, [it] will have to be changed."

35. "Washington Territory—'All's Well that Ends Well,'" *Columbian*, April 30, 1853, 2.

36. "State Government for Oregon," *Weekly Oregonian*, April 1, 1854, 2; "State of Oregon," *Weekly Oregonian*, March 24, 1855, 2; "State Boundaries," *Weekly Oregonian*, May 26, 1855, 2.

37. *Congressional Globe*, 32nd Cong., 2nd sess. (March 2, 1853), 338. This Fort Walla Walla should not be confused with one established later at the present site of Walla Walla, Washington, which stands about thirty miles east of the Columbia River and the original post. See Frazer, *Forts of the West*, 177; Meinig, *The Great Columbia Plain*, 161.

38. *Indians—Oregon and Washington Territories*, 33rd Cong., 1st sess., February 14, 1854, H. Ex. Doc. 55, serial 721.

39. See *Report of the Secretary of War, in Compliance with a Resolution of the Senate of the 21st Ultimo, Calling for Copies of All the Letters of the Governor of Washington Territory, Addressed to Him*

during the Present Year; and Copies of All the Correspondence Relative to the Indian Disturbances in the Territories of Washington and Oregon, 34th Cong., 1st sess., May 12, 1856, S. Ex. Doc. 66, serial 822, 16–17, 23–24.

40. "Indian Difficulties in the North," *Weekly Oregonian*, October 6, 1855, 2; "Indian War," *Weekly Oregonian*, October 20, 1855, 2; *Indian Disturbances in Oregon and Washington*, 34th Congress, 1st sess., March 10, 1856, H. Ex. Doc. 48, serial 853, 3–5.

41. "From the Dalles and Cascades," *Weekly Oregonian*, October 27, 1855, 2.

42. "Latest News from Puget Sound," *Weekly Oregonian*, November 10, 1855, 2; *Appendix to the Congressional Globe* (March 2, 1853), 340.

43. "From the Umatilla Valley," *Weekly Oregonian*, December 8, 1855, 2; "Great Indian Fight near Walla-Walla," *Weekly Oregonian*, December 15, 1855, 2; "The War," *Weekly Oregonian*, December 22, 1855, 2; "How Goes on the War?," *Weekly Oregonian*, December 22, 1855, 2.

44. "Who Killed Umtux?," *Weekly Oregonian*, December 8, 1855, 2; Frazer, *Forts of the West*, 128–29, 168.

45. "From the Seat of War, North," *Weekly Oregonian*, January 19, 1856, 2. See also Scott and DeLorme, *Historical Atlas of Washington*, 28.

46. "Gen. Wool and the Indian War in Oregon," *Weekly Oregonian*, February 2, 1856, 2.

47. *Report of the Secretary of War*, May 12, 1856, 2–3.

48. Ibid., 6, 7, 39.

49. Ibid., 34, 35, 45.

50. See Ficken, *Washington Territory*, 19.

51. *Report of the Secretary of War*, May 12, 1856, 60.

52. Meinig, *The Great Columbia Plain*, 165–66.

53. *Report of the Secretary of War*, May 12, 1856, 58.

54. *Indian Hostilities in Oregon and Washington Territories*, 34th Cong., 1st sess., July 8, 1856, H. Ex. Doc. 118, serial 859, 2–3, 5.

55. *Congressional Globe*, 36th Cong., 1st sess. (May 30, 1860), 2469. See also "Our War Debt," *Washington Standard*, November 23, 1860, 2.

56. *Congressional Globe*, 36th Cong., 2nd sess. (February 21, 1861), 1105. See also ibid., 1100.

57. See Meinig, *The Great Columbia Plain*, 160–68. For a volunteer's version of the Yakima War, see Urban E. Hicks, *Yakima and Clickitat Indian Wars, 1855 and 1856* (Portland, OR: Himes the Printer, 1886).

58. *Appendix to the Congressional Globe*, 36th Cong., 2nd sess. (March 2, 1861), 333.

59. "Admission of Oregon," *Weekly Oregonian*, March 14, 1857, 2.

60. "Proceedings of the Constitutional Convention," *Weekly Oregonian*, September 5, 1857, 1.

61. "Proceedings of the Constitutional Convention," *Weekly Oregonian*, September 12, 1857, 1.

62. *Indian Affairs in the Territories of Oregon and Washington*, 35th Cong., 1st sess., January 25, 1858, H. Ex. Doc. 39, serial 955, 29–31. Three years later the Bureau of Indian Affairs, in a report on the situation of Native peoples in the territories, similarly treated Oregon and Washington Territories as one polity, even though the former received its promotion to statehood in 1859. See "Indian Affairs in Oregon and Washington," *Washington Standard*, January 19, 1861, 2.

63. *Appendix to the Congressional Globe*, 35th Cong., 1st sess. (May 29, 1858), 560.

64. "Proceedings of the Constitutional Convention," *Weekly Oregonian*, September 12, 1857, 1. See also *Report of the Secretary of War*, May 12, 1856, 3; D. W. Meinig, *The Shaping of*

America: A Geographical Perspective on 500 Years of History, vol. 3, *Transcontinental America, 1850–1915* (New Haven: Yale University Press, 1998), 82.

65. *Constitution of Oregon*, 35th Cong., 1st sess., February 1, 1858, H. Misc. Doc. 38, serial 961, 19.

66. "Constitutional Convention," *Weekly Oregonian*, September 26, 1857, 2.

67. Session Laws of Washington, 4th Territorial Legislature, December 15, 1857, 61.

68. "Bill for the Admission of Oregon into the Union," *Weekly Oregonian*, June 25, 1858, 2.

69. *Appendix to the Congressional Globe*, 35th Cong., 2nd sess. (February 14, 1859), 330.

70. See "Admission of Oregon," *Weekly Oregonian*, March 19, 1859, 2; "Admission of Oregon—Its Causes, Effects and Consequences," *Weekly Oregonian*, March 26, 1859, 2.

71. "The Insane Asylum for Oregon and Washington Territory," *Weekly Oregonian*, January 25, 1862, 1.

72. See "The Salmon River Mines," *Daily Alta California*, March 22, 1862, 2; "Later from the North," *Daily Evening Bulletin*, March 31, 1862, 1; "The Salmon River Mines—Letters from Oregon," *Daily Evening Bulletin*, April 9, 1862, 2.

73. "Division of Washington Territory," *Washington Standard*, April 5, 1862, 2. The *Standard* had printed an article from the *Daily Alta California* a few weeks earlier supporting a political split along the Cascades, a notion it vehemently opposed. See "The Importance of a Division in Washington Territory," *Washington Standard*, March 15, 1862, 1.

74. "The Mines of Idaho," *Weekly Oregonian*, June 7, 1862, 2.

75. "Why Stand Ye Here Idle—What a Little Enterprise May Do," *Washington Standard*, May 24, 1862, 2; "The New State," *Washington Standard*, August 23, 1862, 2. See also Ficken, *Washington Territory*, 68–69.

76. "Idaho," *Washington Standard*, November 1, 1862, 2.

77. See "Editorial Correspondence," *Washington Statesman*, January 10, 1863, 2; "The Division Question," *Washington Statesman*, January 10, 1863, 2; "Division of the Territory," *Washington Statesman*, January 17, 1863, 1; "Editorial Correspondence," *Washington Statesman*, January 17, 1863, 2; "The Protest," *Washington Statesman*, January 17, 1863, 2; "By What Right?," *Washington Statesman*, January 31, 1863, 2; "Division—A Voice from Idaho," *Washington Statesman*, February 7, 1863, 4. Sara Jane Richter notes a proposal in the 1861 Washington territorial legislature to support a new Walla Walla Territory on the upper Columbia River, but it met with opposition from Puget Sound members. See Richter, "Washington and Idaho Territories," *Journal of the West* 16, no. 2 (April 1977): 28–29.

78. "The Division Question," *Washington Standard*, February 21, 1863, 2.

79. *Congressional Globe*, 37th Cong., 3rd session (March 3, 1863), 1509; *Appendix to the Congressional Globe*, 37th Cong., 3rd sess. (March 3, 1863), 233. The new line separating Washington and Idaho Territories nearly, but not quite, corresponded to the 117th meridian. See also Merle W. Wells, "The Idaho-Montana Boundary," *Idaho Yesterdays* 12, no. 4 (Winter 1968): 6–9.

80. "Idaho, Otherwise Montana," *Sacramento Daily Union*, March 16, 1863, 1.

81. "The Division of Washington Territory," *Washington Standard*, March 14, 1863, 2.

82. "Under Which King?," *Washington Statesman*, March 21, 1863, 2. See also "Idaho Territorial Convention," *Washington Statesman*, March 28, 1863, 2; *Washington Statesman*, May 2, 1863, 3; ibid., November 28, 1863, 3. A proposal to make Columbia Territory from the upper river valley and northern Idaho simmered in the late 1860s; see Merle W. Wells, "Territorial Government in the Inland Empire: The Movement to Create Columbia Territory," *Pacific Northwest Quarterly* 44, no. 2 (April 1953): 80–87.

83. *Memorial of the Legislature of Oregon, in Favor of Incorporating the County of Walla-Walla, Washington Territory, into the State of Oregon*, 39th Cong., 1st sess., March 14, 1866, S. Misc. Doc. 83, serial 1239.

84. *Eastern Boundary of Oregon*, 41st Cong., 3rd sess., January 9, 1871, H. Misc. Doc. 23, serial 1462; *Memorial of the Legislature of Oregon, Asking a Change of the Northern Boundary of the State of Oregon, So that It Will Conform to the Boundary Described in Its Constitution*, 42nd Cong., 3rd sess., January 8, 1873, S. Misc. Doc. 27, serial 1546; *State Boundary*, 44th Cong., 2nd sess., December 29, 1876, H. Misc. Doc. 23, serial 1762.

85. Paul L. Beckett, *From Wilderness to Enabling Act: The Evolution of a State of Washington* (Pullman: Washington State University Press, 1968), 16–18.

86. *Boundaries of the State of Oregon*, 44th Cong., 1st sess., July 18, 1876, H. Rep. 764, serial 1712, 1, italics original.

87. For a comprehensive discussion of Washington-Idaho boundary disputes in the late nineteenth century, see *In the Senate of the United States*, 50th Cong., 1st sess., March 15, 1888, S. Rep. 585, serial 2520. The following sources offer narrative accounts of the creation, evolution, and debates over the fate of northern Idaho and western Washington: Benjamin E. Thomas, "Boundaries and Internal Problems of Idaho," *Geographical Review* 39, no. 1 (January 1949): 99–109; Wells, "Territorial Government in the Inland Empire," 80–87; Herman J. Deutsch, "The Evolution of Territorial and State Boundaries in the Inland Empire of the Pacific Northwest," *Pacific Northwest Quarterly* 51, no. 3 (July 1960): 129–31; Merle W. Wells, "Idaho's Centennial: How Idaho Was Created in 1863," *Idaho Yesterdays* 7, no. 1 (Spring 1963): 44–58; John J. Peebles, "Retracing a Line: The 1908 Idaho-Washington Boundary Resurvey," *Idaho Yesterdays* 13, no. 3 (Fall 1969), 20–25; E. H. Collins, "Why the Idaho Panhandle?," *Pacific Northwesterner* 18, no. 3 (Summer 1974): 41–46; John R. Wunder, "Tampering with the Northwest Frontier: The Accidental Design of the Washington/Idaho Boundary," *Pacific Northwest Quarterly* 68, no. 1 (January 1977): 1–12; Meinig, *The Shaping of America*, vol. 3, 84, 148; Ficken, *Washington Territory*, 203.

88. Session Laws of Oregon, 49th General Assembly, April 30, 1957.

89. Oregon State Constitution, 2005, Article XVI.

90. Although the initiative was filed with the office of the Washington secretary of state in Olympia, it was not accompanied by any signed petitions and therefore was not eligible for consideration by the state legislature. See "Elections: Initiatives to the Legislature 1914–2005," Washington Secretary of State, accessed http://www.secstate.wa.gov/ (accessed July 18, 2007); "Initiative 289," Washington Secretary of State, http://www.secstate.wa.gov/elections/initiatives/text/i289.pdf (accessed July 16, 2013).

6. A STATE BORDERING UPON ANARCHY

1. *Utah Territory*, 36th Cong., 1st sess., May 2, 1860, H. Ex. Doc. 78, serial 1056, 3.

2. D. W. Meinig, *The Shaping of America: A Geographical Perspective on 500 Years of History*, vol. 3, *Transcontinental America, 1850–1915* (New Haven: Yale University Press, 1998), 46. Many senators and representatives from southern states insisted on an extension of the Missouri Compromise line to divide California in two, making a free state to the north and a slave territory to the south. See *Congressional Globe*, 31st Cong., 1st sess. (February 14, 1850), 367; William Henry Ellison, *A Self-Governing Dominion: California, 1849–1860* (Berkeley: University of California Press, 1950), 167–91; Warren A. Beck and Ynez D. Haase, *Historical Atlas of California* (Norman: University of Oklahoma Press, 1974), 59. California pressed for such an idea again in 1860, proposing the creation of a Colorado Territory from its southern

reaches, presumably for ethnic as well as political reasons, considering the significant Hispano population to the south in contrast to the American-dominated north. The "Crittenden Compromise" of early 1861 made a similar proposal. See *An Act of the Legislature of California, Granting the Consent of Said Legislature to the Formation of a Different Government for the Southern Counties of Said State, with a Statement of the Votes Polled in Pursuance of Said Act*, 36th Cong., 1st sess., January 18, 1860, S. Misc. Doc. 2, serial 1038; "The Territorial Question in a Nutshell," *Daily Alta California*, February 20, 1861, 2; "Modification of the Crittenden Compromise," *Daily Alta California*, March 3, 1861, 2; "Dismemberment of California," *Daily Alta California*, March 21, 1861, 2. As recently as July 2011 a variant on this proposal developed, calling for thirteen eastern and southern California counties to break away as "South California" dominated by Republicans rather than the Democratic-majority remainder of the state. See Andrew Malcolm, "'South California' Proposed as 51st State by Republican Supervisor," *Los Angeles Times*, July 11, 2011, http://latimesblogs.latimes.com/.

3. D. W. Meinig, *Southwest: Three Peoples in Geographical Change, 1600–1970* (New York: Oxford University Press, 1971), 20.

4. Hubert H. Bancroft, *History of Arizona and New Mexico, 1530–1888* (1889; repr., Albuquerque: Horn & Wallace, 1962), 345.

5. Cardinal Goodwin, "The Question of the Eastern Boundary of California in the Convention of 1849," *Southwestern Historical Quarterly* 16, no. 4 (January 1913): 227–29.

6. Beck and Haase, *Historical Atlas of California*, 20–22.

7. David J. Weber, *The Mexican Frontier, 1821–1846: The American Southwest under Mexico* (Albuquerque: University of New Mexico Press, 1982), 23.

8. Malcolm L. Comeaux, "Attempts to Establish and Change a Western Boundary," *Annals of the Association of American Geographers* 72, no. 2 (June 1982), 257.

9. Henry P. Walker and Don Bufkin, *Historical Atlas of Arizona* (1979; repr., Norman: University of Oklahoma Press, 1986), 14.

10. Lansford W. Hastings, *The Emigrant's Guide to Oregon and California, Containing Scenes and Incidents of a Party of Oregon Emigrants; a Description of Oregon; Scenes and Incidents of a Party of California Emigrants; and a Description of California; with a Description of the Different Routes to Those Countries; and All Necessary Information Relative to the Equipment, Supplies, and the Method of Traveling* (1845; repr., Bedford: Applewood Books, 1994), 70. See also Meinig, *The Shaping of America*, vol. 3, 34.

11. John C. Frémont, *Geographical Memoir upon Upper California, in Illustration of His Map of Oregon and California*, 30th Cong., 2nd sess., January 2, 1849, H. Misc. Doc. 5, serial 544, 40. See also J. Ross Browne, *Report of the Debates in the Convention of California, on the Formation of the State Constitution, in September and October, 1849* (Washington, D.C.: J. T. Towers, 1850), 169.

12. *New Mexico and California*, 30th Cong., 1st sess., July 24, 1848, H. Ex. Doc. 70, serial 521.

13. *Congressional Globe*, 30th Cong., 1st sess. (July 26, 1848), 1002–1005.

14. *Congressional Globe*, 30th Cong., 2nd sess. (December 11, 1848), 21.

15. *In the Senate of the United States*, 30th Cong., 2nd sess., January 9, 1849, S. Rep. 256, serial 535, pt. 1, 3–5, italics original.

16. Ibid., pt. 2, 7–8.

17. *Congressional Globe*, 30th Cong., 2nd sess. (January 9, 1849), 193.

18. Ibid., January 22, 1849, 319–20.

19. Ibid., January 29, 1849, 381.

20. See Rodman Wilson Paul and Elliott West, *Mining Frontiers of the Far West, 1848–1880*, revised ed. (Albuquerque: University of New Mexico Press, 2001), 15.

21. Browne, *Report of the Debates in the Convention of California*, 3–5.

22. Ibid., 478–79.

23. Ibid., 4–5, 123–24, 168–70, 417, 426, 454. See also Ellison, *A Self-Governing Dominion*, 33–36.

24. Browne, *Report on the Debates in the Convention of California*, 443, 474.

25. During most of the dispute over the California-Nevada boundary, particularly in the late 1850s and early 1860s, Lake Tahoe was known as Lake Bigler, after California's governor from 1852 to 1856, John Bigler. By the late 1860s most sources referred to the feature as Lake Tahoe.

26. Browne, *Report on the Debates in the Convention of California*, 474–75.

27. See *Almon W. Babbit, Delegate from Deseret*, 31st Cong., 1st sess., April 4, 1850, H. Rep. 219, serial 584; "State of Deseret—Constitution Adopted—Utah," *Sacramento Daily Union*, May 14, 1856, 2; Goodwin, "The Question of the Eastern Boundary of California," 248–50; Warren A. Beck and Ynez D. Haase, *Historical Atlas of the American West* (Norman: University Press of Oklahoma, 1989), 39; Glen M. Leonard, "The Mormon Boundary Question in the 1849–50 Statehood Debates," *Journal of Mormon History* 18, no. 1 (Summer 1992): 114–36.

28. *Congressional Globe*, 31st Cong., 1st sess. (February 15, 1850), 372.

29. Ibid., February 18, 1850, 377.

30. *In the Senate of the United States*, 31st Cong., 1st sess., May 8, 1850, S. Rep. 123, serial 565, 3. See also *Congressional Globe*, 31st Cong., 1st sess. (May 8, 1850), 944.

31. *Congressional Globe*, 31st Cong., 1st sess. (September 9, 1850), 1784. See also *Daily National Intelligencer*, September 10, 1850, 4.

32. *Congressional Globe*, 31st Cong., 1st sess. (May 8, 1850), 947. The compromise bill introduced in Congress in May 1850 separated Utah and New Mexico Territories by the ridge between the Great Basin and the Colorado River valley. By the time of its passage in September, Congress shifted the territories' shared boundary to the thirty-seventh parallel. See *Congressional Globe*, 31st Cong., 1st sess. (September 7, 1850), 1772.

33. For scholarly interpretations of the compromise legislation, see Edwin Charles Rozwenc, *The Compromise of 1850* (Boston: Heath, 1957); Holman Hamilton, *Prologue to Conflict: The Crisis and Compromise of 1850* (Lexington: University of Kentucky Press, 1964); Mark J. Stegmaier, *Texas, New Mexico, and the Compromise of 1850: Boundary Dispute and Sectional Crisis* (Kent, OH: Kent State University Press, 1996); John C. Waugh, *On the Brink of Civil War: The Compromise of 1850 and How It Changed the Course of American History* (Wilmington, DE: Scholarly Resources, 2003); Michael F. Holt, *The Fate of Their Country: Politicians, Slavery Extension, and the Coming of the Civil War* (New York: Hill and Wang, 2004); Robert V. Remini, *At the Edge of the Precipice: Henry Clay and the Compromise that Saved the Union* (New York: Basic Books, 2010).

34. *Senators and Representatives Elect from California*, 31st Cong., 1st sess., March 18, 1850, H. Misc. Doc. 44, serial 581, 11–12.

35. "Latest News from California," *Daily National Intelligencer*, September 7, 1850, 4; Sally Zanjani, *Devils Will Reign: How Nevada Began* (Reno: University of Nevada Press, 2006), 9–13. See also Kent D. Richards, "Washoe Territory: Rudimentary Government in Nevada," *Arizona and the West* 11, no. 3 (Autumn 1969): 214–15.

36. *Joint Resolution of the Legislature of California, in Relation to the Placing of Troops along the Borders, and the Erection of Forts in California for the Protection of the Citizens of That State*, 32nd Cong., 1st sess., December 8, 1851, S. Misc. Doc. 3, serial 629. See also Session Laws of California, 3sr General Assembly, March 6, 1852, 279; Meinig, *The Shaping of America*, vol. 3, 47–49.

37. Beck and Haase, *Historical Atlas of California*, 61.

38. Session Laws of Utah, 3rd Territorial Assembly, January 17, 1854, 19; Richards, "Washoe Territory," 218.

39. James Thomas Butler, *Isaac Roop: Pioneer and Political Leader of Northeastern California* (Janesville, CA: High Desert Press, 1994), 15.

40. See H. L. Wells, "The Sage-Brush Rebellion," *Overland Monthly* 13, no. 75 (March 1889), 254; Charles Curry Aiken, "The Sagebrush War: The California-Nevada Boundary Dispute on the 120th Meridian," *Journal of the Shaw Historical Library* 5, no. 1–2 (1991), 48–49. Aiken's two-volume article offers the most comprehensive interpretation available of the California-Nevada line north of Lake Tahoe.

41. Beck and Haase, *Historical Atlas of California*, 62; Aiken, "The Sagebrush War," 53.

42. *Washington Evening Star*, December 19, 1854, 2.

43. "Washington News and Gossip," *Washington Evening Star*, February 2, 1855.

44. *Resolutions of the Legislature of California, in Favor of Attaching a Part of the Territory of Utah, Carson Valley, to the State of California*, 34th Cong., 1st sess., May 2, 1856, S. Misc. Doc. 48, serial 835.

45. *Carson Valley, Utah—Annexation to the State of California—and Eastern Boundary of California*, 34th Cong., 3rd sess., January 20, 1857, H. Rep. 116, serial 912.

46. William Newell Davis, Jr., "The Territory of Nataqua: An Episode in Pioneer Government East of the Sierra," *California Historical Quarterly* 21, no. 3 (September 1942), 225–27; Butler, *Isaac Roop*, 28–29; Aiken, "The Sagebrush War," 49–51. Davis published the entire declaration creating Nataqua Territory in his article's notes; see 234–35.

47. Butler, *Isaac Roop*, 30.

48. Session Laws of Utah, 7th Sess., January 14, 1857, 13.

49. "Organization of the New Territory," *Sacramento Daily Union*, August 24, 1857, 1. See also "A New Territory in the Great American Basin—Mass Meeting—Great Enthusiasm—Judge Crane Selected Delegate," *Mountain Democrat*, August 22, 1857, 1; *Mountain Democrat*, August 22, 1857, 2.

50. Davis, "The Territory of Nataqua," 230. See also "Proceedings of the Territorial Meeting," *Mountain Democrat*, October 24, 1857, 3.

51. "Carson Valley," *Sacramento Daily Union*, August 24, 1857, 2; "Later from Humboldt Valley," *Sacramento Daily Union*, August 25, 1857, 3.

52. Quotation in Davis, "The Territory of Nataqua," 230.

53. "From California," *Daily National Intelligencer*, October 6, 1857, 3; "Proposed New Territory," *Daily National Intelligencer*, October 10, 1857, 3.

54. "Movement of Mormons," *Daily National Intelligencer*, November 7, 1857, 3. See also "Carson Valley," *Daily National Intelligencer*, November 5, 1857, 3.

55. *Resolution of the Legislature of the State of California, in Favor of the Establishment of a Territorial Government in Carson Valley*, 35th Cong., 1st sess., March 1, 1858, S. Misc. Doc. 181, serial 936; "A Letter from Judge Crane to his Constituents," *Mountain Democrat*, March 27, 1858, 3; "From Washington," *New York Times*, May 13, 1858, 4. See also Session Laws of California, 10th General Assembly, February 9, 1859, 385–86.

56. *Mountain Democrat*, 26 June 1858, 2, 3.

57. *Utah Territory*, 36th Cong., 1st sess., May 2, 1860, H. Ex. Doc. 78, serial 1056, 3, 5; Session Laws of Utah, 8th sess., January 17, 1859, 19–20.

58. Russell R. Elliott, *History of Nevada* (1973; repr., Lincoln: University of Nebraska Press, 1987), 61–68.

59. Horace Greeley, *An Overland Journey from New York to San Francisco in the Summer of 1859*, ed. Jo Ann Manfra (1860; repr., Lincoln: University of Nebraska Press, 1999), 227.

60. "Territorial Convention," *Territorial Enterprise*, June 11, 1859, 2.

61. Greeley, *An Overland Journey*, 228–29. See also *Mountain Democrat*, October 29, 1859, 2; Horace Greeley, *Recollections of a Busy Life*, vol. 2 (1873; repr., Port Washington, NY: Kennikat Press, 1971), 378.

62. *Mountain Democrat*, January 21, 1860, 2.

63. "To the People of Western Utah, Included within the Boundaries of the Proposed Territory of Nevada," *Territorial Enterprise*, December 17, 1859, 1. See also "Honey Lake," *Territorial Enterprise*, April 16, 1859, 2; Aiken, "The Sagebrush War," 67.

64. *Resolution of the Legislature of California in Favor of the Establishment of a New Territory in Western Utah*, 36th Cong., 1st sess., February 23, 1860, S. Misc. Doc. 17, serial 1038.

65. See "Nevada Territory," *Territorial Enterprise*, January 14, 1860, 2; "Nevada Territory—Mormon Legislation," *Territorial Enterprise*, January 28, 1860, 2; "Nevada Territory," *Territorial Enterprise*, March 10, 1860, 1.

66. "The Proposed New Territory in the Great Basin," *Mountain Democrat*, October 31, 1857, 1.

67. *Territorial Enterprise*, March 3, 1860, 1. See also "Honey Lake Valley," *Territorial Enterprise*, March 10, 1860, 1.

68. "Indian Outbreak—Horrid Massacre," *Mountain Democrat*, May 12, 1860, 2.

69. William M. Stewart, *Reminiscences of Senator William M. Stewart of Nevada*, ed. George Rothwell Brown (New York: Neale, 1908), 123–24; Elliott, *History of Nevada*, 92–94.

70. "The Indian Outbreak in Utah Territory," *Mountain Democrat*, May 19, 1860, 2; *In the Senate of the United States*, 52nd Cong., 2nd sess., February 1, 1894, S. Rep. 197, serial 3179, 2. See also "The War," *Mountain Democrat*, June 9, 1860, 2.

71. "Special Correspondence from the Seat of War," *Mountain Democrat*, June 2, 1860, 2, italics original.

72. Stewart, *Reminiscences*, 125.

73. See Helen Hunt Jackson, *A Century of Dishonor: A Sketch of the United States Government's Dealings with Some of the Indian Tribes* (1881; repr., Norman: University of Oklahoma Press, 1995), 447. Ferol Egan's *Sand in a Whirlwind: The Paiute Indian War of 1860* (1972; repr., Reno: University of Nevada Press, 1985) offers a narrative of the conflict; a more concise version can be found in Elliott, *History of Nevada*, 92–94.

74. "Shall We Have Law or Anarchy—Order or Confusion?," *Territorial Enterprise*, August 4, 1860, 2.

75. Quotation in William P. MacKinnon, "'Like Splitting a Man Up His Backbone': The Territorial Dismemberment of Utah, 1850–1896," *Utah Historical Quarterly* 71, no. 2 (Spring 2003): 112.

76. "Our Washoe Correspondence," *Daily Alta California*, February 22, 1861, 1.

77. *In the Senate of the United States*, 36th Cong., 2nd sess., January 16, 1861, S. Misc. Doc. 11, serial 1089. A letter to the *Territorial Enterprise* supported Senator Rice's notion. See "Items from Washoe," *Daily Alta California*, March 11, 1861, 1; "Reasons for the Annexation of Washoe to California," *Daily Evening Bulletin*, March 13, 1861, 3.

78. *Appendix to the Congressional Globe*, 36th Cong., 2nd sess. (March 2, 1861), 337.

79. See "From Honey Lake Valley," *Territorial Enterprise*, August 4, 1860, 1; *Mountain Democrat*, September 29, 1860, 2; "Governor's Message," *Mountain Democrat*, January 26, 1861, 1; "Letter from the Boundary Commission," *Daily Alta California*, March 4, 1861, 1; "Letter from Esmeralda," *Daily Alta California*, March 5, 1861, 1; "Letter from the Boundary Commission," *Daily Alta California*, March 10, 1861, 1; Session Laws of California, 12th sess., March 26, 1861, 683; *Sacramento Daily Union*, April 27, 1861, 3.

80. "The Esmeralda Mining District," *Daily Evening Bulletin*, March 8, 1861, 2. See also Earl W. Kersten, Jr., "The Early Settlement of Aurora, Nevada, and Nearby Mining Camps," *Annals of the Association of American Geographers* 54, no. 4 (December 1964): 490–507.

81. *Daily Bee*, March 20, 1861, 4.

82. Printed in *Daily Bee*, April 16, 1861, 4.

83. *Sacramento Daily Union*, April 6, 1861, 2; ibid., April 11, 1861, 4.

84. "What Shall California Do about Nevada?," *Daily Alta California*, April 2, 1861, 2. See also *Sacramento Daily Union*, April 16, 1861, 2.

85. *Daily Bee*, April 5, 1861, 3.

86. *Daily Evening Bulletin (San Francisco)*, April 5, 1861, 2.

87. See *Mountain Democrat*, September 21, 1861, 2; Session Laws of Nevada, 1st Territorial Assembly, November 29, 1861, 269.

88. Aiken, "The Sagebrush War," 73–74.

89. *Daily Alta California*, March 22, 1862, 2. See also "California Legislature," *Daily Alta California*, March 19, 1862, 1.

90. "The Boundaries with Nevada Territory," *Sacramento Daily Union*, March 24, 1862, 3. See also "Constitutional Objections to Altering the State Boundaries," *Daily Alta California*, March 25, 1862, 2.

91. *Congressional Globe*, 37th Cong., 2nd sess. (May 8, 1862), 2030; *Appendix to the Congressional Globe*, 37th Cong., 2nd sess. (July 14, 1862), 408. Nevada's territorial legislature petitioned Congress to pressure Californians to approve the Sierra Nevada boundary in the fall of 1862. See Session Laws of Nevada, 2nd Territorial Assembly, December 20, 1862, 195.

92. See *Sacramento Daily Union*, March 25, 1862, 3; "Affairs in Washoe," *Daily Evening Bulletin*, March 28, 1862, 1; "Matters in Honey Lake Valley—Trouble with the Indians—Terrible Excitement," *Daily Evening Bulletin* (San Francisco), April 12, 1862, 2.

93. *Appendix to the Congressional Globe*, 37th Cong., 2nd sess., July 1, 1862, 383.

94. *Sacramento Daily Union*, February 21, 1863, 3; "Civil War in the Mountains," *Daily Evening Bulletin*, February 23, 1863, 1.

95. "The Difficulties at Honey Lake," *Sacramento Daily Union*, February 26, 1863, 1. See also "The California Boundary," *New York Herald*, February 24, 1863, 1.

96. Quotation in Aiken, "The Sagebrush War," 96.

97. "Our Eastern Boundary Line—Pretentions of Nevada Territory," *Daily Evening Bulletin*, February 25, 1863, 2; "The Conflict of Jurisdiction in the Mountains," *Daily Evening Bulletin*, February 25, 1863, 2; *Message of the Governor to the Legislature of California, March 30, 1863*, in *Western Americana: Frontier History of the United States* (Woodbridge, CT: Research Publications, 1975), reel 88, doc. 901, 3–12.

98. For the opinions of Mark Twain on the nascent Nevada Territory, see his *Roughing It* (1872, repr., New York: Penguin Books, 1985), 177–209. Twain noted Roop's lingering presence as an arbiter in Nevada Territory in the early 1860s as well (254–57).

99. Aiken, "The Sagebrush War," 102.

100. "Letter from Esmeralda," *Daily Alta California*, March 8, 1863, 1; "First Impressions of Esmeralda," *Daily Alta California*, March 14, 1863, 1.

101. "The Conflict of Jurisdiction in the Mountains," *Daily Evening Bulletin*, February 25, 1863, 2; "Disputed Boundary," *Sacramento Daily Union*, April 4, 1863, 2. Even the California State Assembly factored into the confusion: Speaker Timothy N. Machin hailed from Aurora, although no one knew whether his home actually stood within the boundary of California.

102. "Our Eastern Boundary Line—Pretentions of Nevada Territory," *Daily Evening Bulletin*, February 25, 1863, 2; "The Eastern Line of the State—An Hour among the Snowy

Mountains," *Daily Evening Bulletin*, April 1, 1863, 3; "Legislative Proceedings," *Daily Alta California*, April 1, 1863, 1; "Legislative Proceedings," *Daily Alta California*, April 21, 1863, 1.

103. *Appendix to the Congressional Globe*, 38th Cong., 1st sess. (March 21, 1864), 148. See also *Congressional Globe*, 37th Cong., 3rd sess. (March 3, 1863), 1510; *Congressional Globe*, 38th Cong., 1st sess. (February 24, 1864), 787.

104. Francis S. Landrum, "A Major Monument: Oregon-California Boundary," *Oregon Historical Quarterly* 71, no. 1 (March 1971): 18, 49–53. For narratives of the surveys that marked these lines, see James W. Hulse, "The California-Nevada Boundary: The History of a Conflict, Part I," *Nevada Historical Society Quarterly* 23, no. 2 (Summer 1980): 87–109; James W. Hulse, "The California-Nevada Boundary: The History of a Conflict, Part II," *Nevada Historical Society Quarterly* 23, no. 3 (Fall 1980): 157–78.

105. See *Resolution of the State of Nevada, in Favor of the Passage of a Law Fixing as the Eastern Boundary of the State of Nevada the Thirty-Seventh Degree of Longitude West from Washington*, 38th Cong., 2nd sess., February 24, 1865, S. Misc. Doc. 43, serial 1210; *Congressional Globe*, 29th Cong., 1st sess. (March 14, 1866), 1386; *Congressional Globe*, 39th Cong., 1st sess. (May 3, 1866), 2368–70; *Appendix to the Congressional Globe*, 39th Cong., 1st sess. (May 5, 1866), 321.

106. Session Laws of Nevada, 5th legislative sess., March 2, 1871, 188.

107. "Wheeler Survey," *Daily Evening Bulletin* (San Francisco), November 23, 1876, 1.

108. "The Constitution of the State of Nevada," Nevada Legislature, http://www.leg.state.nv.us/const/nvconst.html#Art15 (accessed 20 March 2011).

7. TWO DISTINCT CIVILIZATIONS

1. "Southward Bound," *Denver Daily Tribune*, January 19, 1878, 4.

2. Thomas J. Noel, Paul F. Mahoney, and Richard E. Stevens, *Historical Atlas of Colorado* (Norman: University of Oklahoma Press, 1994), 10; David J. Weber, *The Mexican Frontier, 1821–1846: The American Southwest under Mexico* (Albuquerque: University of New Mexico Press, 1982), 190–95; María E. Montoya, *Translating Property: The Maxwell Land Grant and the Conflict over Land in the American West, 1840–1900* (2002; repr., Lawrence: University Press of Kansas, 2005).

3. P. David Smith, *Ouray: Chief of the Utes* (1986; repr., Ridgway, CO: Wayfinder Press, 1990), 37; Carl Abbott, Stephen J. Leonard, and Thomas J. Noel, *Colorado: A History of the Centennial State* (Boulder: University Press of Colorado, 2006), 36.

4. Frederic L. Paxson, "The Boundaries of Colorado," *University of Colorado Studies* 2, no. 2 (July 1904): 90–91; P. M. Baldwin, "A Historical Note on the Boundaries of New Mexico," *New Mexico Historical Review* 5, no. 2 (April 1930): 132.

5. Quotation in Glen M. Leonard, "Southwestern Boundaries and the Principles of Statemaking," *Western Historical Quarterly* 8, no. 1 (January 1977), 51. See also Malcolm L. Comeaux, "Attempts to Establish and Change a Western Boundary," *Annals of the Association of American Geographers* 72, no. 2 (June 1982), 255–56.

6. Richard L. Nostrand, *The Hispano Homeland* (Norman: University of Oklahoma Press, 1992), 82–84.

7. Francis T. Cheetham, "The Early Settlements of Southern Colorado," *Colorado Magazine* 5, no. 1 (February 1928): 1.

8. Robert W. Frazer, *Forts of the West: Military Forts and Presidios and Posts Commonly Called Forts West of the Mississippi River to 1898* (1965; repr., Norman: University of Oklahoma Press, 1972), 40; D. W. Meinig, *Southwest: Three Peoples in Geographical Change, 1600–1970* (New York: Oxford University Press, 1971), 32; Abbott, Leonard, and Noel, *Colorado*, 36; William

Wyckoff, *Creating Colorado: The Making of a Western American Landscape, 1860–1940* (New Haven: Yale University Press, 1999), 39–41.

9. Quotation in Robert W. Larson, *New Mexico's Quest for Statehood: 1848–1912* (Albuquerque: University of New Mexico Press, 1968), 73.

10. Jay S. Stowell, *The Near Side of the Mexican Question* (New York: George H. Doran, 1921), 61.

11. W. W. H. Davis, *El Gringo, or, New Mexico and Her People* (1857; repr., Chicago: Rio Grande Press, Inc., 1962), 45, 142–43.

12. Frank C. Spencer, *The Story of the San Luis Valley* (1925; repr., Alamosa, CO: San Luis Valley Historical Society, 1975), 45–49; "Death of Major L. Head," *Rocky Mountain News*, March 9, 1897, 1.

13. Quotation in James H. Baker, ed., *History of Colorado*, vol. 2 (Denver: Linderman, 1927), 485.

14. Ibid., 485–86.

15. Spencer, *The Story of the San Luis Valley*, 80.

16. Luther E. Bean, *Land of the Blue Sky People: A Story of the San Luis Valley* (Alamosa, CO: Ye Olde Print Shoppe, 1975), 59.

17. Paxson, "The Boundaries of Colorado," 92–93.

18. *Congressional Globe*, 36th Cong., 2nd sess. (February 6, 1861), 763.

19. Ibid., 764. See also ibid., February 26, 1861, 1205–1206. See also Howard R. Lamar, *The Far Southwest, 1846–1912: A Territorial History*, revised ed. (Albuquerque: University of New Mexico Press, 2000, 189–90.

20. *Congressional Globe*, 36th Cong., 2nd sess. (February 18, 1861), 1003.

21. Ibid., 1004.

22. Ibid., February 27, 1861, 1248; February 28, 1861, 1274.

23. LeRoy R. Hafen, "Colorado's First Legislative Assembly," *Colorado Magazine* 20, no. 2 (March 1943): 43.

24. "Legislature of Colorado," *Rocky Mountain News*, September 14, 1861, 2. See also Hafen, "Colorado's First Legislative Assembly," 46.

25. Noel, Mahoney, and Stevens, *Historical Atlas of Colorado*, 15.

26. Thomas L. Karnes, *William Gilpin: Western Nationalist* (Austin: University of Texas Press, 1970), 265–66, 305.

27. "Legislature of Colorado," *Rocky Mountain News*, September 25, 1861, 2. See also ibid., September 28, 1861, 2; September 30, 1861, 2; October 2, 1861, 2.

28. *Congressional Globe*, 37th Cong., 2nd sess. (May 8, 1862), 2025.

29. *Boundary Line of New Mexico and Colorado*, 38th Cong., 1st sess., April 11, 1864, H. Misc. Doc. 73, serial 1200.

30. James E. Perkins, *Tom Tobin: Frontiersman* (Pueblo West, CO: Herodotus Press, 1999), 127–70; Thomas T. Tobin, "The Capture of the Espinosas," *Colorado Magazine* 9, no. 2 (March 1932): 61.

31. "Governor's Message," *Rocky Mountain News*, February 4, 1864, 2.

32. Session Laws of Colorado Territory, 4th Legislative Assembly, February 10, 1865, 150.

33. Ibid., 6th Legislative Assembly, January 11, 1867, 47–49, 88.

34. Eugene H. Berwanger, *The Rise of the Centennial State: Colorado Territory, 1861–76* (Urbana: University of Illinois Press, 2007), 42.

35. *Message of the President of the United States, Transmitting a Communication Addressed to Him by John Evans and J. B. Chaffee, as United States Senators Elect from the State of Colorado, and Other*

Information in Relation to the Admission of that State into the Union, 39th Cong., 1st sess., January 12, 1866, S. Ex. Doc. 10, serial 1237, 31–32.

36. Carl Ubbelohde, Maxine Benson, and Duane A. Smith, *A Colorado History* (1965; repr., Boulder, CO: Pruett, 2006), 137–38. Congress also doubted the ability of Coloradans to act intelligently considering their rejection of African American suffrage and their unabashed praise for the ghastly Sand Creek Massacre in November 1864.

37. Tom I. Romero, II, "Wringing Rights Out of the Mountains: Colorado's Centennial Constitution and the Ambivalent Promise of Human Rights and Social Equality," *Albany Law Review* 69, no. 2 (April 2006): 570.

38. John L. Kane, Jr., and Sharon Marks Elfenbein, "Colorado: The Territorial and District Courts," in *The Federal Courts of the Tenth Circuit: A History*, ed. James K. Logan (Denver: US District Court of Appeals for the Tenth Circuit, 1992), 42.

39. "Letter of Hon. Francisco Perea," *Santa Fe Weekly Gazette*, February 18, 1865, 1.

40. "The Conejos," *Santa Fe Weekly Gazette*, February 18, 1865, 2.

41. William J. Convery, "Reckless Men of Both Races: The Trinidad War of 1867–68," *Colorado History* 10 (2004): 19–35.

42. Ibid., 25–26; Warren A. Beck and Ynez D. Haase, *Historical Atlas of New Mexico* (Norman: University of Oklahoma Press, 1969).

43. "The Cimarron Mines," *Rocky Mountain News*, January 20, 1868, 2.

44. *Annexation, New Mexico and Colorado*, 40th Cong., 2nd sess., March 6, 1868, H. Misc. Doc. 96, serial 1350.

45. Quotation in Convery, "Reckless Men of Both Races," 26.

46. Samuel Bowles, *The Parks and Mountains of Colorado: A Summer Vacation in the Switzerland of America, 1868*, ed. James H. Pickering (1869; repr., Norman: University of Oklahoma Press, 1991), 173.

47. *Milwaukee Daily Sentinel*, March 10, 1868, 2.

48. Robert B. Houston, "On the Face of the Earth: Marking Colorado's Boundaries, 1868–1925," *Colorado History* 10 (2004): 90–93.

49. "New Mexico," *Daily National Intelligencer*, October 16, 1868, 3.

50. Spencer, *The Story of the San Luis Valley*, 81.

51. William O'Ryan and Thomas H. Malone, *History of the Catholic Church in Colorado, from the Date of the Arrival of Rt. Rev. J. P. Machebeuf, until the Day of His Death* (Denver: C. J. Kelly, 1889), 35–37.

52. Thomas J. Steele, ed., *Archbishop Lamy: In His Own Words* (Albuquerque: LPD Press, 2000), 10.

53. O'Ryan and Malone, *History of the Catholic Church in Colorado*, 43–45, 49–51, 53–55.

54. Ibid., 57, 94–95, 97; Nancy Hanks, "American Diocesan Boundaries in the Southwest, 1840 to 1912," *Catholic Southwest* 11 (2000): 35–36.

55. "A New State," *Harper's Weekly*, June 6, 1874, 471.

56. "New Mexico," *Harper's Weekly*, April 1, 1876, 262–63.

57. Session Laws of Colorado Territory, 10th Legislative Assembly, February 13, 1874, 56; ibid., 11th Legislative Assembly, February 11, 1876, 14.

58. William B. Taylor and Elliott West, "Patrón Leadership at the Crossroads: Southern Colorado in the Late Nineteenth Century," in *The Chicano*, ed. Norris Hundley, Jr. (Santa Barbara, CA: Clio Books, 1975), 90–91.

59. Colorado State Constitution, 1876, Article V, Sections 46–49; ibid., Article XVIII, Section 8; ibid., Ordinances section; ibid., Address, "Miscellaneous" section. The convention's

secretary, W. W. Coulson, misspelled (and mis-gendered) Casimiro Barela's first name as "Casimira" twice in the printed version of the 1876 Constitution.

60. *Daily New Mexican*, November 7, 1876, 1; *Daily New Mexican*, November 11, 1876, 2.

61. Quotation in "Colorado," *Rocky Mountain News*, December 23, 1876, 1. Colorado's legislature employed interpreters for Hispano members through at least 1887. See *Legislative, Historical, and Biographical Compendium of Colorado* (Denver: C. F. Coleman, 1887), 96, 143.

62. Abbott, Leonard, and Noel, *Colorado*, 114; Ubbelohde, Benson, and Smith, *A Colorado History*, 175–76.

63. Abbott, Leonard, and Noel, *Colorado*, 76, 120.

64. Richard D. Lamm and Duane A. Smith, *Pioneers and Politicians: 10 Colorado Governors in Profile* (Boulder, CO: Pruett , 1984), 34–35.

65. "The Election," *Saguache (CO) Chronicle*, October 14, 1876, 2.

66. *Silver World* (Lake City, CO), January 13, 1877, 2; *Silver World*, February 3, 1877, 2.

67. "United States Senator," *Saguache (CO) Chronicle*, October 28, 1876, 2.

68. "From Southwestern Colorado," *Daily Colorado Chieftain*, April 18, 1877, 2. Alva Adams reconciled his rebellious attitudes later in life. After his time in the legislature Adams served three terms as Colorado's governor, the first member of a Colorado political dynasty that included his younger brother Billy, also elected governor, and his son Alva B. Adams, elected U.S. senator.

69. "Southwest Colorado," *Daily Colorado Chieftain*, May 16, 1877, 2.

70. "Del Norte," *Daily Colorado Chieftain*, April 25, 1877, 2; "Ben Butler vs. San Juan Territory," *Daily Colorado Chieftain*, June 6, 1877, 4.

71. "The Territory of San Juan," *Daily Colorado Chieftain*, June 3, 1877, 4. See also "The San Juan Lunacy," *Daily Colorado Chieftain*, June 20, 1877, 4. Some newspaper accounts argued that women's suffrage marked a key difference between the sections as well. With suffrage, some southern Coloradans feared that the larger female population of the north would help ensure the political dominance of that section, whereas the women of the south, it was thought, would be content to let their husbands' vote represent their family's interests. See "The New Territory," *Daily Colorado Chieftain*, June 21, 1877, 2; "Female Suffrage," *Daily Colorado Chieftain*, August 2, 1877, 2. Colorado voters approved women's suffrage in 1893.

72. *Rocky Mountain News*, April 24, 1877, 2. See also "Secession," *Rocky Mountain News*, April 19, 1877, 4.

73. Quotation in "The New Territory," *Denver Weekly Times*, June 27, 1877, 3.

74. "Dividing the State," *Greeley Tribune*, June 20, 1877, 2.

75. In "Troubles in Colorado," *Daily Press and Dakotaian*, June 15, 1877, 2.

76. "The Election," *Saguache (CO) Chronicle*, October 14, 1876, 2.

77. "From San Juan," *Rocky Mountain News*, June 30, 1877, 4.

78. "We Told You So!," *Colorado Transcript*, June 20, 1877, 2. See also Derek R. Everett, *The Colorado State Capitol: History, Politics, Preservation* (Boulder: University Press of Colorado, 2005), 13–18.

79. "The New Territory of San Juan," *Daily Colorado Chieftain*, June 13, 1877, 2. See also *Daily Colorado Chieftain*, June 19, 1877, 4.

80. Quotation in "The Territory of San Juan," *Daily Colorado Chieftain*, August 7, 1877, 4.

81. "The New Territory," *Albuquerque Review*, July 7, 1877, 2.

82. *Daily New Mexican*, June 20, 1877, 1. This bilingual newspaper printed a Spanish version of its report on the proposed San Juan territory on June 22.

83. "La verdad sin rebozo sobre Nuevo México," *Revista Católica*, July 28, 1877, 354–56; ibid., August 11, 1877, 378–80; August 18, 1877, 391–93; September 1, 1877, 416–17; September 22, 1877, 453–54; September 29, 1877, 464–66.

84. "Southwest Colorado," *Daily Colorado Chieftain*, May 16, 1877, 2.

85. Printed in "The New Territory," *Daily Colorado Chieftain*, June 15, 1877, 4.

86. *Daily Colorado Chieftain*, June 27, 1877, 4.

87. "Fourth of July Carnival," *Daily Colorado Chieftain*, June 28, 1877, 4; "San Juan," *Daily Colorado Chieftain*, July 7, 1877, 4. See also "Fourth of July Celebration," *Daily Colorado Chieftain*, June 23, 1877, 4.

88. "San Juan," *Daily Colorado Chieftain*, July 7, 1877, 4.

89. "Dividing the State," *Rocky Mountain News*, July 3, 1877, 2.

90. See "Good Words for the 'Chronicle,'" *Saguache (CO) Chronicle*, June 30, 1877, 3.

91. Quotation in "Smart Aleck on His Travels," *Daily Colorado Chieftain*, July 31, 1877.

92. See "Ula," *Daily Colorado Chieftain*, July 11, 1877, 2; *Rocky Mountain News*, August 9, 1877, 2; *Saguache (CO) Chronicle*, November 10, 1877, 3. The notion percolated for several months but never gained traction after mid-summer 1877. See *Rocky Mountain News*, August 9, 1877, 2; "The New Territory," *Saguache (CO) Chronicle*, October 13, 1877, 2.

93. See "The New Territory," *Denver Weekly Times*, June 27, 1877.

94. See Robert Athearn, *The Denver and Rio Grande Western Railroad: Rebel of the Rockies* (1962; repr., Lincoln: University of Nebraska Press, 1977), 38–44.

95. Frank Hall, *History of the State of Colorado,* vol. 2 (Chicago: Blakely Printing Company, 1890), 487.

96. From the *San Juan Prospector*, in the *Saguache Chronicle*, July 21, 1877, 2.

97. "Butler and the New Territory," *Silver World* (Lake City, CO), June 16, 1877, 2. See also "The New Territory Twaddle," *Silver World*, June 30, 1877, 2.

98. "From Our Traveling Correspondent," *Daily New Mexican*, September 20, 1877, 1.

99. "Southward Bound," *Denver Daily Tribune*, January 19, 1878, 4.

100. Taylor and West, "Patrón Leadership at the Crossroads," 95; Abbott, Leonard, and Noel, *Colorado*, 40–42; Wyckoff, *Creating Colorado*, 198–201.

101. "The Governor's Message," *Denver Daily Times*, January 3, 1901, 7; *Survey of Boundaries of Colorado, New Mexico, and Oklahoma*, 57th Cong., 1st sess., May 14, 1902, H. Doc. 604, serial 4377, 2; *Resurvey of Boundary between Colorado, New Mexico, and Oklahoma, etc.*, 57th Cong., 2nd sess., December 9, 1902, H. Doc. 120, serial 4489; *Reestablishment of Boundary Line between Colorado and New Mexico and Oklahoma*, 58th Cong., 3rd sess., January 10, 1905, S. Doc. 89, serial 4765; Frank Minitree Johnson, "The Colorado–New Mexico Boundary," *Colorado Magazine* 4, no. 3 (May 1927), 113–14; Houston, "On the Face of the Earth," 100–101. One Denver newspaper suggested that the Democratic majority of Edith's county, Archuleta, sought to foist that Republican-leaning town onto New Mexico to reinforce its local power. See "To Define Boundaries," *Denver Daily Times*, April 12, 1901, 10.

102. *Reestablishment of Boundary Line*, January 10, 1905, 9–10.

103. *Boundary Line between Colorado and Oklahoma and New Mexico*, 60th Cong., 2nd sess., December 19, 1908, S. Doc. 604, serial 5407.

104. For a thorough narrative of the arduous accomplishment of New Mexico statehood in the early twentieth century, see David V. Holtby, *Forty-Seventh Star: New Mexico's Struggle for Statehood* (Norman: University of Oklahoma Press, 2012).

105. "New Mexico Sues Colorado for Land," *Rocky Mountain News*, July 16, 1919, 5; "New Mexico Wants Area in Colorado," *Rocky Mountain News*, October 24, 1915, 1, 4.

106. United States Reports, vol. 267, 26 January 1925, 30–42.

107. "Colorado Wins Boundary Suit from New Mex.," *Denver Post*, January 26, 1925, 1, 11; "Boundary Dispute with New Mexico Won by Colorado," *Rocky Mountain News*, January 27, 1925, 14.

108. "New Mexico Loses Out in Border Dispute," *Santa Fe New Mexican,* January 26, 1925, 1; "N.M. Loses Its Boundary Suit with Colorado," *Albuquerque Morning Journal,* January 27, 1925, 1.

109. "Colorado to Stay Same Size, Says Border Re-survey Chief," *Rocky Mountain News,* September 2, 1949, 5; "State Line Marking Job to Be Completed in '50," *Denver Post,* January 7, 1950, 16.

110. "Colorado–N.M. Boundary Row Nears End," *Rocky Mountain News,* June 28, 1960, 14.

111. "Colorado Wins Boundary Feud," *Rocky Mountain News,* October 26, 1960, 38.

112. Spencer, *The Story of the San Luis Valley,* 60.

113. Virginia McConnell Simmons, *The San Luis Valley: Land of the Six-Armed Cross* (1979; repr. Niwot: University Press of Colorado, 1999), 128. For more information on the heritage and significance of the Hispano population in the American Southwest, see José de Onís, ed., *The Hispanic Contribution to the State of Colorado* (Boulder, CO: Westview Press, 1976); Richard L. Nostrand, "The Hispano Homeland in 1900," *Annals of the Association of American Geographers* 70, no. 3 (September 1980): 382–96; and Nostrand, *The Hispano Homeland.*

8. LET US DIVIDE

1. *Daily Press and Dakotaian,* April 21, 1887, 1.

2. Gary E. Moulton, ed., *The Journals of the Lewis and Clark Expedition,* vol. 3, *August 25, 1804–April 6, 1805* (Lincoln: University of Nebraska Press, 1987), 31–33. The Siouan people recognized three large groups within their common culture: the Dakotas (or Santee Sioux), Lakotas (or Teton Sioux), and Nakotas (or Yankton Sioux). As the Dakota people inhabited the part of the northern plains first occupied by Americans, specifically the region that would become Minnesota, their name became synonymous with the region as a whole even though the Lakotas dominated the modern states of North and South Dakota. The first encounter between the Corps of Discovery and the Sioux—specifically Lakotas—took place in late September 1804 near present-day Pierre, South Dakota.

3. George Catlin, *North American Indians,* ed. Peter Matthiessen (1844; repr., New York: Penguin Books, 1989), 202–206.

4. Francis Newton Thorpe, *The Federal and State Constitutions, Colonial Charters, and Other Organic Laws of the States, Territories, and Colonies Now or Heretofore Forming the United States of America,* vol. 2 (Washington, DC: Government Printing Office, 1909), 1111–12.

5. Ibid., vol. 7, 4065–66.

6. *Missouri Whig and General Advertiser,* March 13, 1841, 2.

7. Francis Newton Thorpe, *The Federal and State Constitutions, Colonial Charters, and Other Organic Laws of the States, Territories, and Colonies Now or Heretofore Forming the United States of America,* vol. 4 (Washington, DC: Government Printing Office, 1909), 1981.

8. *Congressional Globe,* 33rd Cong., 1st sess. (May 30, 1854), 2228–29.

9. See Grant K. Anderson, "The Politics of Land in Dakota Territory: Early Skirmishes, 1857–1861," *South Dakota History* 9, no. 3 (Summer 1979): 210–32; William E. Lass, "The First Attempt to Organize Dakota Territory," in *Centennial West: Essays on the Northern Tier States,* ed. William L. Lang (Seattle: University of Washington Press, 1991), 143–68.

10. *Congressional Globe,* 36th Cong., 1st sess. (May 11, 1860), 2066–77.

11. *Congressional Globe,* 36th Cong., 2nd sess. (February 16, 1861), 946; *Appendix to the Congressional Globe,* 36th Cong., 2nd sess. (March 2, 1861), 346–48.

12. Thorpe, *The Federal and State Constitutions,* vol. 2, 905.

13. *Appendix to the Congressional Globe,* 38th Cong., 1st sess. (May 26, 1864), 165–67.

14. Earl S. Pomeroy, *The Territories and the United States, 1861–1890: Studies in Colonial Administration* (1947; repr., Seattle: University of Washington Press, 1969), 65.

15. Session Laws of Dakota Territory, 7th Legislative Assembly, 1867, 287. See also ibid., 276.

16. Thorpe, *The Federal and State Constitutions*, vol. 7, 4105–11.

17. Moses K. Armstrong, *Centennial Address on Dakota Territory, Giving Its History, Growth, Population, and Resources* (Philadelphia: J. B. Lippincott, 1876), 6.

18. James S. Foster, *Outlines of History of the Territory of Dakota, and Emigrant's Guide to the Free Lands of the Northwest* (Yankton, Dakota Territory: M'Intyre & Foster, 1870), 6.

19. *A Brief History of Dakota, the Wonderland!* (Huron, Dakota Territory: "Huronite" Steam Publishing House, 1882), 5.

20. Armstrong, *Centennial Address on Dakota Territory*, 5.

21. PBS, "Fort Laramie Treaty, 1868," in *Archives of "The West"* (PBS, 2001). http://www.pbs.org/weta/thewest/resources/archives/four/ftlaram.htm (accessed June 7, 2012)

22. Session Laws of Dakota Territory, 9th Legislative Assembly, January 12, 1871, 597–98.

23. Session Laws of Dakota Territory, 10th Legislative Assembly, December 31, 1872, 241.

24. Session Laws of Dakota Territory, 7th Legislative Assembly, 1867, 277.

25. Session Laws of Dakota Territory, 11th Legislative Assembly, December 19, 1874, 347–48.

26. See Herbert S. Schell, *History of South Dakota* (1975; repr., Pierre: South Dakota State Historical Society Press, 2004), 128–40. See also "Ratified," *Daily Press and Dakotaian*, February 16, 1877.

27. "New Territorial Organization Suggested," *Daily Press and Dakotaian*, August 2, 1876, 3.

28. "County Organizations in the Hills," *Daily Press and Dakotaian*, September 24, 1876, 2.

29. In his work on contemporary eastern South Dakota, Jon W. Lauck suggests that the territorial moniker "Lincoln" emerged as a result of the large numbers of Union veterans of the Civil War in the region. While this might have been true for the "east river" communities, the communities east of the Missouri, the prevalence of Union veterans in the dramatically different mining communities is less certain. See Jon W. Lauck, *Prairie Republic: The Political Culture of Dakota Territory, 1879–1889* (Norman: University of Oklahoma Press, 2010), 36.

30. *Daily Press and Dakotaian*, April 11, 1877, 2.

31. Jesse Brown and A. M. Willard, *The Black Hills Trails: A History of the Struggles of the Pioneers in the Winning of the Black Hills*, ed. John T. Milek (Rapid City, SD: Rapid City Journal, 1924), 354.

32. "The Boundary Line," *Daily Press and Dakotaian*, June 16, 1877, 2.

33. *Daily Press and Dakotaian*, June 30, 1877, 2; "Under Which King?," *Daily Colorado Chieftain*, July 19, 1877, 4; *Daily Press and Dakotaian*, July 26, 1877, 2.

34. *Daily Press and Dakotaian*, September 26, 1877, 2.

35. "The New Territory Scheme," *Daily Press and Dakotaian*, November 5, 1877, 2; "Lincoln Territory," *Daily Press and Dakotaian*, November 5, 1877, 3.

36. "A Scheme against Dakota's Interests," *Daily Press and Dakotaian*, February 20, 1878, 2; *Daily Press and Dakotaian*, February 22, 1878, 2; ibid., February 23, 1878, 2; "Dividing Dakota," *Daily Press and Dakotaian*, February 23, 1878, 4; "Bismarck," *Chicago Daily Tribune*, March 2, 1878, 12; *Daily Press and Dakotaian*, March 6, 1878, 2; *Dakota Herald*, March 9, 1878, 2; *Daily Press and Dakotaian*, March 27, 1878, 2.

37. Howard Roberts Lamar, *Dakota Territory, 1861–1889: A Study of Frontier Politics* (New Haven: Yale University Press, 1965), 165.

38. Session Laws of Dakota Territory, 11th Legislative Assembly, December 19, 1874, 348–49.

39. *Daily Press and Dakotaian*, August 10, 1876, 2; "The New Territory," *Daily Press and Dakotaian*, August 12, 1876, 2; *Daily Press and Dakotaian*, February 22, 1877, 2; "State of Dakota," *Daily Press and Dakotaian*, November 23, 1877, 4. For another account of the north-south Dakota division proposals of the 1870s, see Grant K. Anderson, "The First Movement to Divide Dakota Territory, 1871–77," *North Dakota History* 49, no. 1 (Winter 1982): 20–28.

40. "Division and Admission," *Dakota Herald*, December 25, 1880, 2.

41. "The State of Dakota," *Daily Press and Dakotaian*, April 20, 1877, 2. See also *Daily Press and Dakotaian*, April 14, 1877, 2; "The State Organization," *Daily Press and Dakotaian*, April 21, 1877, 2; "Dividing Dakota," *Daily Press and Dakotaian*, December 17, 1877, 4.

42. "The State Movement," *Dakota Herald*, May 5, 1877, 2. See also "The State Movement Again," *Dakota Herald*, May 19, 1877, 2; "The State," *Dakota Herald*, November 23, 1878, 2.

43. *Dakota Herald*, December 6, 1879, 1.

44. "Gov. Ordway," *Dakota Herald*, June 26, 1880, 3.

45. Pomeroy, *The Territories of the United States*, 69.

46. *Dakota Herald*, September 10, 1881, 2.

47. Burleigh F. Spalding, "Constitutional Convention, 1889," *North Dakota History* 31, no. 3 (July 1964): 151–52.

48. *Dakota Herald*, September 24, 1881, 2. See also ibid., October 22, 1881, 2; November 26, 1881, 2.

49. "Dividing Dakota," *Dakota Herald*, October 8, 1881, 2; *Dakota Herald*, December 10, 1881, 2.

50. "The Sioux Falls Meeting," *Dakota Herald*, January 28, 1882, 3.

51. Quotation in "Don't Want Division Without the Offices," *Dakota Herald*, April 1, 1882, 2.

52. *Admission of Dakota into the Union as a State*, 47th Cong., 1st sess., February 16, 1882, H. Rep. 450, serial 2066; *In the Senate of the United States*, 47th Cong., 1st sess., March 20, 1882, S. Rep. 271, serial 2004.

53. *Congressional Record*, 47th Cong., 1st sess. (February 3, 1882), 860–61.

54. Session Laws of Nebraska, 17th Legislature, Extra Session, May 23, 1882, 56–57. See also *Northern Boundary of Nebraska*, 51st Cong., 1st sess., February 14, 1890, H. Ex. Doc. 201, serial 2746. Surveyors marked Nebraska's northern line west from the Keya Paha River in 1874. In 1893 another survey of the northern line, starting at the state's northwestern corner and going eastward past the Keya Paha to the Missouri River, formalized the annexation approved by federal and state officials in 1882. See Rollin C. Curd, *A History of the Boundaries of Nebraska and Indian-Surveyor Stories* (Chadron, NE: Boundaries Publishing, 1999), 49–62.

55. "The Canton Convention!," *Dakota Herald*, June 24, 1882, 2, "The Coming New State!," *Dakota Herald*, July 1, 1882, 1. See also Lamar, *Dakota Territory*, 199–202.

56. In "Dakota as One State," *Dakota Herald*, December 9, 1882, 1.

57. *Congressional Record*, 47th Cong., 2nd sess. (February 5, 1883), 2106, 2108, 2109.

58. "Bismarck," *Chicago Daily Tribune*, February 22, 1878, 2; *Daily Press and Dakotaian*, February 25, 1878, 2.

59. *Dakota Herald*, February 24, 1883, 2.

60. "A Suggestion to Capital Movers," *Dakota Herald*, February 10, 1883, 3.

61. *Dakota Herald*, March 10, 1883, 2.

62. Robert F. Karolevitz, *Yankton: A Pioneer Past* (Aberdeen, SD: North Plains Press, 1972), 108. The *Dakota Herald* excerpted many editorials opposing the capital removal; see March 10, 1883, 3; March 17, 1883, 1–2; March 24, 1883, 1; March 31, 1883, 1; April 14, 1883, 1; May 12, 1883, 2.

63. See L. Martin Perry, "According to Plan: A History of the North Dakota Capitol Grounds," in *The North Dakota State Capitol: Architecture and History*, ed. Larry Remele (Bismarck: State Historical Society of North Dakota, 1989), 31–37.

64. "North Pacific," *Chicago Daily Tribune*, September 6, 1883, 3.

65. "The Huron Resolutions," *Dakota Herald*, July 7, 1883, 2.

66. "The Constitution," *Dakota Herald*, September 22, 1883, 2. For a thorough discussion of the 1883 constitution's provisions and its influence on those drawn up for South Dakota in 1885 and 1889, see Lauck, *Prairie Republic*, 111–35.

67. "The State Schemers' Plight," *Dakota Herald*, November 24, 1883, 1.

68. "The Senate Admission Bill," *Dakota Herald*, March 15, 1884, 1; Schell, *History of South Dakota*, 213–14.

69. *Congressional Record*, 48th Cong., 2nd sess. (December 9, 1884), 107.

70. Ibid., December 10, 1884, 143.

71. Ibid., December 11, 1884, 184.

72. Ibid., December 15, 1884, 234–38; ibid., December 16, 1884, 279–82; "Opponents of Division," *Daily Press and Dakotaian*, September 6, 1887, 1.

73. *In the Senate of the United States*, 49th Cong., 1st sess., January 11, 1886, S. Rep. 15, serial 2355, 1–2, 22–24, 31–32, 40, 68–72.

74. "Dakota Is Assuming Airs," *New York World*, December 19, 1885, 1.

75. Lamar, *Dakota Territory*, 247–48, 255–56. See also *Daily Press and Dakotaian*, June 1, 1887, 1.

76. "Statehood Prospects," *Daily Press and Dakotaian*, May 7, 1886, 3; *Daily Press and Dakotaian*, May 20, 1886, 1; *Admission of Southern Half of Dakota*, 49th Cong., 1st sess., May 25, 1886, H. Rep. 2578, serial 2443; "The Minority," *Daily Press and Dakotaian*, May 26, 1886, 1.

77. *Admission of Dakota as a State*, 49th Cong., 1st sess., May 25, 1886, H. Rep. 2577, serial 2442, 3.

78. *Daily Press and Dakotaian*, December 20, 1886, 1; "Division Discussion," *Daily Press and Dakotaian*, February 16, 1887, 2. See also *Daily Press and Dakotaian*, June 11, 1887, 1.

79. *Daily Press and Dakotaian*, February 1, 1887, 1.

80. Ibid., June 13, 1887, 1.

81. Ibid., April 21, 1887, 1.

82. "The Divisionists," *Bismarck Daily Tribune*, July 14, 1887, 1; "The Division Convention," *Daily Press and Dakotaian*, July 15, 1887, 3.

83. *Bismarck Daily Tribune*, August 13, 1887, 2.

84. Ibid., August 24, 1887, 2.

85. Ibid., September 21, 1887, 2; September 23, 1887, 2.

86. *Daily Press and Dakotaian*, October 6, 1887, 1.

87. "North Dakota Address," *Daily Press and Dakotaian*, October 12, 1887, 4.

88. *Daily Press and Dakotaian*, October 19, 1887, 1; "Looking to Statehood," *Bismarck Daily Tribune*, November 16, 1887, 2.

89. "Officially Reported," *Daily Press and Dakotaian*, December 14, 1887, 3.

90. *Daily Press and Dakotaian*, November 16, 1887, 1; "Dakota's Two Sections," *Daily Press and Dakotaian*, December 16, 1887, 2; "The Convention," *Bismarck Daily Tribune*, December 16, 1887, 1; ibid., December 17, 1887, 2; *Daily Press and Dakotaian*, December 19, 1887, 1;

"Springer's Letter," *Daily Press and Dakotaian*, December 20, 1887, 1; "Statehood for One Dakota," *Bismarck Daily Tribune*, December 28, 1887, 2.

91. "State of Dakota," *Daily Press and Dakotaian*, January 10, 1888, 3; *In the Senate of the United States*, 50th Cong., 1st sess., February 13, 1888, S. Misc. Doc. 55, serial 2516.

92. *In the Senate of the United States*, 50th Cong., 1st sess., January 23, 1888, S. Rep. 75, serial 2519; *Admission of South Dakota into the Union and for the Organization of Territory of North Dakota*, 50th Cong., 1st sess., February 24, 1888, H. Rep. 709, serial 2600; *North Dakota*, 50th Cong., 1st sess., February 24, 1888, H. Rep. 710, serial 2600.

93. "Dakota Statehood," *Daily Press and Dakotaian*, December 22, 1887, 4; "Gifford's Statehood Bills," *Bismarck Daily Tribune*, December 23, 1887, 2; "Division and Admission," *Daily Press and Dakotaian*, February 14, 1888, 4.

94. *Bismarck Daily Tribune*, February 14, 1888, 2; "That Bologna Bill," *Bismarck Daily Tribune*, 17 February 1888, 2.

95. *Bismarck Daily Tribune*, April 21, 1888, 2.

96. "Two States," *Bismarck Daily Tribune*, May 4, 1888, 2; *Daily Press and Dakotaian*, May 8, 1888, 1; *Bismarck Daily Tribune*, May 8, 1888, 2; ibid., May 19, 1888, 2.

97. *Daily Press and Dakotaian*, May 21, 1888, 1; "A Convention Called," *Daily Press and Dakotaian*, May 26, 1888, 1.

98. *Daily Press and Dakotaian*, November 8, 1888, 1.

99. "Good for Dakota," *Bismarck Daily Tribune*, November 16, 1888, 2; *Daily Press and Dakotaian*, November 19, 1888, 1; ibid., November 20, 1888, 1; "Mellette and Harrison," *Daily Press and Dakotaian*, November 21, 1888, 1; "Gifford and Mellette," *Daily Press and Dakotaian*, November 23, 1888, 1; *Daily Press and Dakotaian*, November 28, 1888, 1; "Statehood for Dakota," *Daily Press and Dakotaian*, December 11, 1888, 1.

100. "The Jamestown Meeting," *Bismarck Daily Tribune*, November 29, 1888, 2; *Daily Press and Dakotaian*, December 1, 1888, 1; "Oh, for Statehood," *Bismarck Daily Tribune*, December 6, 1888, 1; "Self Government," *Bismarck Daily Tribune*, December 7, 1888, 1; "North Dakota," *Bismarck Daily Tribune*, January 10, 1889, 4; *In the Senate of the United States*, 50th Cong., 2nd sess., January 16, 1889, S. Misc. Doc. 39, serial 2615.

101. *Daily Press and Dakotaian*, November 27, 1888, 1; *Bismarck Daily Tribune*, November 28, 1888, 2; "Proposed New State," *Daily Press and Dakotaian*, December 7, 1888, 1.

102. *Daily Press and Dakotaian*, January 19, 1889, 1; "The Bill Outlined," *Daily Press and Dakotaian*, January 21, 1889, 3; "As It Passed," *Daily Press and Dakotaian*, January 26, 1889, 1; "Springer's Bill," *Daily Press and Dakotaian*, January 28, 1889, 1; "Admission Sure," *Daily Press and Dakotaian*, February 15, 1889, 3; "Dakota Rejoiceth," *Bismarck Daily Tribune*, February 16, 1889, 1; "The Statehood Law," *Daily Press and Dakotaian*, February 27, 1889, 1–2. For a comprehensive interpretation of congressional debate over Dakota division that winter, see Julie Marvyl Andresen Koch, "The Omnibus Bill: Statehood for Dakota Territory," *North Dakota History* 49, no. 3 (Summer 1982): 18–26.

103. "Signed by Cleveland," *Daily Press and Dakotaian*, February 22, 1889, 3; "North and South," *Bismarck Daily Tribune*, February 23, 1889, 1.

104. "Looks Well in Ink," *Washington Post*, July 31, 1889, 6.

105. Francis Newton Thorpe, *The Federal and State Constitutions, Colonial Charters, and Other Organic Laws of the States, Territories, and Colonies Now or Heretofore Forming the United States of America*, vol. 6 (Washington, DC: Government Printing Office, 1909), 3358; ibid., vol. 5, 2284.

106. A full account of the joint commission's work appeared in the *Bismarck Daily Tribune*, August 6, 1889, 2, 6–7. See also Spalding, "Constitutional Convention," 155–56, 159–60.

107. "Two New States Are In," *Washington Post*, November 3, 1889, 1. Although President Harrison did not indicate which state he admitted first, tradition grants North Dakota the rank of thirty-ninth and South Dakota fortieth in the Union.

108. "North and South Dakota," *Morning Oregonian*, November 15, 1889, 6.

109. "North Dakota's Boundary," *Morning Oregonian*, November 23, 1889, 7.

110. Gordon L. Iseminger, *The Quartzite Border: Surveying and Marking the North Dakota–South Dakota Boundary, 1891–1892* (Sioux Falls, SD: Center for Western Studies, 1988), 18–21, 49–51, 57, 90.

111. *Bismarck Daily Tribune*, September 30, 1892, 2.

112. *The New States: A Sketch of the History and Development of the States of North Dakota, South Dakota, Montana, and Washington, with Map and Illustrations* (New York: Ivison, Blakeman, 1889), 46.

113. "Winner of the NDSU-SDSU Game to Be Presented Dakota Marker Trophy," Great West Conference, www.greatwestconference.org/genrel/042204aaa.html (posted April 22, 2004, accessed November 5, 2011).

114. See James D. McLaird, "From Bib Overalls to Cowboy Boots: The East River/West River Differences in South Dakota," *South Dakota History* 19, no. 4 (Winter 1989): 445–91.

115. Mark Wahlgren Summers, *Party Games: Getting, Keeping, and Using Power in Gilded Age Politics* (Chapel Hill: University of North Carolina Press, 2004), 132–33.

116. Summers, *Party Games*, 132.

117. See John W. Morris and Edwin C. McReynolds, *Historical Atlas of Oklahoma* (Norman: University of Oklahoma Press, 1965), 30, 44; Jon T. Kilpinen, "Land Speculation and the Case of Greer County, Texas," *Southwestern Historical Quarterly* 109, no. 1 (July 2005): 71–98.

118. William A. DeGregorio, *The Complete Book of U.S. Presidents*, 4th ed. (New York: Wings Books, 1993), 346.

CONCLUSION

1. Quoted in *Missouri Republican*, September 3, 1839, 2.

2. Richard Wilbur, "Slap Leather Pard, Yer on My Claim!," *Rocky Mountain News*, March 26, 1962, 8.

3. See John W. Morris and Edwin C. McReynolds, *Historical Atlas of Oklahoma* (Norman: University of Oklahoma Press, 1965), 53.

4. David Sholtz to Edwin C. Johnson, February 29, 1936, Governor Edwin C. Johnson Papers, box 26915, folder 1, Colorado State Archives. The troopers remained in place until the end of Sholtz's term in January 1937. See Joan M. Crouse, *The Homeless Transient in the Great Depression: New York State, 1929–1941* (Albany: SUNY Press, 1986), 226.

5. See H. Mark Wild, "If You Ain't Got that Do-Re-Mi: The Los Angeles Border Patrol and White Migration in Depression-Era California," *Southern California Quarterly* 83, no. 3 (Fall 2001): 317–34.

6. See Richard D. Lamm and Duane A. Smith, *Pioneers and Politicians: 10 Colorado Governors in Profile* (Boulder, CO: Pruett, 1984), 132; Carl Abbott, Stephen J. Leonard, and Thomas J. Noel, *Colorado: A History of the Centennial State* (Boulder: University Press of Colorado, 2005), 357.

7. "Statement by Governor Jan Brewer," April 23, 2010, http://azgovernor.gov/dms/upload/PR_042310_StatementByGovernorOnSB1070.pdf (accessed November 27, 2010).

8. "Anti-Illegal Immigration Laws in States," *New York Times*, April 22, 2012, http://www.nytimes.com/ (accessed April 25, 2012).

9. Bill Mears, "High Court Appears to Lean toward Arizona in Immigration Law Dispute," CNN, April 25, 2012, www.cnn.com/ (accessed April 25, 2012).

10. Tom Cohen and Bill Mears, "Supreme Court Mostly Rejects Arizona Immigration Law; Gov Says 'Heart' Remains," CNN, June 25, 2012, http://www.cnn.com/ (accessed June 25, 2012).

11. Adam Liptak and Adam H. Cushman Jr., "Supreme Court Rejects Part of Arizona Immigration Law," *New York Times*, June 26, 2012, www.nytimes.com/ (accessed June 26, 2012).

12. "Jefferson Is Selected as State Name," *Siskiyou Daily News* (Yreka, CA), November 24, 1941, 1; "Manifesto on Rebellion Is Issued," *Siskiyou Daily News*, November 27, 1941, 1; "Sentinels on Yreka's Roads Spread Word of 'Secession'," *Medford Mail Tribune*, November 28, 1941, 1. See also Richard W. Reinhardt, "The Short, Happy History of the State of Jefferson," *American West* 9, no. 3 (May 1972): 36–42; Thomas K. Worcester, *The State of Jefferson and Other Yarns of the Klamaths, Siskiyous, Southern Cascades and Northern Sierra from an Area Roughly Enclosing the Border Lands of California and Oregon between the 40th and 44th Parallels, Stretching from the Pacific Ocean to the Vast Reaches East of the Mountain Ranges* (Beaverton, OR: TMS Book Service, 1982); Steve Wilson, "Jefferson State of Mind," *American History* 39, no. 6 (February 2005): 20–69; Kevin Dickinson, "Jefferson Call to Start Over," *Siskiyou Daily News*, August 14, 2013, www.siskiyoudaily.com/ (accessed August 21, 2013).

13. Lew Ferguson, "School Plan Would Sock the Southwest," *Garden City (KS) Telegram*, January 22, 1992, 1; Tim Unruh, "Forming a 51st State?," *Garden City Telegram*, March 17, 1992, 1; Tim Unruh, "Leaders Won't Quit 51st State," *Garden City Telegram*, March 18, 1992, 1; Michael Bates, "Secession Leaders Say They're Serious," *Southwest Daily Times* (Liberal, KS), March 18, 1992, 1; Tim Unruh, "Most Vote to Secede," *Garden City Telegram*, April 8, 1992, 1–3; Tim Unruh, "Secession Leaders Set Tax Equality as Their Goal," *Garden City Telegram*, September 9, 1992, 3; Charlie Hayes, "Secessionists Open Convention," *Southwest Daily Times*, September 11, 1992, 1, 2; Tim Unruh, "Defiance Grows at Ulysses," *Garden City Telegram*, September 11, 1992, 1, 3. See also Peter J. McCormick, "The 1992 Secession Movement in Southwest Kansas," *Great Plains Quarterly* 15, no. 4 (Fall 1995): 247–58.

14. Analisa Romano, "Weld County Commissioners Propose Formation of New State, North Colorado," *Greeley Tribune*, June 6, 2013, www.greeleytribune.com (accessed June 6, 2013); Thomas Hendrick and Eli Stokols, "Rural County Commissioners Discuss Seceding from Colorado," Fox 31 Denver, June 6, 2013, http://kdvr.com/ (accessed June 6, 2013); "The 51st State Initiative," Facebook, www.facebook.com/The51stStateInitiative (accessed August 20, 2013).

15. Herbert E. Bolton, *The Spanish Borderlands: A Chronicle of Old Florida and the Southwest* (New Haven: Yale University Press, 1921). Historians Pekka Hämäläinen and Samuel Truett argue that Bolton's editor rather than Bolton himself coined the term "borderlands," yet it remains Bolton's contribution in the eyes of most scholars. See Pekka Hämäläinen and Samuel Truett, "On Borderlands," *Journal of American History* 98, no. 2 (September 2011): 341.

16. Jeremy Adelman and Stephen Aron, "From Borderlands to Borders: Empires, Nation-States, and the Peoples in Between in North American History," *American Historical Review* 104, no. 3 (June 1999): 814–41 (italics original).

17. John R. Wunder and Pekka Hämäläinen, "Of Lethal Places and Lethal Essays," *American Historical Review* 104, no. 4 (October 1999): 1229–34.

18. Hämäläinen and Truett, "On Borderlands," 338–61.

19. Ibid., 348–49.

20. Kelly Lytle Hernández, "Borderlands and the Future History of the American West," *Western Historical Quarterly* 42, no. 3 (Autumn 2011): 325–30.

21. Adelman and Aron, "From Borderlands to Borders," 815.

22. D. W. Meinig, *The Shaping of America: A Geographical Perspective on 500 Years of History*, vol. 2, *Continental America, 1800–1867* (New Haven: Yale University Press, 1993), 128.

23. Glen M. Leonard, "Southwestern Boundaries and the Principles of Statemaking," *Western Historical Quarterly* 8, no. 1 (January 1977): 53.

24. Richard White, *"It's Your Misfortune and None of My Own": A New History of the American West* (Norman: University of Oklahoma Press, 1991), 3.

25. See Patricia Nelson Limerick, "The Adventures of the Frontier in the Twentieth Century," in *The Frontier in American Culture*, ed. James R. Grossman (Berkeley: University of California Press, 1994), 72–80; Kerwin Lee Klein, "Reclaiming the 'F' Word, or Being and Becoming Postwestern," *Pacific Historical Review* 65, no. 2 (May 1996): 179–215. Unable to avoid the scholarly contention, in 1999 Adelman and Aron identified frontiers as zones of interaction without clear limits of the players' authority, whereas borderlands emerged in the disputed area between imperial powers in the Americas. See Adelman and Aron, "From Borderlands to Borders," 815–16. One of their critics, Evan Haefeli, disputes that Adelman and Aron provided and followed a clear difference between the two terms, and he rejected their notion of a steady progression from one condition to the other. See Evan Haefeli, "A Note on the Use of North American Borderlands," *American Historical Review* 104, no. 4 (October 1999), 1222–25. In the vein of Hämäläinen and Truett, I view borderlands as existing between various players in North America, not just Euro-American empires. That interpretation combined with Adelman and Aron's definitions, therefore, suggests to me that "borderlands" and "frontier" are interchangeable terms to describe areas of interaction and competition between two centers of authority, however constituted.

26. *Blazing Saddles* (DVD), directed by Mel Brooks, 1974 (Burbank, CA: Warner Home Video, 1997).

27. Association for Borderlands Studies, www.absborderlands.org/ (accessed November 23, 2011).

28. "Recent Articles," *Western Historical Quarterly* 33, no. 4 (Winter 2002): 521; "Recent Articles," *Western Historical Quarterly* 34, no. 1 (Spring 2003): 117. The shift in borderlands identification corresponded with other updated topic headings, such as the shift from "Women" to "Gender and Sexuality."

29. Greg Kaza, "A Texarkana Tax Tale," *National Review Online*, May 2, 2007, www.nationalreview.com/ (accessed November 21, 2011).

30. *Congressional Globe*, 31st Cong., 1st sess. (August 5, 1850), 1520; Elvis E. Fleming, "The Texas–New Mexico Boundary: Whatever Happened to the 103rd Meridian?," *Southwest Heritage* 5, no. 1 (Spring 1975): 38–39; Ralph H. Brock, "'Perhaps the Most Incorrect of Any Land Line in the United States': Establishing the Texas–New Mexico Boundary along the 103rd Meridian," *Southwestern Historical Quarterly* 109, no. 4 (April 2006): 431, 446–52, 456.

31. Session Laws of New Mexico, 21st Legislature (1953), 670–71.

32. Brock, "Perhaps the Most Incorrect," 461; "New Mexico Seeks to Reclaim Land Lost in Boundary Struggle," *Santa Fe New Mexican*, January 31, 1991, 1; James E. Garcia, "Texas Faces Border Battles," *Austin American-Statesman*, March 24, 1991, B7.

33. Fritz Thompson, "N.M.'s Drawing the Line," *Albuquerque Journal*, April 7, 1991, G1, G7.

34. Garcia, "Texas Faces Border Battles," B1.

35. Brock, "Perhaps the Most Incorrect," 461, 462.

36. Stephen Speckman, "54% Back Nevada Annex of Wendover," *Deseret News*, July 10, 2001 www.deseretnews.com/ (accessed November 21, 2011).

37. *Congressional Record*, 107th Cong., 2nd sess. (June 11, 2002), 9995–99. See also Lee Davidson, "U.S. House OKs 'Secession,'" *Deseret News*, June 12, 2002, www.deseretnews.com/ (accessed November 21, 2011).

38. "Legislative Committee on Public Lands Subcommittee to Study the Feasibility and Desirability of Changing the State Boundary Line along the Border with Utah," Nevada Legislature, August 5, 2004, www.leg.state.nv.us/ (accessed November 21, 2011).

39. "City Merger Looks Dead in Wendover," *Deseret News*, March 6, 2006, www.deseretnews.com/ (accessed November 21, 2011); "West Wendover Puts off Decision," *Deseret News*, March 8, 2006, www.deseretnews.com/ (accessed November 21, 2011).

40. For recent examples of articles that could contribute to the study of intranational borderlands, see Mark Fiege, "The Weedy West: Mobile Nature, Boundaries, and Common Space in the Montana Landscape," *Western Historical Quarterly* 36, no. 1 (Spring 2005): 22–47; and Rob Kuper, "Joining the Great Plains in Space, Place, and Time: Questioning a Time Zone Boundary," *Great Plains Quarterly* 31, no. 3 (Summer 2011): 223–42.

41. Herbert E. Bolton, "The Epic of Greater America," *American Historical Review* 38, no. 3 (April 1933): 473.

42. "Cannot Be Isolated, Harding Declares," *New York Times*, May 24, 1921, 2.

APPENDIX

1. See Floyd Calvin Shoemaker, "Some Historic Lines in Missouri," *Missouri Historical Review* 3, no. 4 (July 1909): 273; Eric McKinley Erickson, "The Honey War," *Palimpsest* 6 (September 1924): 343–44; Duane Meyer, *The Heritage of Missouri—A History*, revised ed. (St. Louis: State Publishing Co., 1973), 185; Thomas M. Spencer, "'Demand Nothing but what is Strictly Right and Submit to Nothing that is Wrong': Governor Lilburn Boggs, Governor Robert Lucas, and the Honey War of 1839," *Missouri Historical Review* 103, no. 1 (October 2008), 31.

Bibliography

ARCHIVAL SOURCES

Association for Borderlands Studies (http://www.absborderlands.org)
Governor Edwin C. Johnson Papers, Colorado State Archives, Denver

GOVERNMENT SOURCES

American State Papers
Annals of the Congress of the United States
Arizona Office of the Governor; http://azgovernor.gov/
Articles of Confederation, 1777
Colorado State Constitution, 1876
Congressional Globe
Congressional Record
The Debates and Proceedings in the Congress of the United States
Gales and Seaton's Register of Debates in Congress
Journals of the Continental Congress
Montana State Constitution, 1972
Nevada Legislature; http://www.leg.state.nv.us/
Oregon State Constitution, 2005
Session Laws of Arkansas, 1905
Session Laws of California, 1852, 1859, 1861
Session Laws of Colorado, 1865, 1867, 1874, 1876
Session Laws of Dakota, 1867, 1871, 1872, 1874, 1877
Session Laws of Iowa, 1838, 1839
Session Laws of Louisiana, 1820
Session Laws of Missouri, 1836, 1839, 1840, 1841
Session Laws of Nebraska, 1882
Session Laws of Nevada, 1861, 1862, 1871
Session Laws of New Mexico, 1953
Session Laws of Oregon, 1957
Session Laws of Utah, 1854, 1857, 1859
Session Laws of Washington, 1857
United States Constitution, 1787
United States Reports
United States Serial Set
United States Statutes at Large, 1905
Washington Secretary of State; http://www.secstate.wa.gov/

BOOKS, ARTICLES, THESES, AND DISSERTATIONS

Abbott, Carl. *Political Terrain: Washington, D.C. from Tidewater Town to Global Metropolis.* Chapel Hill: University of North Carolina Press, 1999.

Abbott, Carl, Stephen J. Leonard, and Thomas J. Noel. *Colorado: A History of the Centennial State.* Boulder: University Press of Colorado, 2005.

Adelman, Jeremy, and Stephen Aron. "From Borderlands to Borders: Empires, Nation-States, and the Peoples in Between in North American History." *American Historical Review* 104, no. 3 (June 1999): 814–41.

Agnew, Brad. "The Cherokee Struggle for Lovely's Purchase." *American Indian Quarterly* 2, no. 4 (Winter 1975): 347–61.

Aiken, Charles Curry. "The Sagebrush War: The California-Nevada Boundary Dispute on the 120th Meridian." *Journal of the Shaw Historical Library* 5, no. 1–2 (1991): 45–112.

Anderson, Fred. *Crucible of War: The Seven Years' War and the Fate of Empire in British North America, 1754–1766.* New York: Vintage Books, 2000.

Anderson, Grant K. "The First Movement to Divide Dakota Territory, 1871–77." *North Dakota History* 49, no. 1 (Winter 1982): 20–28.

———. "The Politics of Land in Dakota Territory: Early Skirmishes, 1857–1861." *South Dakota History* 9, no. 3 (Summer 1979): 210–32.

Armstrong, Moses K. *Centennial Address on Dakota Territory, Giving Its History, Growth, Population, and Resources.* Philadelphia: J. B. Lippincott, 1876.

Aron, Stephen. *American Confluence: The Missouri Frontier from Borderland to Border State.* Bloomington: Indiana University Press, 2006.

———. *How the West Was Lost: The Transformation of Kentucky from Daniel Boone to Henry Clay.* Baltimore: Johns Hopkins University Press, 1996.

Athearn, Robert. *The Denver and Rio Grande Western Railroad: Rebel of the Rockies.* Lincoln: University of Nebraska Press, 1977. First published 1962.

Baker, James H., ed. *History of Colorado.* Five volumes. Denver: Linderman, 1927.

Baird, W. David. "Arkansas's Choctaw Boundary: A Study of Justice Delayed." *Arkansas Historical Quarterly* 28, no. 3 (Autumn 1969): 203–22.

Baldwin, P. M. "A Historical Note on the Boundaries of New Mexico." *New Mexico Historical Review* 5, no. 2 (April 1930): 116–37.

Bancroft, George. *History of the Formation of the Constitution of the United States of America.* Two volumes. New York: D. Appleton, 1882.

Bancroft, Hubert Howe. *History of Arizona and New Mexico, 1530–1888.* Albuquerque: Horn & Wallace, 1962. First published 1889.

———. *The Works of Hubert Howe Bancroft.* Thirty-nine volumes. San Francisco: A. L. Bancroft, 1882–90.

Barrett, John Ira. "The Legal Aspects of the Iowa-Missouri Boundary Dispute." Master's thesis, Drake University, 1959.

Bean, Luther E. *Land of the Blue Sky People: A Story of the San Luis Valley.* Alamosa, CO: Ye Olde Print Shoppe, 1975.

Bearss, Edwin C., and Arrell M. Gibson. *Fort Smith: Little Gibraltar on the Arkansas.* 2nd ed. Norman: University of Oklahoma Press, 2002.

Beck, Warren A., and Ynez D. Haase. *Historical Atlas of California.* Norman: University of Oklahoma Press, 1974.

———. *Historical Atlas of New Mexico.* Norman: University of Oklahoma Press, 1969.

———. *Historical Atlas of the American West.* Norman: University of Oklahoma Press, 1989.

Beckett, Paul L. *From Wilderness to Enabling Act: The Evolution of a State of Washington*. Pullman: Washington State University Press, 1968.

Berkhofer, Robert F., Jr. "Jefferson, the Ordinance of 1784, and the Origins of the American Territorial System." *William and Mary Quarterly* 29, no. 2 (April 1972): 231–62.

Berwanger, Eugene H. *The Rise of the Centennial State: Colorado Territory, 1861–76*. Urbana: University of Illinois Press, 2007.

Bolton, Herbert E. "The Epic of Greater America." *American Historical Review* 38, no. 3 (April 1933): 448–74.

———. *The Spanish Borderlands: A Chronicle of Old Florida and the Southwest*. New Haven: Yale University Press, 1921.

Bolton, S. Charles. "Jeffersonian Indian Removal and the Emergence of Arkansas Territory." In *A Whole Country in Commotion: The Louisiana Purchase and the American Southwest*, edited by Patrick G. Williams, S. Charles Bolton, and Jeannie M. Whayne, 77–90. Fayetteville: University Press of Arkansas, 2005.

———. *Territorial Ambition: Land and Society in Arkansas, 1800–1840*. Fayetteville: University of Arkansas Press, 1993.

Bowen, Catherine Drinker. *Miracle at Philadelphia: The Story of the Constitutional Convention, May to September 1787*. New York: Little, Brown, 1966.

Bowles, Samuel. *The Parks and Mountains of Colorado: A Summer Vacation in the Switzerland of America, 1868*. Edited by James H. Pickering. Norman: University of Oklahoma Press, 1991. First published 1869.

Boyd, Julian P., ed. *The Papers of Thomas Jefferson, Vol. 1, 1760–1776*. Princeton, NJ: Princeton University Press, 1950.

———. *The Papers of Thomas Jefferson, Vol. 6, 21 May 1781–1 March 1784*. Princeton, NJ: Princeton University Press, 1952.

A Brief History of Dakota, the Wonderland! Huron, Dakota Territory: "Huronite" Steam Publishing House, 1882.

Brock, Ralph H. "'Perhaps the Most Incorrect of Any Land Line in the United States': Establishing the Texas-New Mexico Boundary along the 103rd Meridian." *Southwestern Historical Quarterly* 109, no. 4 (April 2006): 431–62.

Brown, Jesse, and A. M. Willard. *The Black Hills Trails: A History of the Struggles of the Pioneers in the Winning of the Black Hills*. Edited by John T. Milek. Rapid City, SD: Rapid City Journal, 1924.

Browne, J. Ross. *Report of the Debates in the Convention of California, on the Formation of the State Constitution, in September and October, 1849*. Washington, DC: J. T. Towers, 1850.

Burrows, J. M. D. "Rumors of War." *Palimpsest* 24, no. 2 (February 1943): 71–72.

Butler, James Thomas. *Isaac Roop: Pioneer and Political Leader in Northeastern California*. Janesville, CA: High Desert Press, 1994.

Byron, Matthew A. "Crime and Punishment: The Impotency of Dueling Laws in the United States." PhD diss., University of Arkansas, 2008.

Calloway, Colin G. *The Scratch of a Pen: 1763 and the Transformation of North America*. Oxford, UK: Oxford University Press, 2006.

Carter, Clarence, ed. *The Territorial Papers of the United States*, vol. 20, *Territorial Papers of Arkansas, 1825–1829*. Washington, DC: Government Printing Office, 1954.

Catlin, George. *North American Indians*. Edited by Peter Matthiessen. New York: Penguin Books, 1989. First published 1844.

Cayton, Andrew R. L. *The Frontier Republic: Ideology and Politics in the Ohio Country, 1780–1825*. Kent, OH: Kent State University Press, 1986.

Cheetham, Francis T. "The Early Settlements of Southern Colorado." *Colorado Magazine* 5, no. 1 (February 1928): 1–8.

Chitwood, Oliver Perry. *A History of Colonial America*. New York: Harper & Brothers, 1931.

Cline, Eric H. and Mark W. Graham. *Ancient Empires: From Mesopotamia to the Rise of Islam*. New York: Cambridge University Press, 2011.

Cohen, Tom, and Bill Mears. "Supreme Court Mostly Rejects Arizona Immigration Law; Gov Says 'Heart' Remains." CNN, June 25, 2012. http://www.cnn.com/2012/06/25/politics/scotus-arizona-law/index.html?hpt=hp_t1 (accessed June 25, 2012).

Collins, E. H. "Why the Idaho Panhandle?" *Pacific Northwesterner* 18, no. 3 (Summer 1974): 41–46.

Combs, H. Jason. "The Platte Purchase and Native American Removal." *Plains Anthropologist* 47, no. 182 (August 2002): 265–74.

Comeaux, Malcolm L. "Attempts to Establish and Change a Western Boundary." *Annals of the Association of American Geographers* 72, no. 2 (June 1982), 254–71.

Convery, William J. "Reckless Men of Both Races: The Trinidad War of 1867–68." *Colorado History* 10 (2004): 19–35.

Corkran, David H. *The Cherokee Frontier: Conflict and Survival, 1740–1762*. Norman: University of Oklahoma Press, 1962.

Cronon, William. *Changes in the Land: Indians, Colonists, and the Ecology of New England*. Revised ed. New York: Hill and Wang, 2003.

Crouse, Joan M. *The Homeless Transient in the Great Depression: New York State, 1929–1941*. Albany: SUNY Press, 1986.

Curd, Rollin C. *A History of the Boundaries of Nebraska and Indian-Surveyor Stories*. Chadron, NE: Boundaries Publishing, 1999.

Danson, Edwin. *Drawing the Line: How Mason and Dixon Surveyed the Most Famous Border in America*. New York: John Wiley & Sons, 2001.

Davis, James E. *Frontier Illinois*. Bloomington: Indiana University Press, 1998.

Davis, William Newell, Jr. "The Territory of Nataqua: An Episode in Pioneer Government East of the Sierra." *California Historical Quarterly* 21, no. 3 (September 1942), 225–38.

Davis, W. W. H. *El Gringo, or, New Mexico and Her People*. Chicago: Rio Grande Press, 1962 First published 1857.

DeGregorio, William A. *The Complete Book of U.S. Presidents*. 4th ed. New York: Wings Books, 1993.

DeRosier, Arthur H. *The Removal of the Choctaw Indians*. Knoxville: University of Tennessee Press, 1970.

Deutsch, Herman J. "The Evolution of Territorial and State Boundaries in the Inland Empire of the Pacific Northwest." *Pacific Northwest Quarterly* 51, no. 3 (July 1960), 115–31.

Drake, James D. *The Nation's Nature: How Continental Presumptions Gave Rise to the United States of America*. Charlottesville: University of Virginia Press, 2011.

DuVal, Kathleen. *The Native Ground: Indians and Colonists in the Heart of the Continent*. Philadelphia: University of Pennsylvania Press, 2006.

Egan, Ferol. *Sand in a Whirlwind: The Paiute Indian War of 1860*. Reno: University of Nevada Press, 1985. First published 1972.

Ehrlich, Daniel Henry. "Problems Arising from Shifts of the Missouri River on the Eastern Border of Nebraska." *Nebraska History* 57, no. 2 (Fall 1973), 341–63.

Elliott, Russell R. *History of Nevada*. Lincoln: University of Nebraska Press, 1987. First published 1973.

Ellison, William Henry. *A Self-Governing Dominion: California, 1849–1860*. Berkeley: University of California Press, 1950.

Erickson, Eric McKinley. "The Honey War." *Palimpsest* 6 (September 1924): 339–50.
Etcheson, Nicole. *Bleeding Kansas: Contested Liberty in the Civil War Era*. Lawrence: University Press of Kansas, 2004.
Everett, Derek R. *The Colorado State Capitol: History, Politics, Preservation*. Boulder: University Press of Colorado, 2005.
———. "Frontiers Within: State Boundaries and Borderlands in the United States." PhD diss., University of Arkansas, 2008.
———. "On the Extreme Frontier: Crafting the Western Arkansas Boundary." *Arkansas Historical Quarterly* 67, no. 1 (Spring 2008): 1–26.
———. "To Shed Our Blood for Our Beloved Territory: The Iowa-Missouri Borderland." *Annals of Iowa* 67, no. 4 (Fall 2008): 269–97.
Ficken, Robert E. "Columbia, Washington or Tacoma: The Naming and Attempted Renaming of Washington Territory." *Columbia Magazine* 17, no. 1 (Spring 2003): 25–30.
———. *Washington Territory*. Pullman: Washington State University Press, 2002.
Fiege, Mark. "The Weedy West: Mobile Nature, Boundaries, and Common Space in the Montana Landscape." *Western Historical Quarterly* 36, no. 1 (Spring 2005): 22–47.
Fleming, Elvis E. "The Texas-New Mexico Boundary: Whatever Happened to the 103rd Meridian?" *Southwest Heritage* 5, no. 1 (Spring 1975): 38–40.
Forbes, Robert Pierce. *The Missouri Compromise and Its Aftermath: Slavery and the Meaning of America*. Chapel Hill: University of North Carolina Press, 2007.
Foreman, Grant, ed. *A Traveler in Indian Territory: The Journal of Ethan Allen Hitchcock, Late Major-General in the United States Army*. Cedar Rapids, IA: Torch Press, 1930.
Foster, Dave. *Tennessee: Territory to Statehood*. 2nd ed. Johnson City, TN: Overmountain Press, 2002.
Foster, James S. *Outlines of History of the Territory of Dakota, and Emigrant's Guide to the Free Lands of the Northwest*. Yankton, Dakota Territory: M'Intyre & Foster, 1870.
Frazer, Robert W. *Forts of the West: Military Forts and Presidios and Posts Commonly Called Forts West of the Mississippi River to 1898*. Norman: University of Oklahoma Press, 1972. First published 1965.
Fulton, A.R. "Van Buren County." *Annals of Iowa* 3 (April 1884): 43–48.
Gabler, Ina. "Lovely's Purchase and Lovely County." *Arkansas Historical Quarterly* 19, no. 1 (Spring 1960): 31–39.
Gannett, Henry. *Boundaries of the United States and of the Several States and Territories with an Outline of the History of All Important Changes of Territory*. Washington, DC: Government Printing Office, 1904.
Gebhard, David, and Gerald Mansheim. *Buildings of Iowa*. New York: Oxford University Press, 1993.
Goodwin, Cardinal. "The Question of the Eastern Boundary of California in the Convention of 1849." *Southwestern Historical Quarterly* 16, no. 4 (January 1913): 227–58.
Greeley, Horace. *An Overland Journey from New York to San Francisco in the Summer of 1859*. Edited by Jo Ann Manfra. Lincoln: University of Nebraska Press, 1999. First published 1860.
———. *Recollections of a Busy Life*, vol. 2. Port Washington, NY: Kennikat Press, 1971. First published 1873.
Guyton, Kathy. *U.S. State Names: The Stories of How Our States Were Named*. Nederland, CO: Mountain Storm Press, 2009.
Haefeli, Evan. "A Note on the Use of North American Borderlands." *American Historical Review* 104, no. 4 (October 1999): 1222–25.
Hafen, LeRoy R. "Colorado's First Legislative Assembly." *Colorado Magazine* 20, no. 2 (March 1943): 41–50.

Hall, Frank. *History of the State of Colorado*, vol. 2. Chicago: Blakely Printing, 1890.
Hämäläinen, Pekka, and Samuel Truett. "On Borderlands." *Journal of American History* 98, no. 2 (September 2011): 338–61.
Hamilton, Holman. *Prologue to Conflict: The Crisis and Compromise of 1850*. Lexington: University of Kentucky Press, 1964.
Hamilton, Stanislaus Murray, ed. *The Writings of James Monroe, Including a Collection of His Public and Private Papers and Correspondence Now for the First Time Published*. New York: AMS Press, 1969. First published 1898.
Hanks, Nancy. "American Diocesan Boundaries in the Southwest, 1840 to 1912." *Catholic Southwest* 11 (2000): 27–43.
Hansen, Tegan, and Jerry Hansen. *State Boundaries of America: How, Why and When American State Lines Were Formed*. Westminster, MD: Heritage Books, 2007.
Hanson, Gerald T., and Carl H. Moneyhon. *Historical Atlas of Arkansas*. Norman: University of Oklahoma Press, 1989.
Harrison, Lowell H. *Kentucky's Road to Statehood*. Lexington: University Press of Kentucky, 1992.
Hastings, Lansford W. *The Emigrant's Guide to Oregon and California, Containing Scenes and Incidents of a Party of Oregon Emigrants; a Description of Oregon; Scenes and Incidents of a Party of California Emigrants; and a Description of California; with a Description of the Different Routes to Those Countries; and All Necessary Information Relative to the Equipment, Supplies, and the Method of Traveling*. Bedford, MA: Applewood Books, 1994. First published 1845.
Hayes, Carlton J. H., and James H. Hanscom. *Ancient Civilizations: Prehistory to the Fall of Rome*. New York: MacMillan, 1983.
Hayes, Troy L. "Missouri/Iowa Boundary Line Investigation." *American Surveyor* 3, no. 2 (March/April 2006): 33–37.
Hebard, Alfred. "The Border War between Iowa and Missouri on the Boundary Question." *Annals of Iowa* 1, no. 8 (January 1895): 651–57.
Hendrick, Thomas, and Eli Stokols. "Rural County Commissioners Discuss Seceding from Colorado." Fox 31 Denver, June 6, 2013. http://kdvr.com/2013/06/06/rural-county-commissioners-discuss-seceding-from-colorado (accessed June 6, 2013).
Hernandéz, Kelly Lytle. "Borderlands and the Future History of the American West." *Western Historical Quarterly* 62, no. 3 (Autumn 2011): 325–30.
Hicks, Urban E. *Yakima and Clickitat Indian Wars, 1855 and 1856*. Portland, OR: Himes the Printer, 1886.
Hill, Craig. "The Honey War." *Pioneer America* 14, no. 2 (July 1982): 81–88.
Hines, Gustavus. *Oregon: Its History, Condition and Prospects*. Buffalo: Geo. H. Derby, 1851.
The History of Van Buren County, Iowa, Containing a History of the County, its Cities, Towns, &c. Chicago: Western Historical, 1878.
Hoig, Stanley. *The Cherokees and Their Chiefs: In the Wake of Empire*. Fayetteville: University of Arkansas Press, 1998.
Holst, Hermann von. *The Constitutional and Political History of the United States*. Seven volumes. Chicago: Callaghan, 1877.
Holt, Michael F. *The Fate of Their Country: Politicians, Slavery Extension, and the Coming of the Civil War*. New York: Hill and Wang, 2004.
Holtby, David V. *Forty-Seventh Star: New Mexico's Struggle for Statehood*. Norman: University of Oklahoma Press, 2012.
Houston, Robert B. "On the Face of the Earth: Marking Colorado's Boundaries, 1868–1925." *Colorado History* 10 (2004): 87–105.

Hubbard, Bill, Jr. *American Boundaries: The Nation, the States, the Rectangular Survey.* Chicago: University of Chicago Press, 2009.

Hulse, James W. "The California-Nevada Boundary: The History of a Conflict, Part I." *Nevada Historical Society Quarterly* 23, no. 2 (Summer 1980): 87–109.

———. "The California-Nevada Boundary: The History of a Conflict, Part II." *Nevada Historical Society Quarterly* 23, no. 3 (Fall 1980): 157–78.

Hunt, Norman Bancroft. *Historical Atlas of Ancient Mesopotamia.* New York: Checkmark Books, 2004.

Hurt, R. Douglas. *The Ohio Frontier: Crucible of the Old Northwest, 1720–1830.* Bloomington: Indiana University Press, 1998. First published 1996.

Iseminger, Gordon L. *The Quartzite Border: Surveying and Marking the North Dakota–South Dakota Boundary, 1891–1892.* Sioux Falls, SD: Center for Western Studies, 1988.

Jackson, Helen Hunt. *A Century of Dishonor: A Sketch of the United States Government's Dealings with Some of the Indian Tribes.* Norman: University of Oklahoma Press, 1995. First published 1881.

Jefferson, Thomas. *Notes on the State of Virginia.* Edited by David Waldstreicher. Boston: Bedford/St. Martin's, 2002. First published 1787.

Jensen, Merrill. *The Articles of Confederation: An Interpretation of the Social-Constitutional History of the American Revolution, 1774–1781.* Madison: University of Wisconsin Press, 1963.

———. *The New Nation: A History of the United States during the Confederation, 1781–1789.* New York: Vintage Books, 1950.

Johnson, Frank Minitree. "The Colorado–New Mexico Boundary." *Colorado Magazine* 4, no. 3 (May 1927): 112–15.

Jones, Landon Y., Jr. "The Council that Changed the West: William Clark at Portage des Sioux." *Gateway Heritage* 24, nos. 2 and 3 (Fall 2003 and Winter 2004): 88–95.

Kane, John L., Jr., and Sharon Marks Elfenbein. "Colorado: The Territorial and District Courts." In *The Federal Courts of the Tenth Circuit: A History,* edited by James K. Logan, 37–78. Denver: U.S. District Court of Appeals for the Tenth Circuit, 1992.

Kappler, Charles J., ed. *Indian Affairs: Laws and Treaties.* Washington, DC: Government Printing Office, 1904.

Karnes, Thomas L. *William Gilpin: Western Nationalist.* Austin: University of Texas Press, 1970.

Karolevitz, Robert F. *Yankton: A Pioneer Past.* Aberdeen, SD: North Plains Press, 1972.

Kastor, Peter J. *The Nation's Crucible: The Louisiana Purchase and the Creation of America.* New Haven: Yale University Press, 2004.

Kaza, Greg. "A Texarkana Tax Tale." *National Review Online,* May 2, 2007. www.nationalreview.com/articles/220802/texarkana-tax-tale/greg-kaza (accessed November 21, 2011).

Kelley, Hall J. *A Geographical Sketch of that Part of North America, Called Oregon.* Boston: J. Howe, 1830.

Kersten, Earl W., Jr. "The Early Settlement of Aurora, Nevada, and Nearby Mining Camps." *Annals of the Association of American Geographers* 54, no. 4 (December 1964): 490–507.

Key, Joseph Patrick. "Indians and Ecological Conflict in Territorial Arkansas." *Arkansas Historical Quarterly* 59, no. 2 (Summer 2000): 127–46.

Kilpinen, Jon T. "Land Speculation and the Case of Greer County, Texas." *Southwestern Historical Quarterly* 109, no. 1 (July 2005): 71–98.

Klein, Kerwin Lee. "Reclaiming the 'F' Word, or Being and Becoming Postwestern." *Pacific Historical Review* 65, no. 2 (May 1996): 179–215.

Koch, Julie Marvyl Andresen. "The Omnibus Bill: Statehood for Dakota Territory." *North Dakota History* 49, no. 3 (Summer 1982): 18–26.

Kochmann, Rachel M. *Presidents: Birthplaces, Homes, and Burial Sites: A Pictorial Guide.* Detroit Lakes, MN: Midwest Printing, 1993. First published 1976.

Kraus, Caroll J. "A Study in Border Confrontation: The Iowa-Missouri Boundary Dispute." *Annals of Iowa* 40, no. 2 (Fall 1969): 81–107.

Kuper, Rob. "Joining the Great Plains in Space, Place, and Time: Questioning a Time Zone Boundary." *Great Plains Quarterly* 31, no. 3 (Summer 2011): 223–42.

Lamar, Howard Roberts. *Dakota Territory, 1861–1889: A Study of Frontier Politics.* New Haven: Yale University Press, 1965.

———. *The Far Southwest, 1846–1912: A Territorial History.* Revised ed. Albuquerque: University of New Mexico Press, 2000.

Lamm, Richard D. and Duane A. Smith. *Pioneers and Politicians: 10 Colorado Governors in Profile.* Boulder, CO: Pruett, 1984.

Landers, Frank E. "The Southern Boundary of Iowa." *Annals of Iowa* 1, no. 8 (January 1895): 641–51.

Landrum, Francis S. "A Major Monument: Oregon-California Boundary." *Oregon Historical Quarterly* 71, no. 1 (March 1971): 5–53.

Larson, Robert W. *New Mexico's Quest for Statehood: 1848–1912.* Albuquerque: University of New Mexico Press, 1968.

Larzelere, Claude S. "Notes and Documents: The Iowa-Missouri Disputed Boundary." *Mississippi Valley Historical Review* 3, no. 1 (June 1916): 77–84.

Lass, William E. "The First Attempt to Organize Dakota Territory." In *Centennial West: Essays on the Northern Tier States,* edited by William L. Lang, 143–68. Seattle: University of Washington Press, 1991.

Lauck, Jon W. *Prairie Republic: The Political Culture of Dakota Territory, 1879–1889.* Norman: University of Oklahoma Press, 2010.

Lavender, David. Fort Vancouver and the Pacific Northwest. In *Fort Vancouver.* National Park Handbook 113. Washington, DC: Government Printing Office, 2001.

Lawson, Gary, and Guy Seidman. *The Constitution of Empire: Territorial Expansion and American Legal History.* New Haven: Yale University Press, 2004.

Legislative, Historical, and Biographical Compendium of Colorado. Denver: C. F. Coleman, 1887.

Leonard, Glen M. "The Mormon Boundary Question in the 1849–50 Statehood Debates." *Journal of Mormon History* 18, no. 1 (Summer 1992): 114–36.

———. "Southwestern Boundaries and the Principles of Statemaking." *Western Historical Quarterly* 8, no. 1 (January 1977): 39–53.

LeSueur, Stephen C. *The 1838 Mormon War in Missouri.* Columbia: University of Missouri Press, 1987.

Limerick, Patricia Nelson. "The Adventures of the Frontier in the Twentieth Century." In *The Frontier in American Culture,* edited by James R. Grossman, 72–80. Berkeley: University of California Press, 1994.

Linklater, Andro. *The Fabric of America: How Our Borders and Boundaries Shaped the Country and Forged Our National Identity.* New York: Walker, 2007.

———. *Measuring America: How the United States Was Shaped by the Greatest Land Sale in History.* New York: Walker, 2002.

MacKinnon, William P. "'Like Splitting a Man Up His Backbone': The Territorial Dismemberment of Utah, 1850–1896." *Utah Historical Quarterly* 71, no. 2 (Spring 2003): 100–24.

Maclay, William. *The Journal of William Maclay, United States Senator from Pennsylvania, 1789–1791.* New York: Frederick Ungar, 1965. First published 1890.

Madison, James. *Journal of the Federal Convention.* Two volumes. Edited by E. H. Scott. Chicago: Albert, Scott, 1893.

Malone, Henry Thompson. *Cherokees of the Old South: A People in Transition.* Athens: University of Georgia Press, 1956.

McClintock, Thomas C. "British Newspapers and the Oregon Treaty of 1846." *Oregon Historical Quarterly* 104, no. 1 (Spring 2003): 96–109.

McCormick, Peter J. "The 1992 Secession Movement in Southwest Kansas." *Great Plains Quarterly* 15, no. 4 (Fall 1995): 247–58.

McCormick, Richard P. "The 'Ordinance' of 1784?" *William and Mary Quarterly* 50, no. 1 (January 1993): 112–22.

McDonald, Forrest. *States' Rights and the Union: Imperium in Imperio, 1776–1876.* Lawrence: University Press of Kansas, 2000.

McKee, Jesse O., and John A. Schlenker. *The Choctaws: Cultural Evolution of a Native American Tribe.* Jackson: University Press of Mississippi, 1980.

McLaird, James D. "From Bib Overalls to Cowboy Boots: The East River/West River Differences in South Dakota." *South Dakota History* 19, no. 4 (Winter 1989): 445–91.

Mears, Bill. "High Court Appears to Lean toward Arizona in Immigration Law Dispute." CNN, April 25, 2012. www.cnn.com/2012/04/25/justice/scotus-arizona-law/index.html?hpt=hp_c1 (accessed April 25, 2012).

Meinig, D. W. *The Great Columbia Plain: A Historical Geography, 1805–1910.* Seattle: University of Washington Press, 1995. First published 1968.

———. *The Shaping of America: A Geographical Perspective on 500 Years of History*, vol. 1: *Atlantic America, 1492–1800.* New Haven: Yale University Press, 1986.

———. *The Shaping of America: A Geographical Perspective on 500 Years of History*, vol. 2: *Continental America, 1800–1867.* New Haven: Yale University Press, 1993.

———. *The Shaping of America: A Geographical Perspective on 500 Years of History*, vol. 3: *Transcontinental America, 1850–1915.* New Haven: Yale University Press, 1998.

———. *Southwest: Three Peoples in Geographical Change, 1600–1970.* New York: Oxford University Press, 1971.

Merk, Frederick. *The Oregon Question: Essays in Anglo-American Diplomacy and Politics.* Cambridge, MA: Harvard University Press, 1967.

Meyer, Duane. *The Heritage of Missouri—A History.* St. Louis: State Publishing, 1973. First published 1963.

Miller, Lee. *Roanoke: Solving the Mystery of the Lost Colony.* New York: Penguin Books, 2000.

Milton, Giles. *Big Chief Elizabeth: How England's Adventurers Gambled and Won the New World.* London: Hodder and Stoughton, 2000.

Monaghan, Jay. *Civil War on the Western Border, 1854–1865.* Lincoln: University of Nebraska Press, 1955.

Montoya, María E. *Translating Property: The Maxwell Land Grant and the Conflict over Land in the American West, 1840–1900.* Lawrence: University Press of Kansas, 2005. First published 2002.

Morris, John W., and Edwin C. McReynolds. *Historical Atlas of Oklahoma.* Norman: University of Oklahoma Press, 1965.

Moulton, Gary E., ed. *The Journals of the Lewis and Clark Expedition*, vol. 3, *August 25, 1804–April 6, 1805.* Lincoln: University of Nebraska Press, 1987.

Myers, Robert A. "Cherokee Pioneers in Arkansas: The St. Francis Years, 1785–1813." *Arkansas Historical Quarterly* 56, no. 2 (Summer 1997): 127–57.

Neely, Jeremy. *The Border between Them: Violence and Reconciliation on the Kansas-Missouri Line.* Columbia: University of Missouri Press, 2007.
Negus, Charles. "The Southern Boundary of Iowa." *Annals of Iowa* 4, no. 4 (October 1866): 743–53.
———. "Southern Boundary of Iowa." *Annals of Iowa* 5, no. 1 (January 1867): 786–93.
The New States: A Sketch of the History and Development of the States of North Dakota, South Dakota, Montana, and Washington, with Map and Illustrations. New York: Ivison, Blakeman, 1889.
Noel, Thomas J., Paul F. Mahoney, and Richard E. Stevens. *Historical Atlas of Colorado.* Norman: University of Oklahoma Press, 1994.
Nostrand, Richard L. *The Hispano Homeland.* Norman: University of Oklahoma Press, 1992.
———. "The Hispano Homeland in 1900." *Annals of the Association of American Geographers* 70, no. 3 (September 1980): 382–96.
Onís, José de, ed. *The Hispanic Contribution to the State of Colorado.* Boulder, CO: Westview Press, 1976.
Onuf, Peter S. *Statehood and Union: A History of the Northwest Ordinance.* Bloomington: Indiana University Press, 1987.
O'Ryan, William, and Thomas H. Malone. *History of the Catholic Church in Colorado, from the Date of the Arrival of Rt. Rev. J. P. Machebeuf, until the Day of His Death.* Denver: C. J. Kelly, 1889.
Paine, Christopher M. "The Platte Earth Controversy: What Didn't Happen in 1836." *Missouri Historical Review* 91, no. 1 (October 1996): 1–23.
Palmer, Joel. *Journal of Travels over the Rocky Mountains, to the Mouth of the Columbia River; Made during the Years 1845 and 1846.* Cincinnati: J. A. and U. P. James, 1847.
Pattison, William D. *Beginnings of the American Rectangular Land Survey System, 1784–1800.* Chicago: Department of Geography, University of Chicago, 1964. First published 1957.
Paul, Rodman Wilson, and Elliott West. *Mining Frontiers of the Far West, 1848–1880.* Albuquerque: University of New Mexico Press, 2001. First published 1963.
Paxson, Frederic L. "The Boundaries of Colorado." *University of Colorado Studies* 2, no. 2 (July 1904): 87–94.
PBS. "Fort Laramie Treaty, 1868." In *Archives of "The West."* PBS, 2001, http://www.pbs.org/weta/thewest/resources/archives/four/ftlaram.htm (accessed June 7, 2012).
Pearcy, G. Etzel. *A 38 State USA.* Redondo Beach, CA: Plycon Press, 1973.
Peebles, John J. "Retracing a Line: The 1908 Idaho-Washington Boundary Resurvey." *Idaho Yesterdays* 13, no. 3 (Fall 1969): 20–25.
Perkins, James E. *Tom Tobin: Frontiersman.* Pueblo West, CO: Herodotus Press, 1999.
Perry, L. Martin. "According to Plan: A History of the North Dakota Capitol Grounds." In *The North Dakota State Capitol: Architecture and History*, edited by Larry Remele, 31–54. Bismarck: State Historical Society of North Dakota, 1989.
Pike, Zebulon M. *An Account of Expeditions to the Sources of the Mississippi, and through the Western parts of Louisiana, to the Sources of the Arkansaw, Kans, La Platte, and Pierre Jaun, Rivers; Performed by Order of the Government of the United States during the Years 1805, 1806, and 1807.* Philadelphia: C. and A. Conrad, 1810.
Plumbe, John, Jr. *Sketches of Iowa and Wisconsin, Embodying the Experience of a Residence of Three Years in Those Territories.* Iowa City: Athens Press, 1948. First published 1839.
Pomeroy, Earl S. *The Territories and the United States, 1861–1890: Studies in Colonial Administration.* Seattle: University of Washington Press, 1969. First published 1947.
Pratt, Joseph Hyde. "American Prime Meridians." *Geographical Review* 32, no. 2 (April 1942): 233–44.

Prosch, Thomas Wickham. *McCarver and Tacoma*. Seattle: Lowman & Hanford, 1906.
Rafferty, Milton D. *Historical Atlas of Missouri*. Norman: University of Oklahoma Press, 1982.
Reeves, Carolyn. *The Choctaw before Removal*. Jackson: University Press of Mississippi, 1985.
Reinhardt, Richard W. "The Short, Happy History of the State of Jefferson." *American West* 9, no. 3 (May 1972): 36–42.
Remini, Robert V. *At the Edge of the Precipice: Henry Clay and the Compromise that Saved the Union*. New York: Basic Books, 2010.
Richards, Kent D. "Washoe Territory: Rudimentary Government in Nevada." *Arizona and the West* 11, no. 3 (Autumn 1969): 213–32.
Richardson, Albert D. *Our New States and Territories, Being Notes of a Recent Tour of Observation through Colorado, Utah, Idaho, Nevada, Oregon, Montana, Washington Territory and California*. New York: Beadle, 1866.
Richter, Sara Jane. "Washington and Idaho Territories." *Journal of the West* 16, no. 2 (April 1977): 26–35.
Riebsame, William E., ed. *Atlas of the New West: Portrait of a Changing Region*. New York: W. W. Norton, 1997.
Rohrbough, Malcolm J. *The Land Office Business: The Settlement and Administration of American Public Lands, 1789–1837*. Belmont, CA: Wadsworth, 1990.
———. *Trans-Appalachian Frontier: People, Societies, and Institutions, 1775–1850*. Bloomington: Indiana University Press, 2008.
Romero, Tom I., II. "Wringing Rights Out of the Mountains: Colorado's Centennial Constitution and the Ambivalent Promise of Human Rights and Social Equality." *Albany Law Review* 69, no. 2 (April 2006): 569–79.
Ross, Alexander. *Adventures of the First Settlers on the Oregon or Columbia River*. London: Smith, Elder, 1849.
Rozwenc, Edwin Charles. *The Compromise of 1850*. Boston: Heath, 1957.
Salter, William. "Iowa in Unorganized Territory of the United States." *Annals of Iowa* 6, no. 3 (July 1903): 185–205.
Schell, Herbert S. *History of South Dakota*. Pierre: South Dakota State Historical Society Press, 2004. First published 1975.
Schulten, Susan. *Mapping the Nation: History and Cartography in Nineteenth-Century America*. Chicago: University of Chicago Press, 2012.
Scott, James C. *Seeing Like a State: How Certain Schemes to Improve the Human Condition Have Failed*. New Haven: Yale University Press, 1998.
Scott, James W., and Roland L. DeLorme. *Historical Atlas of Washington*. Norman: University of Oklahoma Press, 1988.
Shakespeare, William. *The Tragedy of Hamlet, Prince of Denmark*. In *William Shakespeare: Four Tragedies*, edited by T. J. B. Spencer, 3–300. London: Penguin Books, 1994.
Shankle, George Earle. *American Nicknames: Their Origin and Significance*. New York: H. W. Wilson, 1937.
Shoemaker, Floyd Calvin. "Some Historic Lines in Missouri." *Missouri Historical Review* 3, no. 4 (July 1909): 251–74.
Simmons, Virginia McConnell. *The San Luis Valley: Land of the Six-Armed Cross*. Niwot: University Press of Colorado, 1999. First published 1979.
Smith, Gary Alden. *State and National Boundaries of the United States*. Jefferson, NC: McFarland, 2004.
Smith, P. David. *Ouray: Chief of the Utes*. Ridgway, CO: Wayfinder Press, 1990. First published 1986.

Spalding, Burleigh F. "Constitutional Convention, 1889." *North Dakota History* 31, no. 3 (July 1964): 151–61.
Spencer, Frank C. *The Story of the San Luis Valley*. Alamosa, CO: San Luis Valley Historical Society, 1975. First published 1925.
Spencer, Thomas M. "'Demand Nothing but What Is Strictly Right and Submit to Nothing That Is Wrong': Governor Lilburn Boggs, Governor Robert Lucas, and the Honey War of 1839." *Missouri Historical Review* 103, no. 1 (October 2008): 22–40.
Steele, Thomas J., ed. *Archbishop Lamy: In His Own Words*. Albuquerque: LPD Press, 2000.
Stegmaier, Mark J. *Texas, New Mexico, and the Compromise of 1850: Boundary Dispute and Sectional Crisis*. Kent, OH: Kent State University Press, 1996.
Stein, Mark. *How the States Got Their Shapes*. New York: Smithsonian Books/Collins, 2008.
———. *How the States Got Their Shapes Too: The People Behind the Borderlines*. New York: Smithsonian Books/Collins, 2011.
Stevens, Hazard. *The Life of Isaac Ingalls Stevens*, vol. 1. Boston: Houghton, Mifflin, 1900.
Stevens, Walter B. *Centennial History of Missouri (The Center State): One Hundred Years in the Union, 1820–1921*. Chicago: S. J. Clarke, 1921.
Stewart, William M. *Reminiscences of Senator William M. Stewart of Nevada*. Edited by George Rothwell Brown. New York: Neale, 1908.
Stick, David. *Roanoke Island: The Beginnings of English America*. Chapel Hill: University of North Carolina Press, 1983.
Stowell, Jay S. *The Near Side of the Mexican Question*. New York: George H. Doran, 1921.
Sullivan, Buddy. *Georgia: A State History*. Charleston, SC: Arcadia Publishing, 2003.
Summers, Mark Wahlgren. *Party Games: Getting, Keeping, and Using Power in Gilded Age Politics*. Chapel Hill: University of North Carolina Press, 2004.
Taylor, Alan. *American Colonies: The Settling of North America*. New York: Penguin Books, 2001.
Taylor, William B., and Elliott West. "Patrón Leadership at the Crossroads: Southern Colorado in the Late Nineteenth Century." In *The Chicano*, edited by Norris Hundley, Jr., 73–95. Santa Barbara, CA: Clio Books, 1975.
Thomas, Benjamin E. "Boundaries and Internal Problems of Idaho." *Geographical Review* 39, no. 1 (January 1949): 99–109.
Thorpe, Francis Newton. *The Federal and State Constitutions, Colonial Charters, and Other Organic Laws of the States, Territories, and Colonies Now or Heretofore Forming the United States of America*. Seven volumes. Washington, DC: Government Printing Office, 1909.
Tobin, Thomas T. "The Capture of the Espinosas." *Colorado Magazine* 9, no. 2 (March 1932): 59–66.
Townshend, R. B. *A Tenderfoot in Colorado*. Boulder: University Press of Colorado, 2008. First published 1923.
Trinklein, Michael J. *Lost States: True Stories of Texlahoma, Transylvania, and Other States That Never Made It*. Philadelphia: Quirk Books, 2010.
Trotter, Richard L. "For the Defense of the Western Border: Arkansas Volunteers on the Indian Frontier, 1846–1847." *Arkansas Historical Quarterly* 60, no. 4 (Winter 2001): 394–410.
Turner, Frederick Jackson. "The Middle West." In *The Frontier in American History*, 126–56. New York: Dover Publications, 1996. First published 1920.
Twain, Mark. *Roughing It*. New York: Penguin Books, 1985. First published 1872.
Ubbelohde, Carl, Maxine Benson, and Duane A. Smith. *A Colorado History*. Boulder, CO: Pruett, 2006. First published 1965.

Unrau, William E. *The Rise and Fall of Indian Country, 1825–1855*. Lawrence: University Press of Kansas, 2007.
Unruh, John D., Jr. *The Plains Across: The Overland Emigrants and the Trans-Mississippi West, 1840–1860*. Urbana: University of Illinois Press, 1993. First published 1979.
Van Zandt, Franklin K. *Boundaries of the United States and the Several States*. Washington, DC: Government Printing Office, 1966.
Waldie, D. J. *Holy Land: A Suburban Memoir*. New York: W. W. Norton, 2005.
Walker, Henry P., and Don Bufkin. *Historical Atlas of Arizona*. Norman: University of Oklahoma Press, 1986. First published 1979.
Waugh, John C. *On the Brink of Civil War: The Compromise of 1850 and How It Changed the Course of American History*. Wilmington, DE: Scholarly Resources, 2003.
Way, Willard V. *The Facts and Historical Events of the Toledo War of 1835*. Toledo: Daily Commercial Steam Book and Job Printing House, 1869.
Weber, David J. *The Mexican Frontier, 1821–1846: The American Southwest under Mexico*. Albuquerque: University of New Mexico Press, 1982.
Wells, H. L. "The Sage-Brush Rebellion." *Overland Monthly* 13, no. 75 (March 1889): 253–59.
Wells, Merle W. "The Idaho-Montana Boundary." *Idaho Yesterdays* 12, no. 4 (Winter 1968): 6–9.
———. "Idaho's Centennial: How Idaho Was Created in 1863." *Idaho Yesterdays* 7, no. 1 (Spring 1963): 44–58.
———. "Territorial Government in the Inland Empire: The Movement to Create Columbia Territory." *Pacific Northwest Quarterly* 44, no. 2 (April 1953): 80–87.
West, Elliott. "Lewis and Clark: Kidnappers." In *A Whole Country in Commotion: The Louisiana Purchase and the American Southwest*, edited by Patrick G. Williams, S. Charles Bolton, and Jeannie M. Whayne, 3–20. Fayetteville: University of Arkansas Press, 2005.
White, Richard. *"It's Your Misfortune and None of My Own": A New History of the American West*. Norman: University of Oklahoma Press, 1991.
Wild, H. Mark. "If You Ain't Got that Do-Re-Mi: The Los Angeles Border Patrol and White Migration in Depression-Era California." *Southern California Quarterly* 83, no. 3 (Fall 2001): 317–34.
Wilson, Ben Hur. "The Southern Boundary." *Palimpsest* 20 (1938): 413–24.
Wilson, Steve. "Jefferson State of Mind." *American History* 39, no. 6 (February 2005): 20–69.
Wilson, Woodrow. "The Making of the Nation." *Atlantic Monthly* 80, 477 (July 1897): 1–14.
"Winner of the NDSU-SDSU Game to Be Presented Dakota Marker Trophy." Great West Conference, www.greatwestconference.org/genrel/042204aaa.html (posted April 22, 2004, accessed November 5, 2011).
Woodward, Grace Steele. *The Cherokees*. Norman: University of Oklahoma Press, 1963.
Woolworth, James M. *Nebraska in 1857*. New York: A. S. Barnes, 1857.
Worcester, Thomas K. *The State of Jefferson and Other Yarns of the Klamaths, Siskiyous, Southern Cascades and Northern Sierra from an Area Roughly Enclosing the Border Lands of California and Oregon between the 40th and 44th Parallels, Stretching from the Pacific Ocean to the Vast Reaches East of the Mountain Ranges*. Beaverton, OR: TMS Book Service, 1982.
Wunder, John R. "Tampering with the Northwest Frontier: The Accidental Design of the Washington/Idaho Boundary." *Pacific Northwest Quarterly* 68, no. 1 (January 1977): 1–12.
Wunder, John R., and Joann M. Ross, eds. *The Nebraska-Kansas Act of 1854*. Lincoln: University of Nebraska Press, 2008.
Wunder, John R., and Pekka Hämäläinen. "Of Lethal Places and Lethal Essays." *American Historical Review* 104, no. 4 (October 1999): 1229–34.

290 BIBLIOGRAPHY

Wyckoff, William. *Creating Colorado: The Making of a Western American Landscape, 1860–1940*. New Haven: Yale University Press, 1999.

Zanjani, Sally. *Devils Will Reign: How Nevada Began*. Reno: University of Nevada Press, 2006.

FILMS

Blazing Saddles. DVD. Directed by Mel Brooks. 1974. Burbank, CA: Warner Home Video, 1997.

How the West Was Won. DVD. Directed by John Ford, Henry Hathaway, and George Marshall. 1962. Burbank, CA: Warner Home Video, 2008.

NEWSPAPERS AND PERIODICALS

Albuquerque Morning Journal
Albuquerque Review
Arkansas Gazette (*Arkansas Post*) (Little Rock, AR)
Austin American-Statesman (Austin, TX)
Bismarck (ND) Daily Tribune
Burlington (IA) Hawk-Eye
Chicago Daily Tribune
Cherokee Advocate (Tahlequah, OK)
Colorado Transcript (Golden, CO)
Columbian (Olympia, WA)
Daily Alta California (San Francisco)
Daily Bee (Sacramento)
Daily Colorado Chieftain (Pueblo)
Daily Evening Bulletin (San Francisco)
Daily Inter-Ocean (Chicago)
Daily National Intelligencer (Washington, DC)
Daily Press and Dakotaian (Yankton, SD)
Dakota Herald (Yankton, SD)
Denver Daily Tribune
Denver Post
Denver Times
Deseret News (Salt Lake City)
The Far West (Liberty, MO)
The Free Lance-Star (Fredericksburg, VA)
Garden City (KS) Telegram
Greeley (CO) Tribune
Harper's Weekly
Hawkeye and Iowa Patriot (Burlington, IA)
Los Angeles Times
Medford (OR) Mail Tribune
Miles City (MT) Daily Star
Milwaukee Daily Sentinel
The Missoulian (Missoula, MT)
Missouri Argus (St. Louis)
Missouri Courier (St. Louis)

Missouri Republican (St. Louis)
Missouri Whig and General Advertiser (Palmyra, MO)
Mountain Democrat (Placerville, CA)
The News and Courier (Charleston, SC)
New York Herald
New York Times
New York World
North American and Daily Advertiser (Philadelphia)
Oregonian (Portland)
Revista Católica (Las Vegas, NM)
Rocky Mountain News (Denver)
Sacramento Daily Union
Saguache (CO) Chronicle
St. Louis Enquirer
Santa Fe New Mexican
Santa Fe Weekly Gazette
Saturday Evening Post (Philadelphia)
Silver World (Lake City, CO)
Siskiyou Daily News (Yreka, CA)
Southwest Daily Times (Liberal, KS)
Territorial Enterprise (Genoa and Carson City, NV)
Territorial Gazette and Burlington (IA) Advertiser
The Times (London, UK)
Washington Evening Star (Washington, DC)
Washington Post
Washington Standard (Olympia)
Washington Statesman (Walla Walla)
Western Gazette (Bloomfield, IA)
Wyoming State Tribune (Cheyenne)

Index

References to illustrations are in italic type.

Abbott, Carl, 9
Aberdeen, S.Dak., 207
Adair County, Mo., 116
Adams, Alva, 181, 182, 266n68
Adams, John Quincy, 85
Adams-Onís Treaty (1819), 144, 147, 167, 238n20
Adelman, Jeremy, 8, 219–21, 224, 227, 275n25
Alabama, 5, 63, 98, 147, 217, 236n43
Alamosa, Colo., 178
Alaska, 23, 122
Albuquerque, N.Mex., 183, 188
Alta California, 144–46
American Indians, 20, 45, 69, 100, 122, 176, 185, 192, 194, 196, 228, 255n62, 268n2; American attempts to constrain, 47, 56, 59–61, 88–91, 151, 180, 191, 197; and boundary-making, 4, 17, 33, 71–88, 95–99, 117, 195, 201–202, 219, 222–23; conflict with Euro-Americans, 24, 25, 36, 101, 129–33, 144, 156–58, 169, 191, 203, 223. *See also specific cultures*
American meridians, 15, 159, 164, 170, 195, 196, 233n37
American Revolution, 3, 20, 23, 24, 26, 29, 31, 51, 72
Anderson, Thomas L., 112
Appalachian Mountains, 9, 20, 23–26, 29, 31, 33, 34, 36, 40, 45, 50
Arthur, Chester, 202, 203
Arizona, 15, 54, 144, 146, 153, 168, 170, 174, 178, 182, *214*, 216, 217, 221
Arkansas: boundary with Louisiana, 37, 47–50; boundary with Missouri, 57, 64–68, 239n32; boundary with Oklahoma, 71, 92–93; state of, 50, 65, 68, 72, 74, 90, 91–93, 100, 222, 224; territory of, 57, 58, 64, 65, 72, 73, 75, 77, 78, 81, 82, 88–89, 243n36, 243n40, 244n38; western boundary of, *70*, 77–78, 80–84, 86, 222–23
Arkansas Post, Ark., 74
Arkansas River, 47, 52, 71, 72, 74, 77, 78, 81–83, 86, 87, 89, 97, 167, 168, 171, 173, 182
Aron, Stephen, 8, 219–22, 224, 227, 275n25
Articles of Confederation, 26, 41
Association for Borderlands Studies, 222
Astor, John Jacob, 126
Atlantic Ocean, 5, 20, 21, 23–28, 30, 31, 34, 37, 40, 42, 43, 49, 57, 124
Aurora, Calif., 159, 163, 262n101

Baja California, 144, 146
Bancroft, Hubert Howe, 144
Barbour, James, 63–64, 86
Barela, Casimiro, 179, 265n59
Barela, Jesus, 173
Bates, Charles H., 210
Bates, James W., 77
Beck, Warren A., 144
Belford, James B., 7
Belser, James E., 5
Bennet, Hiram P., 174
Benton, Thomas Hart, 51, 52, 80, 114, 168
Big Sioux River, 62, 195, 202
Big Stone Lake, 195
Bighorn River, 198
Bismarck, N.Dak., 203–207, 209
Bitterroot Mountains, 10, 132
Black Hawk, 101, 105
Black Hills, 191, 196, 197–99, 211
Blanchard, Rufus, 6
Blazing Saddles, 222
"Bleeding Kansas," 134

Blount, James, 202
Blue Earth River, 61
Bocock, Thomas, 172
Boggs, Lilburn W., 106–108, 110–14, 229–30
Bolon, Andrew J., 130–31
Bolton, Herbert Eugene, 218–19, 222, 227, 228, 274n15
Bonaparte, Charles S., 187
Borderlands history, 16–17, 218–28
"Border ruffians," 103
Bowen, Catherine Drinker, 38
Bowles, Samuel, 14, 177
Boyd, Julian P., 31
Bradford, William, 78, 80
Breckinridge, John, 68
Brewer, Jan, 217
British Empire, 19–26, 29–30, 35–36, 39, 43, 44, 50, 61, 96, 98, 122–25, 192, 193, 234n17
Brown, Joseph C., 100, 102, 117
Bryce, James, 10
Buchanan, James, 137, 158, 159, 172, 195
Burlington, Iowa, 101, 102, 106, 111, 112, 124
Burrill, James, Jr., 37, 64
Burrows, Julius, 202
Butler, James Thomas, 153
Byers, William N., 173
Byron, Matthew A., 7

Calhoun, John C., 77, 80, 82, 83
California: boundary with Nevada, 10, 12, 15, 71, 142, 159–65, 223, 262n91; boundary with Utah Territory, 148–58, 223; state of, 12, 21 122, 126, 132, 134, 135, 139, 142–65, 168, 216, 218, 223, 257n2
Cal-Neva Casino Resort, *142*, 143, 165
Canada, 23, 25, 29–30, 61, 96, 102, 193
Canadian River, 75, 78
Canton, S.Dak., 202
Cape Girardeau, Mo., 54
Capitol buildings, 11, 28, 117, 146, 185, 201, 204, 209, 223
Carolina (colony), 22, 24. *See also* North Carolina; South Carolina
Carpenter, Howard B., 187–88

Carson City, Nev., 155–61, 223
Carson County, Utah, 151–53, 155, 156
Carson River, 150, 153, 155–59
Cartography, 6
Cascade Mountains, 125–27, 130, 132, 134–36, 138, 139, 141
Catholic Church, 178–79
Catlin, George, 192, 193
Catron, John, 117–18
Cayuses, 127
Central Pacific Railroad, 161
Chambers, John, 114, 116
Cherokee Casino, 92
Cherokees, 72, 74, 76, 78–80, 82, 84–92, 222, 243n36, 243n40
Cherokee Treaty (1828), 85–88, 92
Chicago, Ill., 184, 202, 210
Chickasaws, 91, 92
Chippewa Territory (proposed), 195
Choctaws, 72, 75–80, 82–84, 86, 88–92, 222, 224nn55–56
Choctaw Treaty (1825), 83, 86, 88
Cimarron, N.Mex., 177, 183
Civil War, 4, 41, 68, 91, 138, 139, 161, 174, 196
Clark, William, 96–98, 103, 192
Clark County, Mo., 105, 107, 108, 110, 112
Clarke, James, 116
Clay, Henry, 149, 150
Clearwater River, 125, 138, 139
Clemens, Orion, 162–63
Cleveland, Grover, 208–209, 212
Coastal Range, 125, 127, 134
Colorado: boundary with New Mexico, *166*, 168–69, 171–72, 174, 176–79, 186–89, 216, 223–24; state of, 4, 7, 9, 11, 14, 15, 146, 167–68, 170, 171, 178–89, 216, 217, 218, 223–24; territory of, 159, 171–79, 195
Colorado River, 11, 54, 144, 147, 148, 164, 259n32
Colorado Territory (proposed), 257n2
Columbia (ship), 122
Columbia River, 11, *120*, 122–29, 131–41
Columbia Territory (proposed), 128–29
Colvilles, 131
Comeaux, Malcolm G., 10–11, 144
Compromise of 1850, 67, 150, 225, 259n32

Conejos, Colo., 170, 173, 178
Confederation government, 9, 26–39, 41, 44
Connecticut, *18*, 22, 33, 35, 56, 65, 67, 236n43
Continental Divide, 10, 14, 56, 137, 146, 168, 195
"Continental" identity, 25–26
Convery, William J., 176
Conway, Henry W., 80, 82, 83, 84, 244n36
Conway, James, 83, 91
Costilla, N.Mex., 177–78
Council Bluffs, Iowa, 100
Cowlitz River, 127, 128
Creeks, 98
Crittenden, John J., 133, 257n2
Crittenden, Robert, 79
Cross, Trueman, 90–91
Cumming, Alfred, 155
Curry, George C., 131–34
Custer, George, 197, 198
Custer City, S.Dak., 198

Dakota (culture), 191, 192, 194, 195, 197, 268n2
Dakota Land Company, 194
Dakota Territory: 14–15, 38, 159, 171, 191, 194–209; division of, 182, 191, 196–97, 199–209. *See also* North Dakota; South Dakota
Dakota Trophy, 211
Dalles, The, Ore., 135, 253n17
Dana, Samuel, 56
Darling, Ehud N., 177–78, 187–88
Davidson, John W., 108
Davis, Jefferson, 132
Davis County, Iowa, 116
Dawes, Henry, 201–202
Deadwood, S.Dak., 198
Declaration of Independence, 26, 38
Delaware, 25, 27, 29, 36, 37, 39, 129
Delaware River, 24
Del Norte, Colo., 178, 181, 182, 185, 186
Democratic Party, 114, 134, 138, 194, 200–205, 207, 208, 212, 218, 226
Denver, Colo., 148, 171, 173, 175, 178–80, 182, 186
Deseret (proposed state), 149

Des Moines Rapids, 96, 99, 101–103, 115, 117–19
Des Moines River, 96, 98–100, 102–103, 105, 115–16, 117–19
Dixon, Jeremiah, 24–25
Dodge, Augustus C., 62, 115
Douglas, Stephen, 61, 146–47, 172
Downey, John G., 157, 158
Drainage basins, 11–14, 121
Drake, James D., 25–26, 42
Dred Scott v. Sanford (1857), 67, 134
Dunbar and Hunter Expedition, 51
Dryer, Thomas J., 135, 136
Dueling, 7–8, 87

Edith, Colo., 187
Edmunds, George, 201–202
Edwards, John C., 115
Elizabeth I, 21
Espinosa, Felipe and Vivian, 174–76
Eustis, William, 47
Evans, John, 174, 175
Extralegal polities, 50, 124, 134–35, 151, 153–56, 171, 180–86, 194, 223–24, 253n11. *See also* Secession

Fargo, N.Dak., 200–201, 207
Farmington, Iowa, 105, 110–11
Federalist Party, 42, 43
Fillmore, Millard, 129
Fillmore, Utah, 152
Florida, 9, 25, 29–31, 216
Forsyth, John, 111
Fort, Greenbury L., 140
Fort Clark (Osage), Mo., 72, 97, 98
Fort Des Moines, Iowa, 101
Fort Gibson, Okla., 81, 89, 90
Fort Massachusetts, Colo., 169–70
Fort Smith, Ark., *70*, 74, 76, 78, 80, 81, 83, 86, 89–92
Fort Snelling, Minn., 90
Fort Towson, Okla., 90
Fort Vancouver, Wash., 122–24, 126
Fort Walla Walla (Nez Perces), Wash., 130, 131, 135, 137, 254n37
Four Corners, *2*, 10, 15, 182, 184, 187
Franklin (proposed state), 50
Freeman and Custis Expedition, 51

Frémont, John C., 146, 147
French Empire, 20, 25, 29–30, 45, 69, 96, 192
Fur trade, 101, 122, 124–25

Garcia, Jesus M., 179
Garcia, Victor, 173
Geographic boundaries. *See* State boundaries: geometry versus geography
Geometric boundaries. *See* State boundaries: geometry versus geography
George III, 25
Georgia, 33, 35, 37, 202, 217, 236n43
Gerry, Elbridge, 40
Gifford, Oscar, 207
Gilded Age, 4, 201, 204, 211–13
Gilpin, William, 14, 173–74
Godwin, Cardinal, 144
Grand Forks, N.Dak., 201
Grant, Ulysses S., 204
Graves, Tom, 79, 85, 223
Great Basin, 144, 147, 150, 151, 156, 158, 160, 161, 165
Great Depression, 216–17, 224
Great Lakes, 23, 27, 31, 35, 202
Great Recession, 217
Greeley, Horace, 155
Green, James, 171–72
Greenwich meridians, 15, 134, 159, 164, 171, 196, 225
Gregory, Uriah S., 108, 110–11, 113, 250n82
Grout, William, 202
Grover, Lafayette F., 136
Grow, Galusha, 172
Gulf of Mexico, 25

Haase, Ynez D., 144
Hackett, Cathy, 8
Haefeli, Evan, 275n25
"Hairy Nation," 105, 118, 223
Hallett, Moses, 176
Hämäläinen, Pekka, 219–20, 227, 274n15, 275n25
Hansen, James, 226
Harding, Warren G., 228
Harrison, Benjamin, 15, 204, 208, 209, 212
Hastings, Lansford W., 126, 143

Hayes, Rutherford, 200
Head, Lafayette, 170–71, 174, 179, 181
Hebard, Alfred, 105
Heffleman, Henry, 108, 110–11, 113
Hernández, Kelly Lytle, 220
Hickok, James Butler, 192
Hilliard, Henry, 147
Hinds, Thomas, 75
Hispanos, 145, 147–48, 167–87, 189, 223–24, 257n2
Hitchcock, Ethan Allen, 90
Holly, Charles F., 177
Honey Lake, Calif., 151–54, 156, 157, 159–63, 223
Honey War, 108–13, 119, 229–30
Howell, David, 31
How the West Was Won, 16
Hudson River, 24, 42
Hudson's Bay Company, 122, 124–26, 130
Huron, S.Dak., 204, 206, 207
Huron Territory (proposed), 199

Idaho, 10, 121, 125, 127, 137, 139–41, 171, 195, 212
Illinois, 16, 35, 36, 61, 63, 64, 96, 101, 106, 140, 146, 172, 205, 236n43
Indiana, 15, 36, 37, 49, 63, 188, 204, 217
"Indian country," 47, 50, 59–60, 89–91
Indian Removal Act (1830), 59–60, 88
Indian Territory, 91, 201, 208, 211
Intranational borderlands, 221–28
Iowa: boundary with Missouri, 94, 95, 101–19, 223, 229–30, 248n30, 248n32; state of, 11, 62, 95, 101, 117–19, 194, 195, 202, 211, 223; territory of, 61–62, 102, 103, 105–17, 124, 193, 215, 229–30, 247n17, 248n30, 248n32
Iowa City, Iowa, 115
Iowa River, 99
Ioways, 99
Izard, George, 83, 84, 87

Jackson, Andrew, 28–29, 59, 75, 88, 98, 101
James I, 21
James River, 204
Jamestown (colony), 21
Jamestown, N.Dak., 203, 208
Jefferson (proposed polities), 68, 171–72, 218

Jefferson, Thomas, 24, 31–32, 34–36, 38, 41, 43, 47, 51, 52, 72, 96
Jefferson City, Mo., 106, 110
Johnson, Edwin C., 216
Johnson, Richard M., 64
Journal of Borderlands Studies, 222

Kansas, 54, 60, 65, 91, 105, 134, 152, 153, 171, 178, 194, 218
Kansas City, Mo., 58, 97, 98, 180
Kansas-Nebraska Act (1854), 67, 194
Kansas River, 52, 58, 98
Kastor, Peter, 46, 69
Kelley, Hall Jackson, 125–26
Kentucky, 25, 34, 42–43, 50, 64, 68, 133, 149
Keokuk, Iowa, 101, 102, 119
Keya Paha River, 195, 201, 270n54
Kiamichi River, 82
Kidder, Arthur, 188
King, Rufus, 42
Kinsey, Charles, 66
Klickitats, 131

Lake Tahoe, 142, 143, 149, 159, 163, 164, 259n25
Lamar, Howard Roberts, 199
Lamy, John B., 178
Land Ordinance (1785), 9, 33–35, 197
Land surveys, 33–35, 81, 117, 136, 194, 196, 197, 199, 224
Lane, Joseph, 128
Las Vegas, Nev., 164
Lea, Albert Miller, 102–104, 248n32
Leonard, Glen M., 220–21
Lewis, Meriwether, 96, 97
Lewis and Clark Expedition, 51, 96, 100, 122, 192, 268n2
Lewis County, Mo., 112, 116
Lewiston, Idaho, 139
Lincoln, Abraham, 139, 159, 160, 164, 175
Lincoln Territory (proposed), 198–99, 204, 269n29
Linklater, Andro, 9
Linn, Lewis F., 114
Little Bighorn River, 198
Little Rock, Ark., 74, 75, 81, 85, 86, 91
Long Expedition, 51

Los Angeles, Calif., 147, 216
Louisiana: district/territory of, 37, 47, 49; state of, 7, 23, 37, 47, 50, 57, 65, 76, 242n23
Louisiana Purchase, 3, 16, 23, 37, 43, 45–49, 51, 54–56, 62–65, 68–69, 72, 76, 88, 95–96, 100, 167, 192, 237n64, 238n20
Louisville, Ky., 31, 33, 221
Lovely, William, 74
Lovely's Purchase, 74, 78–82, 84–87, 91–92, 223, 243n40
Lucas, Robert, 106–108, 110–13, 229–30, 251n94
Lyon, Matthew, 78

Machebeuf, Joseph P., 178
Maclay, William, 41–42
Madison, James, 39
Maine, 63–64, 66, 67
Marquette, Jacques, 96
Maryland, 24, 27, 29–31, 66
Mason, Charles, 24–25
Mason-Dixon Line, 24–25, 28, 66, 71
Massachusetts, *18*, 23–24, 33, 35, 40, 47, 63, 65, 67, 177, 201, 234n8
Maury, Matthew Fontaine, 233n37
McCall, Jack, 198–99
McDonald, Forrest, 43
McHenry, James, 30
McKenzie, Alexander, 204
McLoughlin, John, 124
Meigs, Charles R., 135
Meinig, D. W., 11, 133, 143, 220, 227
Mellette, Arthur, 205, 208
Mellin, Prentiss, 66
Meridians. *See* American meridians; Greenwich meridians
Mexican Cession (1848), 67, 143, 144, 146, 149
Mexican land grants, 168–70, 173
Mexico, 3, 20, 38, 67, 68, 78, 143, 144, 146, 148, 168, 170, 217, 224
Michigan, 101, 106, 193, 202, 230, 236n43
Miege, John B., 178
Miller, James, 74, 76
Mining, 14, 103, 138, 146–47, 155, 156, 159, 160, 163, 171, 177, 180, 191, 197–99, 211, 226

Minneapolis, Minn., 90
Minnesota, 68, 158, 194–97, 202, 248n32, 268n2
Mississippi, 47, 63, 75, 77, 84, 236n43
Mississippi River, 25, 44, 47, 49, 55, 57, 63, 75, 90, 96, 98, 102, 119, 124, 167, 192; American settlement west of, 45, 46, 50, 52, 56, 63–64, 101; as an international border, 11, 23, 29–30; and state boundaries, 52, 57, 58, 61, 72, 97, 99, 103, 105, 115, 117–18, 194, 242n23
Missouri: boundary with Arkansas, 57, 62, 64–68, 239n32; boundary with Iowa, *94*, 95, 101–19, 223, 229–30, 248n30, 248n32; state of, 3, 11, 37–38, 50, 56–61, 63–68, 78, 80, 82, 86, 87, 95, 98–103, 105–19, 152, 185, 204, 215, 223, 229–30, 244n58, 348nn32–33; territory of, 37, 54, 57
Missouri Compromise (1820), 37–38, 46, 63–68, 72, 257n2
Missouri River, 11, 52, 53, 57–62, 72, 96–98, 100, 118, 190–98, 202–203, 207, 211, 270n54
Missouri v. Iowa (1849), 117–18, 188
Monoville, Calif., 159, 163
Monroe, James, 35–36, 73, 76, 77, 80, 82, 83, 99
Montana, 8, 10, 15, 137, 139, 141, 178, 195, 196, 198, 199, 208, 209, 212
Monterey, Calif., 147–49
Monticello, Mo., 110
Monticello, Wash., 128, 129, 254n29
Mormon Church, 106, 114, 144, 149–52, 154–56, 184
Mormon Station (Genoa), Nev., 151, 153, 155
Morril, David, 65
Morris, Gouverneur, 40
Mott, Gordon N., 162
Mountain ranges. *See specific ranges*; State boundaries: geometry versus geography
Mulnix, Preston, 116

Nataqua Territory (proposed), 153
Nebraska, 11, 15, 54, 59, 65, 69, 91, 152, 171, 175, 194–96, 199, 201–202, 211, 270n54

Neosho River, 60, 74, 81
Nevada: boundary with California, 10, 12, 15, 71, 142, 143, 159–65, 223, 262n91; state of, 4, 142–44, 146, 153, 158, 164–65, 168, 170, 216; territory of, 154–56, 158–64, 171, 175, 195, 223
New Colorado (proposed state), 218
New Hampshire, 30, 65, 67, 200
New Jersey, 24, 27–30, 39, 66
New Mexico: boundary with Colorado, *166*, 168–69, 171–72, 174, 176–79, 186–89, 222–24; state of, 15, 144, 167, 168, 170, 171, 178, 188–89, *214*, 216, 223–26; territory of, 150, 152, 168–72, 174, 176–80, 182–84, 186–87, 208, 225, 259n32
New York, 24, 27–30, 35, 42, 67, 133, 179, 236n43
New York City, 6, 7, 38, 41, 118, 182, 185, 203, 205, 215, 228
Nez Perces, 130, 131
Niobrara River, 195, 201
North Carolina, 21, 28–29, 33, 35, 41, 42, 236n43. *See also* Carolina (colony)
North Colorado (proposed state), 218
North Dakota: boundary with South Dakota, 10, *190*, 194–212, 224; state of, 4, 15, 23, 195, 210–13, 224, 268n2. *See also* Dakota Territory
Northern Pacific Railroad, 196, 202, 203, 210, 211
Northwest Ordinance (1787), 35–38, 40, 41, 45, 49, 56, 64, 236n47
Northwest Territory, 37, 43, 125
Nye, James W., 160–61, 165

Ohio, 34, 36, 67, 106, 107, 138, 236n43
Ohio River, 29–31, 33–36, 41, 42, 57, 64, 65
Oklahoma, 71, 72, 74, 91–93, 211, 216, 218. *See also* Arkansas: western boundary of; "Indian country"; Indian Territory
Olin, Abram B., 133
Olympia, Wash., 128, 129, 132, 138, 139
Omaha, Nebr., 194
Ordinance of 1784, 31–33, 36, 38, 41, 56
Ordway, Nehemiah G., 200–201, 203–204
Oregon: boundary with Washington, *120*, 122, 127–41, 223; state of, 16, 121, 127,

137–41, 216, 218; territory of, 124, 126–37, 146, 152
Oregon Country, 3, 67, 121–27, 141, 147, 223
Orleans Territory, 47, 49
Ormsby, William M., 157, 158
Osages, 72, 74, 78, 79, 96–98, 223, 244n58
Osage Treaty (1808), 97–98
Otis, Harrison G., 63
Otero, Miguel, 172
Ouachita Mountains, 71, 74
Owyhee River, 134–36
Ozark Mountains, 65, 71, 74, 78

Pacific Ocean, 34, 65, 91, 121, 122, 124, 125, 127–32, 134, 135, 137–39, 146, 147, 150, 159, 161, 164, 223
Paine, Thomas, 39
Paiutes, 153, 156–58, 161, 223
Palmer, Joel, 126–27, 130
Palmer, William A., 64
Palouses, 131
Paterson, William, 39
Patterson, Thomas, 7
Pattison, William D., 9
Pautah County, Calif., 151, 152
Pembina Territory (proposed), 199
Pennsylvania, 14, 24–25, 28, 34, 39–41, 43, 64–66, 172, 202
Perea, Francisco, 176
Pettigrew, Richard F., 201
Pierce, Elisha H., 162, 165
Pierce, Franklin, 132
Pierre, S.Dak., 203, 268n2
Pike Zebulon, 51, 96, 101, 102
Pitkin, Frederick, 185
Placerville, Calif., 155
Platte Purchase (1836), 58–60, 99–100
Platte River (Missouri), 59
Platte River (Nebraska), 52, 54, 59
Plumas County, Calif., 152, 160–62
Plumbe, John, Jr., 101, 103
Plymouth (colony), 21, 22, 35, 234n8
Poinsett, Joel R., 114
Polk, James K., 67, 124, 127, 146
Poncas, 201
Portland, Ore., 128, 129, 132, 135, 210
Poteau River, 74, 82, 92

Potomac River, 24, 42
Powell, John W., 12–14
Pueblo, Colo., 181–86
Puget Sound, 121, 125, 128, 129, 135, 139
Pyramid Lake, Nev., 157, 158

Quapaws, 61, 72, 76, 81, 82, 84
Quebec Act (1774), 235n18
Quincy, Calif., 152, 160

Railroads, 130, 161, 167, 185, 186, 188, 191, 195–97, 202. *See also* Central Pacific Railroad; Northern Pacific Railroad; Union Pacific Railroad
Randall, Samuel, 202
Randolph, Edmund J., 39
Raverdy, John B., 178
Read, George, 39
Red River (north), 192, 196, 200, 203
Red River (south), 75, 78, 80, 82, 83, 86, 90, 211
Reed, Thomas B., 84
Republican Party, 8, 134, 138, 180–81, 194, 200–205, 207, 208, 212, 217, 218, 226
Reynolds, Thomas, 114
Rhode Island, *18*, 31, 37, 64, 67, 129
Rice, Henry M., 68, 158
Richardson, Albert D., 7
Riley, Bennett C., 147
Rio Grande, 146, 167, 170, 173, 180
Rivers. *See specific rivers*; State boundaries: geometry versus geography
Roanoke (colony), 21
Roberts, Jonathan, 64
Rocky Mountains, 14, 51, 53, 56, 121, 124, 126, 127, 137–39, 145–46, 155, 167, 168, 171, 178, 183, 192, 195
Rogue River, 130, 133
Romney, Mitt, 217
Roop, Isaac, 151–53, 156, 157, 160–62, 165, 223
Roop County, Nev., 161–62
Roosevelt, Theodore, 187, 188
Ross, Alexander, 126
Routt, John, 180–81
Royal Proclamation Line (1763), 25, 90, 235n18
Russian Empire, 122

Sabine River, 7, 65, 242n23
Sac and Foxes, 99, 101
Sacramento, Calif., 10, 139, 151, 154, 156, 159–61, 163
Salem, Ore., 131, 134, 136, 140
Salmon River, 138
Salt Lake City, Utah, 149, 153, 154, 158
San Diego, Calif., 147
Sanford, Edward T., 188
San Francisco, Calif., 147, 154, 157, 158, 160, 162
Sangre de Cristo Mountains, 168, 173, 176, 177
San Juan (proposed state), 181–86, 224
San Juan Mountains, 168, 180–82, 185, 187
San Juan River, 182, 184
San Luis, Colo., 168, 173
San Luis Valley, 166–68, 171–74, 176–82, 185–87, 189
Santa Fe, N.Mex., 168, 172, 178, 184, 186, 188
Saunders, Alvin, 199
Scalia, Antonin, 217
Schulten, Susan, 6
Schuyler County, Mo., 116
"Sea to Sea" charters, 21, *22*, 23, 27, 35, 234n8
Secession, 17, 156, 158, 167, 171, 180–86, 189, 195, 217–18, 224, 257n2. *See also* Extralegal polities
Senecas, 61
Sevier, Ambrose H., 86–88
Shakespeare, William, 19
Sholtz, David, 216
Shoshone Territory (proposed), 139
Sierra Nevada Mountains, 10, 12, 143, 144, 146–64, 171, 262n91
Sierra Nevada Territory (proposed), 153–54
Siloam Springs, Ark., 92
Sinatra, Frank, 143, 165
Sioux. *See* Dakota (culture)
Sioux Falls, S.Dak., 194, 201, 204, 205, 208, 209
Sitting Bull, 209
Slavery, 33, 37, 41, 46, 56, 62–68, 74, 75, 102, 108, 150, 172
Smith, Delazon, 135, 136
Snake River, 11, 124, 125, 127, 131, 134, 136–39
South California (proposed state), 257n2
South Carolina, 28–29, 33, 68, 217. *See also* Carolina (colony)
South Dakota: boundary with North Dakota, 10, *190*, 194–212, 224; state of, 15, 62, 194–96, 198, 210–13, 224, 268n2. *See also* Dakota Territory
South Platte River, 171, 182
Southwest Ordinance (1790), 41, 45
Spanish Empire, 20, 21, 25, 29, 31, 45, 69, 76, 77, 96, 122, 144, 147, 167, 219, 238n20
Spanish language, 170, 173–76, 179–80, 184
Spokanes, 131
Spring River, 60
Springer, William, 205, 208
Stanford, Leland, 162–63
State boundaries: and Americanization, 4, 17, 46, 47, 50–51, 68–69, 127, 170, 171, 178, 183, 215, 228; as borderlands, 17, 220–28; geometry versus geography, 9–16, 23–24, 26, 31, 33–36, 47, 51–62, 121, 125–29, 134–41, 144, 148–49, 152–53, 159–61, 192, 202, 205; and identity, 8–9, 95–96, 119; and law enforcement, 8, 11–14, 17, 79, 105, 118–19, 152, 154–55, 160–64, 174–76, 187, 223; modifications to, 58–62, 91–92, 99–100, 139–41, 151, 152, 161, 164, 176, 177, 180–86, 191–92, 201–202, 211–12, 216, 218, 256n77, 256n82, 257n2; and the press, 5–6, 51–57, 62, 100, 105–108, 113, 116, 118, 128, 129, 138–39, 152, 154, 160, 176, 177, 182–86, 198–201, 206–10, 215, 216, 218; redrawing original, 39, 41–42; surveys of, 12–13, 87, 89, 91, 95, 98, 100–102, 114–15, 162, 163, 174, 177–78, 187–88, 198–99, 210–11, 270n54
Statehood, 30–33, 35–38, 40–41, 61, 146, 175–76, 179, 200. *See also specific states*
Stevens, Isaac I., 130–34
Stevens, James, 65
St. Francis River, 58, 72
Stimson, Eugene K., 184–85
St. Louis, Mo., 50, 57, 96, 98, 100, 105, 202

St. Paul, Minn., 195, 197, 202
St. Peter (Minnesota) River, 52, 61
Sullivan, John C., 98–100, 102–103, 116–19
Summers, Mark Wahlgren, 211
Susanville, Calif., 152, 156, 161–62

Tahlequah, Okla., 91
Taylor, Alan, 20, 220, 227
Telegraph, 139, 161, 203
Tennessee, 25, 27, 34, 41–43, 50, 56, 68, 236n43, 248n32
Territory status. *See* Northwest Ordinance (1787)
Texarkana, Ark./Tex., 224, *225*, 226
Texas, 3, 7, 9, 23, 34, 61, 65, 67, 68, 71, 77, 82, 90, 147, 150, 185, 194, 211, 224–26, 243n29
Thomas, Jesse B., 64, 66
Toledo Strip, 106
Townley, Jeremiah, 31
Transylvania (proposed state), 50
Treaty of Doak's Stand (1820), 75–78, 82, 83
Treaty of Fort Laramie (1868), 196
Treaty of Guadalupe Hidalgo (1848), 38, 144, 168
Treaty of Paris (1783), 30, 43
Treaty of Portage de Sioux (1815), 98
Trinidad War, 176–77
Truett, Samuel, 219–20, 227, 274n15, 275n25
Turner, Frederick Jackson, 4, 12, 219, 227
Twain, Mark, 162
Tweed, William M., 203

Umatillas, 127
Union Pacific Railroad, 161, 196
United States Army, 80, 101, 115, 130–34, 146, 157, 168, 169, 174, 203
United States borders, 3–4, 29–30, 56, 167–68
United States Constitution, 38–41, 146, 228, 237n64
United States House of Representatives, 12, 63, 65, 66, 114, 115, 124, 128, 139, 140, 146, 149, 152, 172, 174, 176, 194–95, 201, 202, 204, 205, 207, 212, 226
United States Senate, 7, 41–42, 63–64, 66, 76, 77, 137, 139, 146–47, 149–50, 171–72, 201, 204, 205, 207, 212, 226, 237n69
United States Supreme Court, 117–18, 188, 212, 217
Utah: state of, 15, 146, 153, 158, 164, 168, 171, 178, 216, 217, 226; territorial boundary with California, 148–58, 223; territory of, 54, 134, 150–58, 161, 168, 171, 173, 178, 182, 184, 223, 255n32
Utes, 168, 169, 174, 180

van Buren, Martin, 103, 106, 107
Van Buren County, Iowa, 103, 106–108, 111, 250n82
van Dyke, Nicholas, 37
Verdigris River, 82
Vermont, 30, 42, 64, 201, 202
Vest, George, 204
Vigil, Agapeta, 179
Virginia, 21–24, 27–31, 33, 35, 39, 42–43, 63, 64, 66, 68, 201, 202, 236n43
Virginia City, Nev., 157

Waldie, D. J., 34
Walker, Freeman, 37
Walker, John Hardeman, 57–58
Walla Walla, Wash., 133, 136–41, 254n37
Walla Wallas, 131
War of 1812, 50, 51, 63, 64, 98, 122
Washington: boundary with Oregon, *120*, 122, 127–41, 223; state of, 23, 121, 125, 127, 137, 140, 212; territory of, 121, 129–40, 208, 209
Washington, D.C., 5, 15, 45, 55, 67, 82, 85, 88, 91, 123, 128, 132, 134, 158, 175, 191, 203, 205, 208, 210, 224
Washington, George, 41, 42
Washington (proposed state), 68
Waterloo, Mo., 108, 111, 112
Water rights, 11–12, 215
Watts, John S., 174
Weaver, James, 212
Weber, David J., 144, 221–22, 227
Wendover, Utah, 226
Western Historical Quarterly, 222, 227
Western History Association, 221
West Kansas (proposed state), 218

West Virginia, 30, 201, 202
West Wendover, Nev., 226
Whig Party, 114
White, Richard, 221
White Earth River, 193–94
White River, 72
Whitman Massacre (1847), 127
Willamette River, 121–28, 134, 135
Willard, Emma, 6
Willock, David, 112, 229
Wilmot Proviso, 68
Wilson, Woodrow, 4, 69, 227, 228
Windom, William, 201
Winnemucca, 156
Wirt, William, 88
Wisconsin, 101–103, 105, 193, 236n43, 247n17
Wool, John E., 132–34
Wounded Knee Massacre (1890), 191
Wunder, John, 219
Wyeth, Nathaniel J., 125–26
Wyoming, 4, 15, 67, 137, 144, 146, 168, 171, 178, 195–96, 198, 199, 211, 212

Yakima War, 130–34
Yakimas, 131
Yankton, S.Dak., 195–200, 203, 204, 206, 210
Yellowstone River, 52
Young, Brigham, 152, 154